THE FRONTIERS COLLECTION

Series Editors

Avshalom C. Elitzur
Unit of Interdisciplinary Studies, Bar-Ilan University, 52900, Ramat-Gan, Israel
e-mail: avshalom.elitzur@weizmann.ac.il

Laura Mersini-Houghton
Department of Physics, University of North Carolina, Chapel Hill, NC 27599-3255, USA
e-mail: mersini@physics.unc.edu

Maximilian Schlosshauer
Department of Physics, University of Portland, 5000 North Willamette Boulevard, Portland, OR 97203, USA
e-mail: schlossh@up.edu

Mark P. Silverman
Department of Physics, Trinity College, Hartford, CT 06106, USA
e-mail: mark.silverman@trincoll.edu

Jack A. Tuszynski
Department of Physics, University of Alberta, Edmonton, AB T6G 1Z2, Canada
e-mail: jtus@phys.ualberta.ca

Rudy Vaas
Center for Philosophy and Foundations of Science, University of Giessen, 35394 Giessen, Germany
e-mail: ruediger.vaas@t-online.de

H. Dieter Zeh
Gaiberger Straße 38, 69151 Waldhilsbach, Germany
e-mail: zeh@uni-heidelberg.de

For further volumes:
http://www.springer.com/series/5342

THE FRONTIERS COLLECTION

Series Editors
A. C. Elitzur   L. Mersini-Houghton   M. Schlosshauer
M. P. Silverman   J. A. Tuszynski   R. Vaas   H. D. Zeh

The books in this collection are devoted to challenging and open problems at the forefront of modern science, including related philosophical debates. In contrast to typical research monographs, however, they strive to present their topics in a manner accessible also to scientifically literate non-specialists wishing to gain insight into the deeper implications and fascinating questions involved. Taken as a whole, the series reflects the need for a fundamental and interdisciplinary approach to modern science. Furthermore, it is intended to encourage active scientists in all areas to ponder over important and perhaps controversial issues beyond their own speciality. Extending from quantum physics and relativity to entropy, consciousness and complex systems—the Frontiers Collection will inspire readers to push back the frontiers of their own knowledge.

For a full list of published titles, please see back of book or springer.com/series/5342

Douglas A. Vakoch
Editor

EXTRATERRESTRIAL ALTRUISM

Evolution and Ethics in the Cosmos

Springer

Editor
Douglas A. Vakoch
SETI Institute
Mountain View, CA
USA

ISSN 1612-3018
ISBN 978-3-642-37749-5   ISBN 978-3-642-37750-1   (eBook)
DOI 10.1007/978-3-642-37750-1
Springer Heidelberg New York Dordrecht London

Library of Congress Control Number: 2013940087

Springer is part of Springer Science+Business Media (www.springer.com)

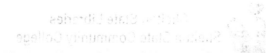

To Gerry Harp, for innovative leadership in the Search for Extraterrestrial Intelligence

Foreword

Extraterrestrial altruism. What can be said about it? Is it important in any way beyond being a good exercise for our logical abilities, and to excite our curiosity? Most people would answer, off-hand, that the answers to these questions are "very little" and just plain "no." The readers of this book may come away with surprisingly revised and positive answers to these questions!

With regard to what can be said about extraterrestrial altruism, we are seriously inhibited by having no information whatsoever about the nature, cultures, and histories of any actual extraterrestrial beings, either non-intelligent or intelligent. To arrive at some ideas about the possible existence and abundance of extraterrestrial altruism, we can, for now, depend only on knowledge of the one intelligent, altruistic, species we know of, ourselves. An example of one! In science, it is always very dangerous and even probably fallacious to develop conclusions, based on a sample of one, about anything complicated. This is not the case in reaching understandings of many matters related to extraterrestrial life, such as the origins of life, the nature and evolution of planets, and the possible chemistry on other worlds. We have plenty of data from the planets of our own solar system to guide us in those studies. But with regard to altruism in an extraterrestrial intelligent species, there is just the one example, us, on which to depend.

Of course, we do have many examples on Earth of altruism in complex creatures other than ourselves—for example, in the marine mammals, and in many terrestrial animals, such as the wolves, the simians, and the birds. Indeed, in the milieu of the higher complex animals, altruism is ubiquitous. Think of the penguins sheltering their young in the brutal Antarctic winter, or the mother bird who fights off predatory birds to bring food to her chicks. Indeed, the ubiquity of altruism in terrestrial animals is so extensive that it would appear to support strongly the idea, discussed later in this book, that altruism is not a choice, or an accident, or even a vanity, as it is sometimes with humans, but is of Darwinian origin. It works very well to enhance survival. If evolution has discovered altruism so widely as an important contributor to survival on Earth, might it not be similarly widespread on other worlds?

As an aside to this thought, as an astronomer I would point out that the driver that causes altruism to flourish is one that is somewhere between trivial and profound. It is that all planets large enough to support life are round (well, or almost round)! Planets that are most suitable for life as we know it must be massive enough to have a gravitational field strong enough to retain an atmosphere and probably liquid water. A simple calculation based on Newton's Law of Gravitation shows that the lowest energy state of the accumulated planetesimals that eventually merge to become a planet is a sphere. If there is rotation, the shape will deviate slightly from being a sphere, as in the planet Jupiter. So planets are near spheres. Never something like a sheet of paper! The consequence of this is that there always will be a finite surface area, and therefore, importantly, finite resources to nourish the growth and livelihood of living things. Indeed, the surface area of a sphere is the smallest surface area of a volume of any shape. Finite resources means there will be competition by living things for them, powering evolution, and in turn, at some point, altruism. This is explored in this book. Hopefully, despite our lack of extraterrestrial empirical evidence, we can use this model to make some credible progress in predicting the nature and extent of extraterrestrial altruism.

Now, are there some benefits of importance to us that we can, with some credibility, predict will derive from extraterrestrial altruism? There are probably many possibilities. Here I will write about one kind that comes to my mind. It is guidance in the search for extraterrestrial intelligence, SETI. Success in such endeavors is a major goal in astrobiology, and altruism may play a major role in them. Indeed, the existence and conduct of extraterrestrial altruism is becoming a central issue in SETI, and may play a crucial role in organizing our SETI programs for success.

How does this come about? In planning SETI programs, the planners must make assumptions as to what detectable technology the ETs may be using. What might those technologies be? Here we are faced, again, with the problem of dealing with a statistic of one. We have only our technological history and capabilities for guidance. Historically, SETI programs on Earth have looked to the status of contemporary technologies to point to the most promising search approaches. Fifty years ago the answer seemed to be to search for radio signals. Nothing else we were doing seemed reasonable as a detectable signal. The paradigm, based on our history, was that radio signals would remain a main sign of our (and their?) existence, and that, indeed, the power of such signals would continue to increase as more powerful transmitters were developed. Then in recent years, the development of very high power optical and infra-red lasers has suggested that, at least, we should search for signals at optical and infra-red wavelengths. Accordingly, we have added searches for such signals to our "classical" radio searches. We had a clear belief that signals at all electromagnetic wavelengths would forever be present and probably growing in intensity. We had no other examples to improve that belief.

Now, only 50 years after the first scientific SETI programs, we see unforeseen changes in the paradigm taking place rapidly. The radio signals from Earth are rapidly becoming fainter and more difficult to detect. The once strongest evidence of our existence, our traditional television transmitters, is fast being replaced by cable television and direct-to-home transmission of television and other media from orbiting satellites. The sensitivity of affordable signal receivers has improved, and signals are more efficiently transmitted using digital techniques. The result, for now, is that the one million watt power of traditional television transmitters is fast being replaced by cable television, which transmits no power into space, and the roughly 40 watt power of satellite-transmitted television, almost all of which power does not go into the extraterrestrial space. The exact numbers are not important of course; the real point is that the power levels to space are decreasing by enormous amounts. Similar reductions, but not so dramatic, in the power released to space by other means, particularly by the military, are taking place. It seems obvious that within a very short time on the cosmic timescale, perhaps only decades, the "brightness" of our civilization will be reduced by a huge factor. The great bulk of the electromagnetic transmissions of human civilization are fast fading to black! Not long from now the remaining signs of our existence may be the lights of cities at night, and at interstellar distances these are very hard to detect. We have no technology that can come close to making such detections. Of course, the city lights that are visible from space are wasted light, a consequence of inefficient light fixtures. They are, to astronomers, light pollution. Already on Earth there are efforts to develop light fixtures that will not waste light in this way, and at optical wavelengths, too, we will fade to black.

All of this adds up to bad news for SETI, and for our chances of learning of other civilizations, and being enriched by the flood of useful information we might well obtain from other civilizations. Our old paradigm was probably very wrong. We need some realistic new thinking.

It is plausible that other civilizations have traced this same history in their technology development. They, like us, have seen their visibility grow fantastically, but then, in a short time on the cosmic timescale, decrease as a result of their own growing expertise. They will be curious as we are, and will know that new, young, civilizations like ours are out there in space, searching and dreaming as we do. They will know that those civilizations have been confronted with the same eventual sobering recognition that intelligent civilizations may be very hard to find.

This is where altruism moves to center stage. Will they practice altruism, and expend some resources to make themselves detectable? Maybe through electromagnetic signals of great power? Is this the crucial step that will allow civilizations to share information, after all? We have tried to think of alternatives to this, but know of none.

If this is the case, we need to change our search strategies. The altruistic beacons of other worlds may come not from the planets of nearby stars, but from rare altruistic systems far away. It is a fact, rarely emphasized, that most of the brightest stars in the sky are not nearby, but are intrinsically very bright very distant stars. The same is true of cosmic radio sources. If altruism is motivating the technology choices for interstellar communication, the same could be true of signs of intelligent life. We should search for signals that are designed to be easily detectable at great distances. We already do this. We should search where we can test as many stars as possible at once for the presence of signals. It could well be that the first civilization we detect will be hosted by a star that is not in any catalog, or not even visible.

The nature and abundance of extraterrestrial altruism is not just an interesting and challenging exercise in logic; it may be very important to the discovery and understanding of extraterrestrial civilizations. This book will serve as the first major step in making progress with these questions, whose answers may be of far more importance than we might have imagined.

Mountain View, CA, USA   Frank Drake
                                                                 SETI Institute
                                                                 fdrake@seti.org

Acknowledgments

To the contributors of the chapters that appear in this volume, I especially appreciate the depth and innovation of the work they share here. I am grateful to Joe Castrovinci for so capably copyediting each chapter, enhancing the consistency across chapters while still encouraging authors to keep their own voices.

Over the past 15 years, many colleagues from the SETI Institute have shared with me their insights into the universality of altruism. I especially thank Shannon Atkinson, Molly Bentley, Linda Bernardi, Anu Bhagat, John Billingham, Leslie Bisno, James Brewster, Steve Brockbank, Edna DeVore, Frank Drake, Sophie Essen, Andrew Fraknoi, John Gertz, Gerry Harp, Jane Jordan, Ly Ly, Chris Munson, Chris Neller, Tom Pierson, Karen Randall, Jon Richards, Pierre Schwob, Seth Shostak, and Jill Tarter.

More recently, I warmly acknowledge the administration, faculty, staff, and students of the California Institute of Integral Studies (CIIS), especially for support from Joseph Subbiondo, Judie Wexler, and Tanya Wilkinson. In addition, I thank Harry and Joyce Letaw, as well as Jamie Baswell, for their support of this work.

For shepherding the book through the editorial process at Springer, I am indebted to Ramon Khanna. Also at Springer, Tamara Schineller and Charlotte Fladt have my gratitude for helping to move the book swiftly and efficiently through the review process and into production. My appreciation for faithfully overseeing all aspects of production goes to S. A. Shine David of Scientific Publishing Services. Moreover, I thank the editors of the Frontiers Collection for including this volume in their series.

Finally and most importantly, to my wife Julie Bayless, I am grateful in more ways that I can or will share here. Thank you, forever.

Mountain View and San Francisco, CA, USA   Douglas A. Vakoch

Contents

Part I Introduction

Cosmic Evolution, Reciprocity, and Interstellar Tit for Tat   3
Albert A. Harrison

Part II The Evolution of Altruism: Is Transmitting Dangerous?

Extraterrestrial Intelligence: Friends, Foes, or Just Curious?   25
George Michael

Eliciting Altruism While Avoiding Xenophobia:
A Thought Experiment   37
Jerome H. Barkow

Predator—Prey Models and Contact Considerations   49
Douglas Raybeck

Harmful ETI Hypothesis Denied: Visiting ETIs Likely Altruists   65
Harold A. Geller

Altruism Toward Non-Humans: Lessons for Interstellar
Communication   79
Abhik Gupta

Caring Capacity and Cosmocultural Evolution: Potential
Mechanisms for Advanced Altruism   93
Mark L. Lupisella

The Precautionary Principle: Egoism, Altruism, and the Active
SETI Debate   111
Adam Korbitz

Part III   Inferring Altruism

The Accidental Altruist: Inferring Altruism from an
Extraterrestrial Signal.   131
Mark C. Langston

Interstellar Intersubjectivity: The Significance of Shared
Cognition for Communication, Empathy, and Altruism in Space.   141
David Dunér

Other Minds, Empathy, and Interstellar Communication   169
Tomislav Janović

Interspecies Altruism: Learning from Species on Earth   191
Denise L. Herzing

Part IV   Universal Ethics and Law

Terrestrial and Extraterrestrial Altruism.   211
Holmes Rolston III

Kenotic Ethics and SETI: A Present-Day View.   223
George F. R. Ellis

Altruism, Metalaw, and Celegistics: An Extraterrestrial
Perspective on Universal Law-Making   231
Adam Korbitz

Part V   Representing Altruism

A Logic-Based Approach to Characterizing Altruism in Interstellar
Messages.   251
Alexander Ollongren

Equity and Democracy: Seeking the Common Good as a Common
Ground for Interstellar Communication.   261
Yvan Dutil

Patterns of Extraterrestrial Culture 275
William Sims Bainbridge

Evolutionary Perspectives on Interstellar Communication:
Images of Altruism 295
Alfred Kracher

About the Editor 309

About the Authors 311

Index 317

Part I
Introduction

Cosmic Evolution, Reciprocity, and Interstellar Tit for Tat

Albert A. Harrison

Abstract Drawing on many fields in the physical, biological and social sciences, the present book explores altruism and related topics such as predation, conflict, and reciprocity within the broadest possible context: the cosmos. This introductory chapter compares two opposing themes in discussions of how societies achieve great age. The paranoid theme holds that authoritarian, egotistical, self-serving societies that rely on threat, bluff and force in interstate relations will prevail. The pronoid theme holds that democratic, inclusive, and cooperative societies that are reluctant to use force achieve greater longevity. Research shows that over time, societies on Earth have accorded increasingly larger groups of people favored insider status; that authoritarian governments are giving way to liberal democracies; that democracies do not go to war with one another but rather enter into effective defensive pacts; and that across many different timescales and locations, violence of all kinds, including warfare, is on the decline. Reciprocity and win-win thinking form powerful social glue that binds diverse societies and people together. These results, which are more consistent with pronoid than paranoid expectations, hint that our own civilization may be working its way through a period of "technological adolescence" where the availability of dangerous technology outpaces cultural adaptations to use such technology with wisdom and restraint. After considering a few possible implications of these discoveries for the search for extraterrestrial intelligence, this chapter concludes with an overview of the rest of this book.

Keywords Altruism · Astrotheology · Authoritarian states · Cold War · Democratic states · Democratic peace · Golden Rule · Interstellar negotiation · Prisoner's Dilemma · Rights Movement · Security pacts · Tit for Tat · Violence

A. A. Harrison (✉)
University of California, Davis, 324 Anza Avenue, Davis, CA 95616-0402, USA
e-mail: aaharrison@ucdavis.edu

D. A. Vakoch (ed.), *Extraterrestrial Altruism*, The Frontiers Collection,
DOI: 10.1007/978-3-642-37750-1_1, © Springer-Verlag Berlin Heidelberg 2014

1 Introduction

As we survey life on Earth we find abundant examples of selfishness, predation, conflict and war along with equally abundant examples of altruism, cooperation, conflict resolution and peace. Altruism is generally defined as a concern for the welfare of others coupled with a willingness to make concessions of benefit to others without direct personal gain. Interest in moral and immoral attitudes and behaviors extends to the distant past and commands the attention of philosophers, theologians, politicians and diplomats. Now, as Michael Shermer (2004) points out, science adds to the discussion.

Issues regarding war and peace, competition and cooperation, self-interest and altruism on Earth are vexing enough without concerning ourselves about the moral status of unknown and perhaps nonexistent beings. Yet, try as we may, we are strongly tempted to speculate about life beyond Earth (Finney 1990; Harrison 1997; 2000; 2007; Michaud 2007; Peters 2009; 2011). This book explores altruism in the largest of all possible frameworks, the cosmos. How did altruism evolve on Earth and what are its implications for the evolution of life on other planets? What about conflict and predation? Are there universal ethics? Can we encode altruism and signal friendly intent so that they can be understood by extraterrestrial civilizations? This introductory chapter touches on each of these themes, traces possible implications for our own future and for the search for extraterrestrial intelligence (SETI), and concludes with an overview of the remaining chapters.

2 Stability, Peace, and Longevity

How can societies achieve great age? This question is crucial for ascertaining our own prospects for the future and for helping us understand the development and course of extraterrestrial civilizations. Any society that SETI is likely to detect is expected to be far older than our own, perhaps by many millions or billions of years (Shostak 2009). What governmental forms and policies help societies prosper over long periods of time? What other governmental forms and policies lead societies to stagnation, decay, and collapse? If we could answer such questions, we could make educated guesses as to the relative proportions of belligerent and friendly societies "out there" (Harrison 2000).

This chapter explores evidence that bears on paranoid and pronoid themes in speculation about life in the universe (Finney 1990). The paranoid theme holds that hostile, aggressive, and self-serving societies survive the longest and thus predominate in the universe. Based on simplistic notions of "survival of the fittest," successful extraterrestrials live in highly regimented, authoritarian, and egotistical societies that use threat, bluff, and force to vanquish all opposition. Borrowing terms from Steven Pinker (2011), predation, dominance and revenge are such societies' rallying cries. The pronoid theme maintains that rigid authoritarian regimes contain the seeds of their own destruction. Their harsh repressive practices make them ripe for rebellion from within, and their belligerent posture invites containment from without. Empathy, self-control, morality and reason have the upper hand (Pinker 2011). From the pronoid perspective it is the inclusive, democratic and peaceful societies that endure.

In the present context, paranoia and pronoia are terms of convenience and neither implies mental impairment or an adverse psychiatric state. Paranoids worry that the universe is packed with predatory societies that could do great damage to us, while pronoids expect wonderful things from newfound friends. To some extent these views project our fears and hopes, respectively, on to imaginary civilizations (Michaud 2007). These contrasting views pervade both highbrow and lowbrow discussions of extraterrestrial life (Harrison 2007) and are evident in science fiction (Wilcox 1991). They are reflected in realist and idealist views of international relations (Wilcox 1991) and have religious overtones (Peters 2009; 2011).

Of course, right now the only place we can look for evidence is on Earth. Earth may be only one case, but it contains millions of species, thousands of cultures, hundreds of nations, and written history stretching back over 5,000 years. Does our own history provide more support for the paranoid or pronoid views? Can we identify shifts towards or away from positive social behaviors over time? Are there any reasons to suspect that evolutionary and developmental processes on Earth will be replicated in distant star systems and galaxies? This extension of my previous work (Harrison 1997; 2000; 2010) reaffirms that research on social inclusiveness, democracies, declining violence, non-zero sum bargaining, and reciprocity may have important implications for our own future status and what might await us if SETI succeeds.

3 Expanding Inclusiveness

Cosmic evolution—change in the universe over time—is history on the largest of all possible scales (Chaisson 2001; Christian 2004; Dick 2009). So far, cosmic evolution has been a continuing shift in the direction of localized order, organization and complexity including greater mind or consciousness, larger societies, and more elaborate cultures. As James Grier Miller (1977) explained years ago, since life formed on Earth 3.7 billion years ago, there has been a steady progression in the direction of larger biological and social entities: cell, organ, individual, group, organization, community, society, and organizations of societies. Each level encompasses all preceding levels, and at certain junctures, phenomena such as consciousness and culture appear. In a similar vein, Alexander Wendt (2003) traces political organization from states through five increasingly encompassing stages: systems of states, societies of states, world society, collective security, and world state. Each stage remedies the flaws in the stage that precedes it.

These shifts in the direction of larger communal and sociopolitical units have been accompanied by an increase in the numbers of people who are defined as "like us" and hence accorded "insider" or favored treatment (Shermer 2004). For the greatest part of human history, biological factors controlled moral sentiments and the overriding goal was to protect one's family. About 35,000 years ago, when communities formed, cultural factors came into play, and favored treatment was extended to all members of the community. The division of labor, which increased people's dependency on one another, was integral to this process.

Modern society further expanded the size of the in-group. The end of the Cold War sparked a new wave of liberal internationalism, whose adherents view national boundaries as obstacles to peace, and seek to hold states accountable for high standards of conduct. This new wave looks beyond sovereign rights to human rights, beyond national security to human security, and beyond national interest to planetary interest (Michalak 2004). Perhaps after we treat our entire species as kin we will raise our sites to bioaltruism—honoring life as a whole (Shermer 2004; Lupisella 2013).

Steven Pinker (2011) identifies trends, some beginning thousands of years ago and some dating back only decades, that have contributed to expanding inclusiveness. The Pacification Process marked the beginning of the transition from hunting-gathering societies to agricultural communities with an extended sense of kinship beyond the immediate family. This stage introduced the rule of law. Next the Civilizing Process consolidated feudal societies into large kingdoms and eventually into modern nation states. The Humanitarian Revolution dates to the Age of Reason and European Enlightenment, and reduced despotism, public executions, torture, slavery, dueling, and other forms of cruelty that are considered unacceptable, even silly, today. The Rights Movement had been around a century or so before it gained momentum in the 1960s. The extension of rights—to women, to members of different ethnic and religious groups, to gays, children and just about everyone else—has continued to accelerate since that time.

Pinker mentions another important factor: integration of the workplace and travel bring people into contact with members of groups that in earlier days they would never meet firsthand. The net effect has been a reduction in prejudice. Moreover, in their massive study of American religion, Robert D. Putnam and David E. Campbell (2010) found that, although there are some divisive forces within religion in America today (primarily on a liberal-conservative dimension), there are compensatory unifying forces as well. One of the most important is a level of religious intermarriage that would have been unthinkable fifty years ago. How much of a Baptist hard-liner can you be if your brother is an atheist, your daughter is a Catholic nun, and your son married a Jew? Today the Internet helps erase barriers between groups (Barkow 2013).

4 The Rise of Liberal Democracies and the Democratic Peace

Authoritarian regimes and liberal democracies (defined as states whose citizens elect their leaders and are guaranteed human rights in return) differ in many significant ways. Timothy Ferris (2010) notes a very close link between democracy and science. Unlike citizens beholden to totalitarian regimes, those who enjoy democracy are free to associate with one another, experiment, hash-out disagreements, and change directions in midstream. Democracies are typified by disagreement, conflict, and a never-ending quest to make things "better." Science also requires freedom of association, opportunities to experiment, disagreement and conflict, and changes of opinions as successive waves of scientists triangulate on empirical truths. Authoritarian regimes that do not allow freedom to associate, are committed (sometimes forcibly) to a particular ideology, and discourage the pursuit of new ideas, get lousy grades for discovery and invention. Joseph Stalin's and Chairman Mao's ventures into science had terrible consequences. Stalin pushed Lysenko's scientifically wrong but politically correct theory of evolution, which contended that artificial modifications (such as chopping off a mouse's tail) could be passed along to subsequent generations. Mao's "experiments" led to widespread famine and death.

Events in Nazi Germany may appear to run counter to the link between democracy and science, but a close look at the record does not bear this out (Ferris 2010). Germany had good science in the 1930s. At first, Nazi scientists were allowed to attend international meetings where they could share ideas, but soon this freedom was lost. Many talented scientists who could have helped develop atomic weapons left Germany when Hitler ascended to power, and many other scientists, as political dissidents, ended up in concentration camps. Nazi Germany was able to turn pre-war rocket technology into functional weapons, and this (along with contributions from American Robert H. Goddard) provided a basis for postwar rocketry. But during the war, German rocket scientists were expressly prohibited from discussing new scientific directions and were kept in line by the Gestapo and SS.

According to Ferris (2010, 1) the scientific and democratic revolutions are proceeding hand in hand:

> [Science] has already revealed more about the universe than had been learned in all prior history, while technological applications of scientific knowledge have rescued billions from poverty, ignorance, fear, and an early grave. The democratic revolution has spread freedom and equal rights to nearly half the world's inhabitants, making democracy the preference of informed people everywhere.

If the link between freedom and science reflects underlying principles, shouldn't friendly extraterrestrial civilizations have the upper hand? As David Dunér points out, successful extraterrestrials will be egalitarian rather than hierarchical, and they will be democratic rather than fascistic (Dunér 2013).

Totalitarian regimes are being replaced by liberal democracies (Ray 1988), and this is important because totalitarian and democratic societies have different propensities to go to war. Bruce Russett (1993) found that at the end of the nineteenth century only 12–15 states qualified as democracies, but by 1990 two-thirds of the nations included in the sample—126 out of 183—were either democracies (91 nations) or moving in that direction. George Modelski and Gardner Perry, III (2002) studied transformations in relatively large societies over the past 160 years. They defined democracies as multiparty systems with elected executives who are subject to checks and balances, and they charted both the increase in the number of democratic societies and the percentage of people living within them.

In 1840, there were two democracies populated by 44 million people, a mere 4 % of the world's population. In 1890, an even dozen democracies served 176 million people, or 11.9 % of the world's population. By 1990 there were 60 democracies totaling 2.367 billion people, that is, 44.8 % of the global population, and only ten years later the halfway mark had been passed with 3.469 billion or 57 % of all people living in democracies. By extrapolation it will take a total of 228 years (1840–2068) to shift from 10 to 90 % democracies. These findings are consistent with those of other researchers who have used different time scales, samples of nation-states, and definitions of democracy.

In contrast to people in totalitarian states, citizens of democracies are less likely to be subject to repression and political violence, and more likely to have their basic needs for food and shelter satisfied, and to develop economic surpluses for trade. Most importantly, as already mentioned, democracies are less likely to get embroiled in wars. This is illustrated by the work of Bruce Russett (1993), who asks us to assume that each society has the opportunity to go to war with every other society. His research shows that at each level of belligerence (threat of force, display of force, war) pairs consisting of two democracies were less likely to escalate a conflict than were pairs that included at least one autocracy. In his sample, none of the pairs of democracies actually went to war; all wars involved at least one totalitarian government. The generality of these results led one scholar to conclude that the link between democracy and peace is the strongest empirical law in international relations, and another to propose that other research on conditions of war and peace might as well be consigned to the scrap heap (Ray 1988). Efforts to overturn this principle or to invoke a third-factor explanation have not withstood close scrutiny, leading Allan Dafoe (2011, 247) to restate "the democratic peace remains one of the most robust empirical associations" in international relations.

Democracies are aware that other democracies share their preferences for peaceful, negotiated solutions to problems, and it is these strategies that they favor in international affairs. Democracies also know that totalitarian governments do not share their principles, and democracies will go to war with totalitarian states if necessary. Democracies do join together in defensive pacts, that is, gang up on aggressors. Mathematical models show that nations that practice collective security, that is, do not initiate war but do go to one another's defense, survive longer than states that follow other foreign policies. The greater the number of states that join a collective security pact, the longer the international system survives. "Our results," write Thomas Cusack and Richard A. Stoll (1994, 56), "suggest that practitioners of collective security are ecologically superior to states following more self-interested…strategies." In a later quantitative study of collective-security behavior in militarized interstate disputes, Stoll (1998) concluded that nice states finish pretty well.

5 The Decline of Violence

Several decades ago, a few historians foresaw, within maybe 200 years, a world without war (Keegan 1994; Mueller 1988). This trend prevails despite the recent carnage of two world wars. The percentage of people who have died from warfare (as compared to all other causes) has decreased in recent centuries. Michael Shermer's (2004) analysis of representative societies shows that pre-modern political states had the highest death rates from war (up to 30 %) while the death rate in Western Europe during the seventeenth century was about 2.5 % and about half of that in the US and Europe during the twentieth century. Steven Pinker (2011) points to statistics that can be described only as dumfounding given what people learn from the news. This discrepancy is understandable given that war and violence make the front pages, while successfully defusing a problem or effecting reconciliation do not. People are more moved by images and stories than by numbers, and it is the scariest stories that are the hardest to forget. Which of the following do you think most influences people's views of the world as a hostile and dangerous place: the terrible mass murder that took place in December 2012 at the Sandy Hook Elementary School in Newtown, Connecticut, or long-term FBI statistics showing that violent crime is decreasing?

Archeological excavations allow estimates of violence when they unearth bodies with smashed or decapitated heads, spearheads embedded in bones, and projectiles inside rib cages (Pinker 2011). The percentage of deaths due to war steadily declines as we move from pre-historical archaeological sites, through hunter-gatherer and horticultural groups, hitting all-time lows in modern states. Pinker's list of the 21 worst wars and atrocities in history—adjusted for population size—shows that four of the deadliest events were the Au Lushan revolt in eighth century China (casualties equivalent to 429,000,000 mid-twentieth century casualties), the Mongol conquests of the eleventh century (equivalent to 278,000,000 deaths), the Mideast Slave Trade (132,000,000 fatalities) and the Fall of the Ming Dynasty (112,000,000 fatalities). Adjusted for mid-twentieth century population size, World War II is in ninth place (55,000,000 casualties) and World War I in the sixteenth (15,000,000 deaths) (Pinker 2011, 195).

All of this has taken place against a backdrop of a reduction in violence as a whole. This includes dramatic decreases in homicides, lynching and other hate crimes, executions, torture and other kinds of corporeal punishment, rape and other forms of violence against women, gay bashing, child abuse, and cruelty to animals.

Pinker (2011, 48) summarizes his findings as "five dozen graphs that plot violence over time and display a line that meanders from the top left to the bottom right. Not a single category of violence has been pinned to a fixed rate over the course of history. Whatever the cause of violence it is not a perennial urge like hunger, sex, or the need to sleep."

Joshua Goldstein (2011) has a somewhat less sweeping view of the decline of war. Yes, he agrees, there has been a reduction in war over the years, and no, some earlier centuries were not as peaceful as most people think. Good statistics, he stresses, are hard to find. Some researchers have ignored wars in Asia because statistics were too hard to assemble, discarded cases that ran counter to their theses, and cut up the pie of history in convenient rather than logical ways. Unlike Pinker, whose rankings of war are based on mid-twentieth century populations, Goldstein looks at absolute numbers and accepts two world wars as dramatic exceptions to an otherwise encouraging trends towards peace. His focus is on 1945 to present, decades that reveal a somewhat uneven but evident decline in war.

Most importantly we are stepping back from the nuclear abyss. Goldstein reports that from the 1950s to the 1980s, the number of strategic nuclear weapons held by the US and USSR was 30,000 to 40,000 each, with about ten thousand deployed on each side. By 2010, the US had about 5,000 weapons and Russia had about 9,000 (it takes time and money to dismantle these safely) and each side had about 2,500 deployed. A treaty ratified in 2011 reduces the number of warheads to a somewhat less ridiculous 1,550 per side. Goldstein finds that large wars between states are giving way to civil wars within states (although occasionally aided and abetted by an outside power). These are fought with cheap, light weapons, not the kinds of tools that would make one master of the universe.

Both Pinker and Goldstein credit this change to recently-evolved overarching social systems such as NATO, the UN, and other terrestrial versions of the Galactic Club. Goldstein praises the United Nation's efforts since 1946 to promote peace. Credible diplomacy, peace-keeping missions and peace-building missions, aided by humanitarian and relief organizations, are curtailing conflict and reducing the likelihood that, once resolved, a conflict will re-ignite.

6 Getting Along Better: Reciprocal Altruism and the Golden Rule

Although biology, social science and religion view the same behavior through different lenses, they converge on the law of reciprocity (Turner 2005). This law states: "Do unto others as you would have them do unto you," "I'll scratch your back then you scratch mine," "Turnabout is fair play," and "What goes around comes around." This law's failure is revealed in "Where were you when I needed you?" Harm, also, is reciprocated: "An eye for an eye and a tooth for a tooth."

Few, if any theories approach the theory of biological evolution in terms of power, scope, and applicability. It is a "selfish" theory in that self-interest (defined in terms of assuring the survival of one's genes) is paramount. An altruistic act involves making a concession or sacrifice that benefits someone else. Simple examples are giving pocket change to strangers, mounting enormously expensive and widespread relief efforts for devastated countries, and suicidal acts to save one's comrades. Since altruistic behavior implies improving someone else's chances for survival at the expense of one's own, it is one of the few activities that was not easily reconciled with evolutionary biology. After all, the whole point of existence is to pass one's genes on to subsequent generations, so why should one forego this to help someone else? The answer is inclusive fitness—self-sacrifice that indirectly promotes continuation of one's genes. A woman who is past child-bearing age may give a fortune to her children, and in the course of improving their life circumstances also increase their ability to support a large family, thus giving her many healthy grandchildren. More detailed discussions of altruism transcending biological self-interest are provided by Mark Lupisella (2013) and other authors in later chapters of this book.

Social sciences, including economics and psychology, offer their own analyses of generous behaviors, as well as alternatives such as competition and spite. People have choices, and these choices yield "outcomes" in the form of rewards and costs. The working assumption is that people are rational, in the sense that they strive to maximize their rewards and minimize their costs. In social interaction the various participants or "players" (whether individuals or societies) make choices that affect each other's outcomes, and they realize that their selections will influence how they are treated in return. Social-exchange theory casts social interaction in terms of economic transactions with the pattern of outcomes determining the quality and duration of the relationship. Here, too, reciprocity is the key. The expected payback may be immediate (splitting the profits right after a sale) or may come due at some future time. The latter is illustrated by a slogan on T-shirts worn by members of the Gold Beach, Oregon, Fire Department: "Come to our breakfast, we'll come to your fire" (Putnam 2000, 21).

Equity theory contends that psychological forces and social norms encourage us to seek fairness or justice. People feel most comfortable with situations that are fair and equitable, where everyone (themselves included) receives what they deserve. The best sales person deserves the biggest bonus, and the murderer who pulls the trigger should get a longer sentence than does an accessory after the fact. We are disquieted when an honorable, long-term suitor gets dumped in favor of a one-night stand, an outstanding employee of many years is passed over for a promotion, or a boss usurps credit for a subordinate's hard work. We are motivated to reduce inequity, if not in fact then at least psychologically. Workers who feel underpaid demand raises, and if this fails may slack off so that their productivity sinks to the level of their pay. Less reliably, workers who feel overpaid relative to their peers pick up the pace, work more carefully, stay late, or find other ways to raise their contributions to their level of reward. Because of his family connections, U.S. Senator John McCain, an admiral's son, was offered early repatriation from a North

Vietnamese POW camp. He declined on the basis that he had done neither more nor less than his fellow internees, and voluntarily suffered the same fate.

Preferences for fairness extend to procedures. It's perfectly acceptable if a young woman wins first prize at the science fair because of the technical excellence of her project, but not okay if she got the medal because her mother is a member of the school board. And equity theory posits minimum standards for treating other people, a norm of human-heartedness. No matter how inept or incompetent an employee, he or she should not be publicly berated and humiliated by the boss. Of course equity is not always achieved, but the pressures for equity outweigh pressures, say, to create an unprincipled and unjust world.

One of religion's foremost principles squares with studies of evolution and social behavior. The Golden Rule—"Treat others as you want to be treated"—is predicated on reciprocity. The Golden Rule is a moral guideline that is found in all cultures and represents the very foundation of universal morality (Shermer 2004). Its behavioral expressions include empathy, hospitality, cooperation, and generosity. The law of reciprocity is widespread because it provides a useful prescription for behavior, a blueprint for action that has proven effective in many cultures and at many times. Even more generous than the Golden Rule is the prescription to do unto others as they would have you do unto them (Korbitz 2013).

The norm of specific reciprocity—"I'll scratch your back if you scratch mine"—is now supplemented by a norm of generalized reciprocity—"I'll scratch your back confident that someone else will do something for me down the road" (Putnam 2000). Communities that adhere to this norm of generalized reciprocity (also known as indirect reciprocity) tend to be better off than communities that rely on specific reciprocity alone (Putnam 2000). People who make concessions or do favors for one another develop good will or "social capital." This is akin to money in the bank, a reservoir that can be drawn upon as needed. Individuals build social capital with altruistic or pro-social acts: doing volunteer work at the local library, feeding bereaved neighbors, visiting friends in the hospital, or trouble-shooting an acquaintance's computer. Societies accrue social capital by entering into defensive pacts, limiting activities that harm the environment, reducing tariffs, easing border restrictions, and offering foreign aid. All of this leads to trust, which increases one's proclivity to be the first to offer a concession. In addition to the kind of "thick trust" that is based on years of give and take, we develop "thin trust," a generalized but not tenuous form that encourages giving strangers the benefit of the doubt.

7 Game Theory

Experimental games are widely employed for controlled studies of interactive behavior. As will become evident in later chapters, the Prisoner's Dilemma Game (PDG) is the best known. This is used for studying processes of mutual accommodation in non-zero sum situations, that is, where one party's gains are not offset by the other party's losses. Each of two "players" is allowed to choose one of two responses, and it is the combination of these responses that determines each player's outcomes. The cooperative choice holds the lure of small but equal gains for each player. The competitive choice offers each player the prospect of a larger gain, but at the other party's expense. If both players "defect," that is, make the greedy choice, then each sustains a loss. Although at first glance, it may appear that taking advantage of the other player offers the highest personal payoff, long-term success requires that over repeated iterations each player choose the cooperative response. Enlightened self-interest demands minor concessions for long-term benefits.

The most effective strategy for locking into win-win choices is conditional cooperation, or Tit for Tat. Start cooperatively, but if the other player defects then retaliate. PDG players who are unconditionally cooperative run the risk of being taken advantage of, and thanks to retaliation, people who are unconditionally selfish get locked into a long string of mutual losses. Hundreds of studies involving the PDG and other non-zero sum situations show that conditionally-cooperative strategies gain toeholds even in competitive environments, rapidly spread, and, once entrenched, protect themselves from invasion by exploitative strategies. Studies conducted across species, cultures, and historical epochs reveal advantages for reciprocity and turn-taking (Axelrod 1984). Is it possible that Tit for Tat is, in the broadest and most literal sense, universally recognized?

One challenge for initiating interstellar Tit for Tat is that "outcomes" are expressed in terms of utilities. In an experimental game, the win-win strategy may deliver so many pennies to each participant, whereas the win-lose strategy gives more small change to the defecting player than to hoodwinked opponent. We have little problem fantasizing that ETs have what we want, such as scientific insights and practical information that could help us erase poverty, disease, and war. If we ignore claims from UFOlogy that ET wants our water to power their circular spacecraft, or human sperm and ova for a hybrid fertilization program, what can we offer them? Trading artifacts seems very unlikely: any bargaining that takes place will be based on information exchange. We may suspect that our science is running behind theirs, but perhaps like our field anthropologists they would like insider views of our culture, or have some interest in our literature and arts.

Communications challenges fall into two categories: signaling our intent to cooperate, and swapping information within a meaningful time frame. If we can communicate at all, it might not be that difficult to convey our interest in win-win strategies (Dutil 2013). After all, experimental games such as the PDG are defined by their formal properties. Relative, not absolute numbers define the payoffs, meaning that the essence of the game can be expressed in terms of equalities and inequalities. Through formal notation, we might be able to signal our interest in extending the hand of friendship but with the intention of either breaking off communication or retaliating in the absence of a reciprocal response. Furthermore, as Douglas Vakoch (2009) has suggested, we may be able to express our understanding of altruism and reciprocity by telling the story of evolution, or by means of carefully-devised graphic representations of people helping one another (Kracher 2013).

Another problem is that even at the speed of light, one-way communication could take many years. Participants in experimental-games research may undergo scores, even hundreds, of iterations. In interstellar communication, each "exchange" could take several lifetimes. Normally, in social interaction, it does not take very long to separate people who are deserving of trust from freeloaders. In interstellar Tit for Tat, iterations may be so far apart that each trial seems separate and unique, and a consistent pattern is hard to discern.

If neither party expects a continuing relationship, there will be little room for cooperation to evolve. When it is extremely unlikely that the parties will meet again, then there is a strong temptation to defect. Nineteenth century Yankee peddlers could sell carved wooden nutmegs with impunity because they never expected to encounter their victims again. Still, there are instances where people treat others generously despite the absence of an ongoing relationship. Tipping a server at a restaurant where you will never return is a good example of this. Unlike the server at a favorite haunt, a waiter at a restaurant you visit only once cannot retaliate.

Our knowledge of other people's reputations is a powerful determinant of our behavior toward them, as is recognition that our own reputations will affect how they will treat us. In the case of interstellar negotiations, each side may be of unknown reputation, thereby putting a heavy burden on trust. At first it is unlikely that we will find any reputable third parties, willing to put in a good word for either side or serve as a helpful mediator or coach.

If it is up to Earth to make the first move, it is difficult to argue for anything other than a Tit for Tat strategy and "doing unto others as they would have done unto themselves" is a promising gambit. If we are insecure, anxious, xenophobic, riddled with fear, interested in domination and prestige or tempted to lie, we should not enter the arena of interstellar relations. If we have confidence in ourselves and enough trust to extend the benefit of the doubt, we should start with the hand of friendship, communicate or at least signal our intentions, try to build a good reputation, and reciprocate both cooperation and defection. If we blunder, we can take heart that reciprocity has many variations. Even small concessions can de-escalate international conflicts, and eventually transform lose-lose into neutral then win-win situations (Osgood 1962).

8 Discussion

When it comes to culture, society, and behavior, it is possible to find examples that support almost any point of view. To assess our current status and identify trends over time we have to look beyond the blare of current news, images that flit across our television screens, and engaging stories. When we look at data rather than listen to anecdote we find that that growing social units are erasing differences between in-groups and out-groups; that oppressive and warlike authoritarian regimes are giving way to peaceful liberal democracies; and that on time scales ranging from millennia to decades, violence and war are on the decline. These trends away from predation and conflict and towards cooperation and altruism may accelerate, decelerate, plateau, or reverse. Still, the overall pattern of results is consistent with optimistic views of our future. At the very least these findings provide scant support for paranoid models of extraterrestrial civilizations.

As Kathryn Denning (2009; 2011) and Seth Shostak (2009) point out, longevity is important to SETI because the only civilizations that we can detect are high-tech societies that overlap ours in time. Estimates based on the age of the universe, birth and deaths of stars and similar factors suggest that the average extraterrestrial civilization could be millions or billions of years older than our own (Shostak 2009). One of many possible explanations for continuing silence following fifty years of listening is that technologically adept civilizations simply do not last that long (Shostak 2009). Of course there are many other reasons that a civilization could grind to a halt: a gamma-ray burst that sterilizes a huge chunk of the galaxy, a supernova that extinguishes life within a few star systems, a giant asteroid that extinguishes a one-planet civilization, and fantastic science projects that go dreadfully wrong (Darling and Schulz-Makuch 2012). One threat that has received considerable attention is all-out nuclear war (Shostak 2009). This is understandable given that atomic weapons and SETI search technology became available at the same time.

The fear is that societies gain dangerous new technology before they have the political stability, emotional control and negotiations skills needed to keep this technology under control (Shostak 2009; Ellis 2013). It is common for technology to outrun the knowledge and skills of its human users, a phenomenon we call culture lag. In some civilizations, culture is unlikely to catch up with destructive technology before that civilization self-destructs. In many discussions of SETI, "technological adolescence" refers to that period between the availability of potentially devastating technology and the cultural adaptations to use it wisely (Shostak 2009). If a civilization can work its way through its period of technological adolescence, then it may survive for millions or billions of years, particularly if through spacefaring it becomes widely dispersed and cannot disappear in a single-planet catastrophe. The studies discussed in this chapter give hope that we are working our way through our technological adolescence.

Do the data presented in this chapter tell us anything about extraterrestrial civilizations? Perhaps even as there are fundamental laws of physics and biology, there are fundamental laws of society and behavior that are likely to apply on many worlds (Harrison 1993; 1997; 2010). Democracies, inclusiveness, international alliances, and the Golden Rule are likely to be widespread because they work. They produce better results than authoritarian, egotistical, war-mongering and unabashedly self-serving regimes. This does not suggest that every society out there is friendly, or that we should immediately embrace every civilization that we can find. It does mean that based on our experience on Earth, the odds may be stacked against war and in favor of peace. SETI is like walking down a dark street and night not knowing what lurks in the shadow. Whereas it is possible to encounter an occasional ruffian, the neighborhood itself is pretty good. The paranoid model has not deterred the search, nor has it prevented people from using high-powered transmitters to send messages, sometimes frivolous ones, on interstellar journeys.

What if we find a civilization that is post-biological; that is, dominated by artificial intelligence and robots? Could electric brains and exotic gizmos qualify as empathic moral agents, ready to extend the hook of friendship? A recent article in *The Economist* discusses efforts to imbue robots with ethical principles and prepare them to resolve ethical dilemmas (Anonymous 2012). For example, should a driverless car swerve to avoid a pedestrian even though it poses a risk to the passengers? Should a robot inform a person that he or she is dying, or reveal the true nature of a situation even though this might trigger panic? Machine ethics brings together engineers, ethicists, lawyers, policy makers (and we should hope many others) to develop ethical systems that, embedded into robots, load the dice in favor of judgments that seem right to most people. Presumably, biological organisms come first and develop the machines that eventually supplant them. There's time yet, but already we are starting to develop artificial morality.

9 Plan of This Book

In 2010, Stephen Hawking warned of dire consequences if our radio transmissions caught the attention of an extraterrestrial species. In his dystopian view, this would be tantamount to inviting an alien invasion. Hawking's claim is subjected to close scrutiny in "Extraterrestrial Intelligence: Friends, Foes, or Just Curious?," a broad overview written by George Michael, and in "Harmful ETI Hypothesis Denied: Visiting ETIs Likely Altruists," authored by Harold Geller. Their criticism of Hawking is incisive. Alien civilizations may be very old, and their technology far advanced in comparison to our own. However, we need to question the widespread claim that contact between a more technologically-advanced civilization and a less technologically-advanced civilizations is catastrophic for the latter. The analogues presented in support of this type of claim (for example, Columbus encountering Native Americans) tend to be based on widely-accepted views of history (as compared to what actually happened) and in any case tend to break down when viewed in the context of human-extraterrestrial contact. Each author points out that despite the ease with which fictional aliens flit around the universe, actual interstellar travel is extremely difficult. Furthermore, it is difficult to find compelling motives for an extraterrestrial civilization to covet Earth's resources. Geller points out biological differences are likely to disqualify humans as potential foodstuff, mating partners, or as likely to succumb to extraterrestrial disease.

"Eliciting Altruism While Avoiding Xenophobia: A Thought Experiment" and "Predator–Prey Models and Contact Considerations" continue to explore whether extraterrestrial civilizations are likely to be predatory or friendly. In the first of these, Jerome H. Barkow conducts a thought experiment on extraterrestrial psychology. What is the psychology of creatures that can produce high technology that we can recognize? He points out that they must be a social species, capable of cooperating and pooling knowledge. There are two types of altruism, nepotism and reciprocal. It is reciprocal altruism that is most likely to fit with our sense of justice and our sense of morality, and it is reciprocal altruism that we hope to cultivate. An extraterrestrial species may have had to compete with another extraterrestrial species, and this could lead to levels of distrust and hostility towards humans. If we choose to enter into communication we should avoid sending information that could trigger a xenophobic response such as pictures of ourselves or descriptions of our political beliefs. Instead, we should convey that we are generous and good candidates for collaboration. In "Predator–Prey Models and Contact Considerations," Douglas Raybeck, while guardedly optimistic, weighs in on the other side. He observes that although pronoid visions are alluring, we should always remember that predation is a path to intelligence. Predators must find ways to locate and trap their prey, which are oftentimes environmentally challenged. Interspecies behavior tends to be competitive, aggressive, and self-serving. Predation is only one path to intelligence—for example, tools set in motion pressures for better tool users and the complexities of social life requires cooperation and cohesion—but it is an unmistakable and important path. The potential benefits of contact may be too alluring to ignore, but we should not forget that their intelligence may have been driven by predation.

In "Altruism Toward Non-Humans: Lessons for Interstellar Communication," Abhik Gupta points out that altruism is widely dispersed across Earth, and that human caring extends not just to other humans and animals but also to nonliving forms of nature, including mountains, oceans and rivers, and forests. Nature, the religion of our ancestors, lives on in human minds. Since extraterrestrials are unrelated to humans, we cannot expect kin relationships to govern their relations with us, and considerations of distance and time would make it difficult to develop reciprocity. However our love of life or biophilia, which leads us to treat animals decently and take care of our environment, shows that we can be caring to other species. The theme of caring is carried forward by Mark Lupisella in "Caring Capacity and Cosmocultural Evolution: Potential Mechanisms for Advanced Altruism." He discusses three types of altruism: biological, biocultural, and advanced. Lupisella argues that both technology and social norms have reduced the costs of caring, with the result that biological altruism, which is motivated by the selfish gene and results in favored treatment of kin, has been supplemented by biocultural altruism, which encompasses more people. The product of cosmocultural evolution, advanced altruism, involves a broad-based respect for the universe and all beings within it. Based on principles, rather than circumstance, advanced altruism constitutes a framework of avoiding harm and viewing others as entitled to respect and rights. Then, in "The Precautionary Principle: Egoism, Altruism, and the Active SETI Debate," Adam Korbitz discusses the Precautionary Principle. Applied to SETI, this principle suggests that since we really know nothing of other civilizations that might be "out there" the wisest strategy is to protect ourselves by remaining inconspicuous. We should avoid transmissions to distant stars, and, if we should receive a message, we should not respond. Korbitz questions the value of the Precautionary Principle, at least in its strong form. The Precautionary Principle is flawed: it is applied inconsistently in different societies at different times and in different contexts, tends to draw attention to worst-case scenarios, confuses possibilities and probabilities, and whereas it may reduce one risk (entanglement with a xenophobic civilization) it may increase another (lost opportunity for friendly interaction). The mere possibility that something could go wrong should not hamper our scientific, technological, or intellectual activity.

How may we infer or demonstrate altruism? In "The Accidental Altruist: Inferring Altruism from an Extraterrestrial Signal," Mark Langston asks if we could infer the sender's intention from a message. This is a multifaceted and knotty problem, especially given that the message may take a long time to decode (if it is understandable at all) and that messages that take years to reach their destination do not promote lively discussion. Further, the beacon that attracts our attention and the contents of the message are not necessarily congruent. For example, a warm and welcoming greeting could be followed by a "Trojan Horse" recipe that, if followed, would lead to our self-destruction. In "Interstellar Intersubjectivity: The Significance of Shared Cognition for Communication, Empathy, and Altruism in Space," David Dunér points out that one of the fundamental requirements for communication is each side showing that it is alive, intelligent, self-conscious, and aware of the possibility of other sentient beings in the universe. This will require some degree of intersubjectivity or shared cognition. This will be crucial not only for sharing about the real world out there (objects and events) but perhaps more importantly experience (perceptions, thoughts, and feelings). On Earth, intersubjectivity was essential for developing civilization, culture, and technology, and we expect it to be a pre-requisite to cultures that develop elsewhere. The question is how to build bridges between different mindsets.

In "Other Minds, Empathy, and Interstellar Communication," Tomislav Janović weighs in with a number of assumptions about the mental and social characteristics of aliens. He proposes that we need to separate the physically possible from the biologically probable. We can study minds that we know of, and consider the extent to which findings can be generalized to other species, and electronic minds. Conscious awareness and empathy are essential. The ability to recognize and express emotions and intentions—along with the capacity for altruistic behaviors—are prerequisites for the development of cooperation.

Cross-cultural interactions provide many examples that may help us communicate with extraterrestrial intelligence. However, such communication will require transmitting information and affect across species. In "Interspecies Altruism: Learning from Species on Earth," Denise Herzing shows how we learn from interspecies communication and altruism on Earth. She points out that there are abundant examples of cross-species altruism on Earth, including examples of one-way (generous or non-reciprocal) altruism. We are capable of understanding other species' distress and have empathy, at least for social animals such as dolphins, elephants, and primates. Although our interpretation of the behavior of other species is anthropocentric, our recognition of this fact helps us think broadly and critically. Herzing covers a range of topics, with an extended discussion of human-dolphin communication and altruism.

The next three chapters raise questions of universal ethics and law. In "Terrestrial and Extraterrestrial Altruism," Holmes Rolston, III points out that because of our deeply-engrained anthropocentric perspective we believe that we will be able to recognize their science. But will we be able to recognize their ethics—their ability to tell the truth, keep promises, and be just? Many animals show helping behavior, but in Rolston's view these are "pre-ethical" because they are not considered praiseworthy or just. Ethics is more than acts that we see as kindly: it involves a larger framework which includes evaluative and emotional components and a conception of self as a moral agent. We may never become free of nature, but we are free to develop ethical frameworks within nature. Ethics is part of a larger, always evolving worldview, but across time and cultures we find the same underlying intent.

Next, in "Kenotic Ethics and SETI: A Present-Day View," George F. R. Ellis advances the thesis that whereas movies often stress unfriendly, hostile and destructive extraterrestrials in reality such civilizations may be few. Over time, morality will win the race between destructive technology and the ability to keep such technology under control. Earlier fears of the other can evolve into an appreciation of their natures and capacities. An ethic that truly serves the interests of tolerance and peace must include forgiveness and reconciliation, abandoning the need for revenge in favor of the greater good. Ellis suggests that everywhere that ethics evolves it would proceed along similar lines. Extraterrestrials could have religious views that are very different from ours but ethical views that are recognizable.

In "Altruism, Metalaw, and Celegistics: An Extraterrestrial Perspective on Universal Law-Making," his second contribution to this volume, Adam Korbitz focuses on extraterrestrial altruism as a philosophical topic related to human conceptions of law and morality, including the work of Ernst Fasan, Robert A. Freitas and G. Harry Stine. He addresses the history of Metalaw—the fundamental legal precepts of universal applicability to all intelligences—and uses celegistics as the reference term for theories of extraterrestrial legal systems. Metalaw is intended to help us put our best foot forward, but we cannot be assured that all extraterrestrial civilizations would do the same. Korbitz explores the possibility that rather than being human creations, laws are imminent in nature and waiting to be discovered, which suggests universality. We need to be aware of what we don't know, and move beyond the Golden Rule, treating others not as we would hope to be treated but as they hope to be treated.

How do we represent altruism to distant, unknown audiences? This is the focal point for the last group of chapters in this volume. In "A Logic-Based Approach to Characterizing Altruism in Interstellar Messages," Alexander Ollongren reviews Lingua Cosmica, Freudenthal's early and comprehensive attempt to develop an interstellar language based on math and logic. He then proposes a New Lingua Cosmica, based on computer-implementable intuitionistic (constructive) logic. Although New Lingua Cosmica may take some time to learn, it is conceptually simple and should facilitate communicating ideas of moral behavior and altruism. Then, in "Equity and Democracy: Seeking the Common Good as a Common Ground for Interstellar Communication," Yvan Dutil tackles the problem of very different civilizations finding common grounds for cooperation. He finds promise in equity theory, with its goal of fair distribution of resources, and social-choice theory, which involves resolving conflicting individual choices into a common good and represents a direct path to democracy. Both equity theory and social choice theory are good candidates for interstellar messages because they can be expressed mathematically. Dutil discusses both sharing rules and voting rules. Good sharing rules are equitable, efficient, and do not create envy at the end of the process.

In "Patterns of Extraterrestrial Culture," William Sims Bainbridge uses examples from science fiction (especially from the virtual world of *Star Wars: The Old Republic*) to illustrate human representations of extraterrestrial personality and culture. These representations, many traceable to a small number of highly-respected science fiction authors, are based on known human and cultural characteristics that are highly exaggerated, combined into unusual patterns or considered pathological by us but make sense in light of radically different extraterrestrial physical and social environments. Bainbridge, who has created his own avatars and participated in virtual worlds, suggests that science fiction may lead to principles that apply in actual extraterrestrial cultures.

To conclude this book, Alfred Kracher describes how images may be useful for communicating altruism. The preparation of these images should involve cross-cultural collaboration of representatives of science, the arts, and humanities. His discussion is informed by studies of religious icons on Earth exemplified by Madonna images that portray benevolent adult-child interactions. His emphasis is not on what images to send, but on how to think about the kinds of images that are likely to promote mutual understanding. Images are inherently ambiguous, but sending multiple and diverse images, accompanied by text, may speed comprehension. Kracher views altruism as a virtue that provides moral guidance and is in some sense universal, but civilizations that more closely resemble our own should more easily grasp our message.

We do not know if extraterrestrial civilizations exist and, if they exist in large numbers, the proportions that would become friends, foes, or disinterested bystanders. The possibility of contact excites the imagination and generates spirited discussions across a range of interdisciplinary topics, in this book the complex topics of extraterrestrial ethics and altruism. The chapters in this book reflect a mixture of scholarship, imagination, and educated speculation. If we never discover an extraterrestrial civilization are we wasting our time? The contributors to this volume do not seem to think so. As we ponder interstellar altruism we learn more about ourselves, develop new understandings of our place in the universe, and prepare ourselves for possible discoveries to come.

References

Anonymous. 2012. "Morals and the Machine." *The Economist*, 2 June, p. 18.

Axelrod, Robert. 1984. *The Evolution of Cooperation*. New York: Basic Books.

Barkow, Jerome H. 2013. "Eliciting Altruism While Avoiding Xenophobia: A Thought Experiment." In *Extraterrestrial Altruism: Evolution and Ethics in the Cosmos*, edited by Douglas A. Vakoch, 37–48. Heidelberg: Springer.

Chaisson, Eric. 2001. *Cosmic Evolution: The Rise of Complexity in Nature*. Cambridge, MA: Harvard University Press.

Christian, David. 2004. *Maps of Time: An Introduction to Big History*. Berkeley: University of California Press.

Cusack, Thomas R., and Richard A. Stoll. 1994. "Collective Security and State Survival in the Interstate System." *International Studies Quarterly* 38:33–59.

Dafoe, Allan. 2011. "Statistical Critiques of the Democratic Peace: Caveat Emptor." *American Journal of Political Science* 55:247–262.

Darling, David, and Dirk Schulze-Makuch. 2012. *Megacatastrophes!: Nine Strange Ways the Earth Could End*. London: Oneworld Books.

Denning, Kathryn. 2009. "The Evolution of Culture." In *Cosmos and Culture: Cultural Evolution in a Cosmic Context*, edited by Steven J. Dick and Mark Lupisella, 63–124. Washington, DC: National Aeronautics and Space Administration.

Denning, Kathryn. 2011. "'L' on Earth." In *Civilizations Beyond Earth: Extraterrestrial Life and Society*, edited by Douglas A. Vakoch and Albert A. Harrison, 74–86. New York: Berghahn Books.

Dick, Steven J. 2009. "Cosmic Evolution: History, Culture, and Human Destiny," In *Cosmos and Culture: Cultural Evolution in a Cosmic Context,* edited by Steven J. Dick and Mark Lupisella, 25–62. Washington, DC: National Aeronautics and Space Administration.

Dunér, David. 2013. "Interstellar Intersubjectivity: The Significance of Shared Cognition for Communication, Empathy, and Altruism in Space." In *Extraterrestrial Altruism: Evolution and Ethics in the Cosmos*, edited by Douglas A. Vakoch, 141–167. Heidelberg: Springer.

Dutil, Yvan. 2013. "Equity and Democracy: Seeking the Common Good as a Common Ground for Interstellar Communication." In *Extraterrestrial Altruism: Evolution and Ethics in the Cosmos*, edited by Douglas A. Vakoch, 261–273. Heidelberg: Springer.

Ellis, George F. R. 2013. "Kenotic Ethics and SETI: A Present-Day View." In *Extraterrestrial Altruism: Evolution and Ethics in the Cosmos*, edited by Douglas A. Vakoch, 223–229. Heidelberg: Springer.

Ferris, Timothy. 2010. *The Science of Liberty: Democracy, Reason, and the Laws of Nature*. New York: Harper Perennial.

Finney, Ben R. 1990. "The Impact of Contact." *Acta Astronautica* 21:117–121.

Goldstein, Joshua S. 2011. *Winning the War on War: The Decline of Armed Conflict World Wide*. New York: Dutton.

Harrison, Albert A. 1993. "Thinking Intelligently about Extraterrestrial Intelligence: An Application of Living Systems Theory." *Behavioral Science* 38:189–217.

Harrison, Albert A. 1997. *After Contact: The Human Response to Extraterrestrial Life*. New York: Plenum.

Harrison, Albert A. 2000. "The Relative Stability of Belligerent and Peaceful Societies: Implications for SETI." *Acta Astronautica* 46:707–712.

Harrison, Albert A. 2007. *Starstruck: Cosmic Visions in Science, Religion, and Folklore*. New York: Berghahn Books.

Harrison, Albert A. 2010. "The ETI Myth: Idolatrous Fantasy or Plausible Inference?" *Theology and Science* 8(1):51–67.

Keegan, John. 1994. *A History of Warfare*. New York: Vintage Books.

Korbitz, Adam. 2013. "Altruism, Metalaw, and Celegistics: An Extraterrestrial Perspective on Universal Law-Making." In *Extraterrestrial Altruism: Evolution and Ethics in the Cosmos*, edited by Douglas A. Vakoch, 231–247. Heidelberg: Springer.

Kracher, Alfred. 2013. "Evolutionary Perspectives on Interstellar Communication: Images of Altruism." In *Extraterrestrial Altruism: Evolution and Ethics in the Cosmos*, edited by Douglas A. Vakoch, 295–308. Heidelberg: Springer.

Lupisella, Mark L. 2013. "Caring Capacity and Cosmocultural Evolution: Potential Mechanisms for Advanced Altruism." In *Extraterrestrial Altruism: Evolution and Ethics in the Cosmos*, edited by Douglas A. Vakoch, 93–109. Heidelberg: Springer.

Michalak, Stanley. 2004. "Post-Democratic Cosmopolitans: The Second Wave of Liberal Internationalism." *Orbis*, Fall:593–607.

Michaud, Michael A. G. 2007. *Contact with Alien Civilizations: Our Hopes and Fears about Encountering Extraterrestrials*. New York: Copernicus Books.

Miller, James Grier. 1977. *Living Systems*. New York: McGraw-Hill.

Modelski, George, and Gardner Perry, III. 2002. "Democratization in Long Perspective Revisited." *Technological Forecasting and Social Change* 69:359–376.

Mueller John. 1988. *Retreat from Doomsday: The Obsolescence of Modern War*. New York: Basic Books.

Osgood Charles E. 1962. *An Alternative to War or Surrender*. Chicago: University of Chicago Press.

Peters, Ted. 2009. "Astrotheology and the ETI Myth." *Theology and Science* 6(4):3–29.

Peters, Ted. 2011. *UFO's—God's Chariots?* Everett, WA: Wittenberg Workshop Publication.

Pinker, Steven. 2011. *The Better Angels of Our Nature: Why Violence Has Declined*. New York: Viking.

Putnam, Robert D. 2000. *Bowling Alone: The Collapse and Revival of American Society*. New York: Simon and Schuster.

Putnam, Robert D., and David E. Campbell. 2010. *American Grace: How Religion Divides and Unites Us*. New York: Simon and Schuster.

Ray, James Lee. 1988. "Does Democracy Cause Peace?" *Annual Review of Political Science* 1:27–46.

Russett, Bruce. 1993. *Grasping the Democratic Peace: Principles for a Post-Cold War World*. Princeton, NJ: Princeton University Press.

Shermer, Michael. 2004. *The Science of Good and Evil*. New York: Times Books.

Shostak, Seth. 2009. "The Value of 'L' and the Cosmic Bottleneck." In *Cosmos and Culture: Cultural Evolution in a Cosmic Context,* edited by Steven J. Dick and Mark Lupisella, 399–414. Washington, DC: National Aeronautics and Space Administration.

Stoll, Richard J. 1998. "Nice States Finish… Pretty Well: Collective Security Behavior in Militarized Interstate Disputes, 1816–1992." *International Interactions: Empirical and Theoretical Research in Intercultural Relations* 24(3):287–313.

Turner, Derek D. 2005. "Altruism: Is It Still an Anomaly?" *Trends in Cognitive Science* 9(7):317–318.

Vakoch, Douglas A. 2009. "Encoding Our Origins: Communicating the Evolutionary Epic in Interstellar Messages." In *Cosmos and Culture: Cultural Evolution in a Cosmic Context*, edited by Steven J. Dick and Mark Lupisella, 415–440. Washington, DC: National Aeronautics and Space Administration.

Wendt, Alexander. 2003. "Why a World State is Inevitable." *European Journal of International Relations* 9(4):491–532.

Wilcox, Clyde. 1991. "Governing Galactic Civilization: Locke and Hobbes in Outer Space." *Extrapolation* 32(2):111–124.

Part II
The Evolution of Altruism: Is Transmitting Dangerous?

Extraterrestrial Intelligence: Friends, Foes, or Just Curious?

George Michael

Abstract For over 50 years, scientists involved in the Search for Extraterrestrial Intelligence (SETI) have searched for evidence of alien civilizations, and several messages have been transmitted into space with the intention of communicating with intelligent extraterrestrial beings. In April 2010, the esteemed British astrophysicist Stephen Hawking warned that such attempts could be potentially dangerous. He based his conclusion on historical analogies on Earth, such as the conflict between Native Americans and European settlers, and he reasoned that when a more technologically advanced civilization encountered a less advanced one, the results have often been catastrophic for the weaker party. Although this argument has intuitive appeal, upon closer examination, it appears misguided. Even for technologically-advanced civilizations, interstellar voyages would probably be justified only for major purposes, and plundering the Earth for its resources would be neither practical nor desirable.

Keywords Alien invasion · Analogy · Christopher Columbus · Extraterrestrial intelligence · Extraterrestrial motivations · Fermi Paradox · Interstellar travel · Kardashev scale · SETI · Stephen Hawking · Wormhole

---

This chapter is an adaptation of Michael, George. 2011. "Extraterrestrial Aliens: Friends, Foes, or Just Curious?" *Skeptic* 16(3):46–53 and is published here with permission of Michael Shermer, publisher and editor-in-chief of *Skeptic*.

G. Michael (✉)
Criminal Justice, Westfield State University, 333 Western Avenue, Westfield, MA 01085-2560, USA
e-mail: gmichael@westfield.ma.edu

D. A. Vakoch (ed.), *Extraterrestrial Altruism*, The Frontiers Collection, DOI: 10.1007/978-3-642-37750-1_2, © Springer-Verlag Berlin Heidelberg 2014

1 Alone in a Crowded Universe

Popular culture continues to reflect our fascination with extraterrestrial aliens. In May 2012 a new science fiction film—*Battleship*—was released which depicted a confrontation between the U.S. Navy and a fleet of extraterrestrial invaders in the Pacific Ocean. The story begins in 2005, when NASA scientists discover an exoplanet (a planet outside of our solar system) with characteristics similar to Earth. Assuming that there could be intelligent life on the planet, NASA transmits a powerful signal from communications array in Hawaii, which is boosted by a satellite in orbit. Remarkably, the alien recipients of the message are able to mount an interstellar invasion a mere seven years later.

The above scenario, though implausible because it ignores the cosmic speed limit (i.e., the speed of light, which according to Einstein's special theory of relativity, nothing can exceed), nevertheless illustrates the potential perils of attempting to communicate with extraterrestrial aliens.

In 1974, Carl Sagan and Frank Drake created the Arecibo Message, which was beamed into space aimed at the globular star cluster M13 some 25,000 light years away. The first intentional message sent into space, it consisted of 1,679 binary digits that collectively formed an image of the Earth. The message was controversial, as some commentators feared that it was potentially dangerous to announce our position in the cosmos to extraterrestrial aliens who might harbor malicious intentions toward us.

In April 2010, the Discovery Channel broadcasted a documentary in which Stephen Hawking speculated on the existence of extraterrestrial life. In his mind, the multitude of billions of stars and galaxies suggests that life in other solar systems is almost a certainty. Hawking pondered an important question: Could visitors from extraterrestrial civilizations pose a threat to Earth? In the documentary, an armada of massive space ships roams the galaxy as interstellar nomads who, having exhausted all the resources on their home planet, search for other planets to conquer and colonize. He concluded that making contact with aliens is "a little too risky." Moreover, drawing upon the experience on Earth, he mused that "if aliens ever visit us, I think the outcome would be much as when Christopher Columbus first landed in America, which didn't turn out very well for the Native Americans" (Leake 2010).

Hawking is not the first scientist to warn of the potential hazards of alien contact. In his book, *The Third Chimpanzee*, the noted geographer and evolutionary theorist Jared Diamond (2006) offered an essay on the perils of attempting to contact alien civilizations, observing that whenever a more advanced civilization encountered a less advanced one, or species that have evolved from different ecosystems, came into contact with each other, the results have often been catastrophic for the weaker party, including slavery, colonialism or extinction (Brin 2006). He once described the 1974 Arecibo message as suicidal folly, comparing it to the Incan emperor describing the wealth of his kingdom to the gold-crazed Spaniards. Sending out radio signals, he concluded, was "naïve, even dangerous"

(Michaud 2007, 246). Even the late astronomer, Carl Sagan, who was generally sanguine about the prospects of interstellar comity, once counseled that a relatively young civilization such as Earth's should listen quietly, "before shouting into an unknown jungle that we do not understand" (Brin 2006).

While this argument seems reasonable at first blush, upon further reflection I believe such fears are greatly overstated for a straightforward reason related to the physics of interstellar travel.

2 The Methods and Feasibility of Interstellar Space Travel

When considering interstellar travel, two important factors to keep in mind are the vast distances between solar systems and the enormous energy requirements that would be necessary to fuel space vessels that could traverse such distances. The closest solar system, Proxima Centauri, is about 4.2 light years away from Earth. To put that in perspective, if a vessel from Earth could somehow travel at the speed of light, it would take over four years to reach that solar system. According to Einstein's Special Theory of Relativity, interstellar travel is seemingly impossible because no vessel could travel faster than the speed of light, thus it would take centuries or millennia to travel the distances between solar systems. For example, at the speed of the Voyager spacecraft, it would take over 70,000 years to get to Proxima Centauri. However, Einstein's General Theory of Relativity indicates that faster than speed of light travel is theoretically possible by using large amounts of energy to continuously stretch space and time. Theoretically, empty space could warp space faster than light. Inasmuch as only empty space is contracting or expanding, one could exceed the speed of light by this fashion; however, this approach would require enormous amounts of energy and would be feasible only for a very advanced civilization (Kaku 2002).

Another way to avoid the speed of light limit is through the use of a wormhole through which an extraterrestrial civilization could create a shortcut across space and time. A functioning wormhole could serve as a bridge between two different regions of the universe. One could enter the wormhole at one end and emerge out of it moments later in a place thousands of light years from the starting point (Webb 2000; Sagan 1985). Carl Sagan depicted this device as a means of interstellar travel in his novel *Contact* (Davies 2010). The Russian physicist, Sergei Krasnikov once demonstrated that a certain class of wormholes could be created using positive mass-energy matter (Webb 2000). The energy requirements for such a system, though, would be, in a word, massive, and well beyond the scope of our capabilities. Assumedly, practical interstellar space travel could be viable only for very advanced civilizations that have learned how to harness enormous amounts of energy.

3 Kardashev's Civilization Scale

In 1964, the Russian astrophysicist and director of the Russian Space Research Institute, Nikolai Kardashev, first proposed a classification of alien civilization based on their methods of energy extraction. His scale has three categories. A Type I civilization can harness all available energy sources on its planet. A Type II civilization utilizes all the energy from its star. A Type III civilization is able to harness the power not of only its own star, but other stars in its galaxy (Kardashev 1985; Kaku 1994). Such a civilization, as the futurist and string theorist Michio Kaku once mused, would be immortal.

According to Kardashev (1997), energy consumption determines civilizational progress and could one day enable interstellar travel. As Michio Kaku (1997, 18) explained, Kardashev's system of classification is reasonable because it relies upon available supplies of energy. "Any advanced civilization in space will eventually find three sources of energy at their disposal: their planet, their star, and their galaxy. There is no other choice."

How would a civilization advance on this scale? A civilization might advance to Type I status by applying fusion power or by producing antimatter to be used as an energy source. Alternatively, one might be able to harness zero-point energy (Webb 2000). To advance beyond Type I status, Freeman Dyson (1960) theorized that a hypothetical megastructure could be employed to encompass a star as a system of orbiting solar power satellites to capture most of a star's energy output. Constructing such a device—a Dyson sphere—would be a gargantuan engineering undertaking, but theoretically possible. Dyson conjectured that an alien civilization could tear apart planets and asteroids to use as the material to build the necessary structures (Davies 2010).

A Type III civilization might also be able to harness energy by building a device around a spinning black hole, which would release far more energy than a star can through nuclear fusion. By exploiting the law of conservation of angular momentum, the prodigious power of its rotation could be used to extract energy (Davies 2010). Furthermore, an advanced extraterrestrial civilization might be able to tap into the energy released from massive black holes that reside at the center of some galaxies.

Dyson observed that human civilizations have a tendency to constantly increase their energy consumption. Based on this reasoning, eventually civilizations would be compelled to search for more ways to harness energy (Dyson 1960). Currently, our civilization occupies a position that has not quite attained Type 1 status. In 1973, Carl Sagan computed it as 0.7 on the Kardashev scale (Sagan 1973). In order for an extraterrestrial civilization to conduct interstellar travel, most probably it would have to obtain at least a Type II status.

4 Alien Motivations

Why would aliens visit the Earth? As Hawking mused, alien civilization might exhaust their home planet and search for other planets from which to extract resources. Would an alien civilization make the long trip to Earth to plunder our resources? Conceivably, one could envisage a resource war in the sense of trying to grab up all the planets in a galaxy, a recurrent theme in science fiction (the blockbuster 1996 film, *Independence Day*, for example).

Upon closer examination, though, this scenario seems unlikely. It stretches the bounds of credulity to believe that an advanced alien civilization would come all the way to Earth for energy products. It's a safe bet that any civilization capable of traveling such long distances either by way of spacecraft and/or wormholes would not be using oil and other pre-Type I civilization energy sources. What is more, the transportation costs to bring the energy products back to the mother planet would not be economical. Possibly, fusion reactors, or even anti-matter reactors, could be used to fuel such space vessels, in which case hydrogen, or some isotope thereof, would be required (Crawford 1990). However, hydrogen is one of the most abundant chemical elements in the universe and would thus not require interstellar travel to obtain. Possibly, an alien civilization might want to extract minerals from other planets. Yet it would not be practical to come all the way to Earth for minerals. After all, they could more than likely be found on planets in much greater quantities in their own solar system or in nearby solar systems. And it would be far more practical to conduct strip mining on planets on which they would not have to deal with restive denizens, such as earthlings. In short, the rarity of advanced life and the tremendous distances between civilizations suggest that there would be plenty of planets and stars for all those that were capable of exploiting such methods of resource extraction.

Still, others point out that the way in which humans treat other animals is not reassuring. Higher-level mammals, such as dolphins and even chimpanzees, are often mistreated for commercial or experimental reasons (Michaud 2007). Moreover, evolutionary biologists point out that altruism occurs with decreasing intensity as individuals grow more distantly related (Michaud 2007). Perhaps, aliens would crave us as food. There is a good possibility that alien visitors would be descended from carnivores (Kaku 2008). Thus some observers, such as the biologist Michael Archer, fear that aliens would more likely be predatory than benevolent (Michaud 2007). The food scenario has some intuitive appeal. After all, many humans consume animal flesh. In a classic episode of the science fiction program *The Twilight Zone*—"To Serve Man"—seemingly magnanimous aliens called the Kanamits come to Earth and share their advanced technology, which solves the planet's greatest woes, including eradicating hunger, disease, and the need for warfare. Humans are encouraged to take trips to the Kanamits' home planet, which is supposedly a paradise. However, a female code breaker eventually deciphers that a Kanamit book ironically titled *To Serve Man* is actually a cookbook on how to serve humans as meals.

When examined more closely, however, this scenario lacks credibility as well. Presumably a Type II or Type III civilization capable of interstellar travel would also have mastered agricultural engineering and would thus have solved problems in farming and livestock a long time ago. Even on Earth, which has not yet attained Type I civilizational status, there have been marked improvements in nutrition on a global scale. A portion of the human population still suffers from hunger, but there has been a steady reduction in that segment over the years. In fact, more people are obese today than undernourished ("Overweight…" 2006). Although it is certainly reasonable to assume that aliens would consume animal flesh, creating a wormhole just to go to Kentucky Fried Chicken seems rather far-fetched.

Might aliens use humans to help biologically propagate their species? Some scientists, including the co-discoverer of the structure of DNA Francis Crick, have speculated that aliens may have actually "seeded" planets in the universe by sending microbe-laden probes out into space in a process called "directed panspermia" as a way in which to spread the building blocks of life (Webb 2000). In the film *Mars Needs Women*, Martians suffer a genetic deficiency that produces only male offspring. However, a civilization that could travel great distances would almost certainly have mastered bioengineering as well and could ameliorate such a deficiency. Moreover, inasmuch as alien life-forms would probably be based on entirely different DNA and protein molecules, they would probably have no interest in either eating or mating with us (Kaku 2008). This scenario is humorous, but aliens would probably not be so desperate to travel light years to meet species of the opposite sex. That's an awful long way to travel to get a date.

Perhaps an advanced martial extraterrestrial civilization might use Earth as a type of training ground for their warriors. This was the plot of the *Predator* film series, in which alien creatures visit various hot spots and war zones on Earth to hone their martial skills. This could be exciting for members of some alien civilizations. After all, at one time, big game hunting was the province of distinguished gentlemen in the West. The scenario has some plausibility until one considers the frivolousness of such a trip. Would aliens be willing to spend a thousand years in suspended animation for such an excursion? Would aliens create a wormhole just to go big game hunting?

More seriously, territorial motives could inform interstellar colonization. Based on assumptions of terrestrial life, we would assume that life has a natural tendency to expand. Should extraterrestrial life be any different (Webb 2000)? In H.G. Wells' classic novel, *The War of the Worlds*, Martians invade Earth to take over the planet. The novel was written at the height of the British Empire when power was often measured in land (Davies 2010). Eventually, an advanced extraterrestrial civilization would be forced to embark on interstellar travel if it wanted to survive insofar as its sun would have a limited life (Webb 2000). In such a scenario, an extraterrestrial civilization might want to colonize so-called Goldilocks planets, that is, those that that falls within a star's habitable zone and would be roughly in size of the planet Earth so that it could have an atmosphere. Such a planet would avoid overly hot or cold temperatures so that it could retain liquid water on its surface, assumedly a sine qua non for the emergence of life. The so-called "Rare

Earth hypothesis" posits that the existence of such planets is extremely uncommon insofar as a number of unlikely events and conditions would be necessary in order to give rise to such planets. Thus such prime real estate, so the argument goes, would be highly coveted by aliens, not unlike the American continents were for European settlers centuries ago.

The Columbus analogy, though, would probably be inapplicable for alien encounters. Judged by contemporary standards, the Spanish conquistadors were not much more technologically superior to their Native American hosts. The Spaniards coveted the gold of the latter and eventually their territory, both of which were limited commodities that could only be obtained on Earth for both cultures. Moreover, it was foreign diseases for which Native Americans had no immunity which decimated that population, rather than a systematic plan of conquest and genocide. Rather than travel to Earth and subdue humans, it would probably be more feasible for a Type III civilization to artificially create the conditions supportable of life on a much closer planet. This was the scenario of the 1990 film *Total Recall*, starring Arnold Schwarzenegger, in which an alien artifact—a terraforming machine—has the ability to create an oxygen-bearing atmosphere on Mars. Conceivably, a Type III civilization might even be able to reconfigure a planetary system so that more planets orbit inside the Goldilocks zone (Davies 2010). Perhaps out of a sense of magnanimity, an alien civilization would leave our sun and solar system alone and choose to colonize other solar systems. In that sense, settling other planets would not really be imperialism in the classic sense of the term because, as Frank Tipler and John Barrow opined, the planets would just be "dead rocks and gas" (Michaud 2007, 312–312). A technologically sophisticated extraterrestrial civilization may also be accompanied with an advanced ethical development that values other life forms and decide to leave them unmolested (Ghirardi 2010). Finally, if a Type III civilization sought more stars from which to extract energy, it could find a great abundance of more desirable and larger stars instead of our Sun, which is relatively small (Geller 2010).

Some have argued that superior civilizations would seek out and destroy less-advanced civilizations for their own self-defense. The analogue of the Cold War is instructive, as neither superpower was principally concerned about territory, but rather feared the prospect of a preemptive nuclear strike. Preemptive galactic warfare, the cosmologist Edward Harrison argued, would be prudent insofar as a species that has overcome its own self-destructive qualities may view other species bent on galactic expansion as a sort of virus (Archer 1989). Ironically, the 1959 science fiction film *Plan 9 from Outer Space*—often cited as the worst movie ever made—contained such a plot that logically explained hostile alien intentions. In the film, extraterrestrial beings seek to stop humans from developing a doomsday weapon that could destroy the universe. One is reminded of the first test of the atomic bomb at the Trinity site at Los Alamos in New Mexico on July 16, 1945. The Manhattan Project physicist Enrico Fermi offered to take wagers on the probability that the test would destroy either the entire state of New Mexico or wipe out all life on Earth (Rhodes 1986). More recently, some observers of projects such as the Relativistic Heavy Ion Collider (RHIC) on Long Island feared that experiments to create a miniature black hole could devour our planet (Webb 2000). Conceivably, an alien civilization might seek to cripple another civilization before it developed the capability to retaliate or wreak galactic havoc. The Manhattan Project scientist John von Neumann and the electrical engineer Ronald Bracewell conjectured that extraterrestrial civilizations might send self-replicating robotic probes to explore other solar systems. Some scientists feared that such artificially-intelligent automatons might be programmed to destroy other civilizations (Webb 2000). However, as Gerald O'Neil pointed out, the fact that Earth has yet to be attacked suggests that nobody out there is hostile to us (Michaud 2007). Furthermore, optimists contend that the vast distances between solar systems insulate our planet.

Historical analogies, Carl Sagan once argued, were probably inapplicable to alien and human contact as he found it unlikely that we would face "colonial barbarity" from a technologically more advanced alien civilization. Quarrelsome extraterrestrials, he reasoned, would probably not last long in interstellar space because they would be eliminated by a more powerful species. According to this line of reasoning, those civilizations that lived long enough to perform significant colonization of the Galaxy would be those least likely to engage in aggressive galactic imperialism (Michaud 2007). Moreover, extraterrestrials visiting Earth would assuredly be far more technologically advanced than we, so that the former would have nothing to fear from us. And Earth would not likely have anything that they would need (Michaud 2007).

At our current technological development, we would not be a threat to a Type II or Type III civilization. Moreover, even a military invasion could be problematical for an advanced extraterrestrial civilization. If interstellar travel is exceedingly difficult, then interstellar invasion would even be more challenging (Webb 2000). Keep in mind that Operation Overlord, the Allied invasion at Normandy in World War II—a mere 100 miles across the English Channel—required considerable planning and was not launched until the Allies had already attained near complete superiority in the air and in the sea.

5 Fermi's Paradox: Where Are They?

In 1950, while having lunch with his colleagues at Los Alamos, New Mexico and after a brief discussion about the prospects of extraterrestrial life, Enrico Fermi blurted out, "Where is everybody?" As he reasoned, if technologically advanced civilizations were common in the universe, and assuming many of them had preceded Earth by many tens of thousands of years, then it follows that we should have been visited by now (Webb 2000). This is called the Fermi Paradox, for which there are at least 50 different answers (Crawford 2000). Theoretical physicist and futurist Michio Kaku points out that the transition from a Type 0 to Type I civilization carries a strong risk of self-destruction. Such civilizations may face a Malthusian catastrophe that precludes transition to a higher type. Thus, a short-lived civilization would be unlikely to establish contact. Yet, despite the eons necessary for a civilization to attain Type III status, when measured by the estimated life of the galaxy, the time necessary for colonization would be relatively short (Webb 2000). According to some estimates, it would take anywhere from 5 million to 60 million years for an advanced extraterrestrial civilization to colonize a galaxy (Webb 2000).

Alas, so far, the SETI project has not yielded any tangible evidence of alien civilizations (Webb 2000).[1] SETI uses scientific methods to scan the galaxy for electromagnetic transmissions from civilizations on other planets. Supposedly, a Type II or Type III civilization would leave a detectable footprint in the galaxy, such as radio transmissions or infrared emissions. According to the Second Law of Thermodynamics, an advanced civilization would generate large quantities of waste heat that could be detected by our instruments (Kaku 1994). Almost certainly a device such as a Dyson sphere would dramatically alter the light spectrum of the enclosed star, and in doing so create a noticeable infrared glow that could be identified by peering astronomers, even on the far side of the galaxy (Davies 2010). Nevertheless, no telltale signs have yet to be detected. Kardashev (1985) countered that alien civilizations have not been found because they have not been properly searched for.

Although we have not detected aliens, they may have detected us. Why then don't they make an effort to communicate? Perhaps they are not interested in us. Michio Kaku (1994) uses the analogy of human contact with ants. When we come upon an anthill, he explains, we do not request to see their leader or bring trinkets to them and offer unparalleled prosperity through the fruits of our technology. Because of astronomical time scales, a civilization capable of visiting Earth could be many millions of years ahead of us and thus might find us uninteresting. Moreover, it would be unlikely that any such advanced civilization would find any resources on Earth that could not be found in numerous other star systems closer to their civilization. As Kaku (2008, 17) points out, the main danger ants would face is not that humans want to invade them or eradicate them. Rather, we might simply "pave them over because they are in the way." In this scenario, the danger would be if Earth got in the way of the aliens' highway.

Others, however, are more sanguine. An advanced civilization would probably be interested in encountering another civilization. Kardashev (1997) speculated that alien civilizations might find it advantageous to unite with others in order to reduce the time delay for communications and other types of activities. The SETI astronomer Jill Tarter (2010) speculated that "rather than exploiting us, they might value and support the natural biodiversity of the galaxy."

---

[1] That is with the possible exception of the "Wow signal," which after numerous attempts to rediscover, has not been found again. On August 15, 1977, Jerry Ehman, a SETI volunteer at Ohio State University, observed a startling strong signal received by telescope. After circling the indication on a printout, he scribbled "Wow!" in the margin. Some observers consider the message to be the most likely candidate from an extraterrestrial source ever discovered. However, subsequent searches have failed to detect it again. The powerful narrowband spike most likely emanated from a man-made satellite.

Why would aliens visit us? In a word, curiosity. As Frank Drake opined, "Many human societies developed science independently through a combination of curiosity and trying to create a better life, and I think those same motivations would exist in other [alien] creatures" (Kaku 1994, 283). Presumably, members of an extraterrestrial civilization capable of reaching Earth would be highly curious—a quality that would be necessary to practice the science that led to their advanced technological development (Webb 2000). After all, curiosity is a sine qua non of science. On Earth, human scientists are intensely interested in lower life forms, such as insects; it would logically seem to follow that aliens would be interested in other life forms as well. In fact, some scientists have even speculated that our solar system has been cordoned off and functions as a "zoo" in the natural development of our planet is allowed to occur (Webb 2000). Certainly, an alien civilization traveling to Earth would have knowledge of physics that far exceeded our own. Therefore, the aliens might be more interested in areas such as ethics, religion, and art (Webb 2000). One reason they might visit Earth is to practice cosmological anthropology (Kardashev 1985). If that is the case, then alien intentions would probably lean towards altruism rather than hostility.

If extraterrestrial civilizations do exist, their aspirations could differ markedly from our own. Nevertheless, as Robert Forward pointed out, even for technologically-advanced civilizations, interstellar voyages would probably be justified only for major purposes (Michaud 2007). From the perspective of representatives of alien civilization, plundering the Earth for its resources would be neither practical nor desirable. To prepare for the eventuality of alien contact, we must consider alien motives. To that end, the former chairman of the Transmissions from Earth Working Group, Michael Michaud, recommended that the Committee on SETI be broadened beyond astronomers to include philosophers, historians, theologians, etc. (Brin 2006).

New hopes of life in the cosmos emerged in late 2010 when astronomers discovered an exoplanet in the star system Gliese 581, a mere 20 light years away, that is believed to be of similar size to Earth and that may reside in a habitable zone in its solar system. Contact with intelligent sentient beings would be of monumental significance and could have numerous scientific, political, and theological implications (Michaud 2007). Understandably, based on certain episodes of human history, some people view this proposition with trepidation. Despite the potential dangers, however, more than likely we need not worry.

References

Archer, Michael. 1989. "Slime Monsters Will Be Human Too," *Australian Natural History* 22:546–547.

Brin, David. 2006. "Shouting at the Cosmos…Or How SETI Has Taken a Worrisome Turn into Dangerous Territory," September. Accessed December 28, 2012. http://lifeboat.com/ex/shouting.at.the.cosmos.

Crawford, I. A. 1990. "Interstellar Travel: A Review for Astronomers," *Quarterly Journal of the Royal Astronomical Society* 31:377–400 cited in Harold A. Geller, 2010. "Stephen Hawking Is Wrong: Earth Would Not Be a Target for Alien Conquest." *The Journal of Cosmology* 7:1790. Accessed December 28, 2012. http://journalofcosmology.com/Aliens111.html.

Crawford, I. A. 2000. "Where Are They?: Maybe We Are Alone in the Galaxy After All," *Scientific American* 283(1):38–43.

Davies, Paul. 2010. *The Eerie Silence: Renewing Our Search for Alien Intelligence.* Boston and New York: Houghton Mifflin Harcourt.

Diamond, Jared M. 2006. *The Third Chimpanzee: The Evolution and Future of the Human Animal.* New York: Harper Perennial.

Dyson, Freeman J. 1960. "Search for Artificial Stellar Sources of Infra-Red Radiation." *Science* 131:1667–1668.

Geller, Harold A. 2010. "Stephen Hawking Is Wrong: Earth Would Not Be a Target for Alien Conquest," *The Journal of Cosmology* 7:1790. Accessed December 28, 2012. http://journalofcosmology.com/Aliens111.html.

Ghirardi, GianCarlo. 2010. "Why Should Hawking's Aliens Wish to Destroy?" *The Journal of Cosmology* 7:1785. Accessed December 28, 2012. http://journalofcosmology.com/Aliens108.html.

Kaku, Michio. 1994. *Hyperspace: A Scientific Odyssey Through Parallel Universes, Time Warps, and the 10th Dimension.* New York: Doubleday.

Kaku, Michio. 1997. *Visions: How Science Will Revolutionize the 21st Century.* New York: Anchor Books.

Kaku, Michio. 2002. "The Physics of Interstellar Travel," November 29. Accessed December 28, 2012. mkaku.org/home/?page_id=250.

Kaku, Michio. 2008. *Physics of the Impossible: A Scientific Exploration into the World of Phasers, Force Fields, Teleportation, and Time Travel.* New York: Anchor Book.

Kardashev, Nikolai S. 1985. "On the Inevitability and the Possible Structures of Supercivilizations," In *The Search for Extraterrestrial Life: Recent Developments*; Proceedings of the Symposium, Boston, MA, June 18–21, 1984 (A86-38126 17-88), 497–504. Dordrecht: D. Reidel Publishing Co.

Kardashev, Nikolai S. 1997. "Cosmology and Civilizations" *Astrophysics and Space Science* 252:25–40.

Leake, Jonathan. 2010. "Don't Talk to Aliens, Warns Stephen Hawking." *The Sunday Times* (London), April 25. Accessed December 28, 2012. http://www.thesundaytimes.co.uk/sto/news/uk_news/Science/article272392.ece.

Michaud, Michael A. G. 2007. *Contact with Alien Civilizations: Our Hopes and Fears about Encountering Extraterrestrials.* New York: Copernicus Books.

"Overweight 'Top World's Hungry.'" 2006. *BBC News*, August 15. Accessed December 28, 2012. http://news.bbc.co.uk/2/hi/health/4793455.stm.

Rhodes, Richard. 1986. *The Making of the Atomic Bomb.* New York: Simon & Schuster.

Sagan, Carl. 1973. *Cosmic Connection: An Extraterrestrial Perspective.* Cambridge, UK: Cambridge University Press.

Sagan, Carl. 1985. *Contact.* New York: Simon & Schuster.

Tarter, Jill. 2010. "Should We Fear Space Aliens?" CNN.COM, April 27. Accessed December 28, 2012. http://www.cnn.com/2010/OPINION/04/27/tarter.space.life.fears/index.html.

Webb, Stephen. 2000. *Where is Everybody?: Fifty Solutions to the Fermi Paradox and the Problem of Extraterrestrial Life.* New York: Copernicus Books.

Eliciting Altruism While Avoiding Xenophobia: A Thought Experiment

Jerome H. Barkow

Abstract How do we convey our own altruism to extraterrestrials and how do we trigger altruism from them while avoiding their possible ethnocentrism or even xenophobia? The answer depends on the evolutionary processes that produced a species with which we can communicate. Only a social, internally cooperative, culture-bearing species is likely to have the necessary high technology. Many species use tools but this does not guarantee them a capacity for advanced technology without an evolutionary amplification process. The three candidate processes are: (1) self-predation; (2) co-predation, that is, rivalry with one or more other tool-using species; and (3) sexual selection. Self-predation, in which competing bands cull one another of the dull and uncooperative, likely featured in our own evolutionary history. It leads to a species capable of altruism, alliance, and ethnocentrism and which is likely to share our ideas of social exchange and fairness. Co-predation, in which competing species engage in comparable culling of one another, could lead to a tendency towards xenophobia. These extraterrestrials may be scanning for enemies rather than looking for friends. In sexual selection, males and females mate preferentially with the most intelligent, cooperative, and successful. In such sexually selected species, art functions as a display of superior health and genes in a potential mate. Species with a history of sexual selection are likely to produce art and to recognize and respect our own. The practical implication of this analysis is that, though the risk is no doubt very low, it could be unwise to communicate information that would demonstrate to xenophobes that we are an alien species. Instead, we could communicate our art and music, along with scientific knowledge, because doing so is safe, conveys

J. H. Barkow
Department of Sociology and Social Anthropology, Dalhousie University,
1459 Oxford Street, Halifax, NS B3H 4R2, Canada

J. H. Barkow (✉)
Institute of Cognition and Culture, Queen's University Belfast, Belfast, UK
e-mail: barkow@dal.ca

D. A. Vakoch (ed.), *Extraterrestrial Altruism*, The Frontiers Collection,
DOI: 10.1007/978-3-642-37750-1_3, © Springer-Verlag Berlin Heidelberg 2014

similarity, and is likely to elicit an altruistic response from any species whose intelligence was produced by either self-predation or by sexual selection.

Keywords Culture · Evolution · Altruism · Art · Predation · Intelligence · Extraterrestrials · Risk reduction · Precautionary Principle

1 The Problem

How do we convey our own altruism to extraterrestrials and how do we elicit altruism from them? Can we recognize their encodings of altruism and increase the likelihood that they will understand ours? The answers to these questions depend on what we think will be the psychology of the extraterrestrials. The first step is therefore to determine what extraterrestrial psychologies evolution is likely to produce. This task is simplified because we can only communicate with high-technology extraterrestrials. The issue therefore is not "what kinds of psychology could conceivably evolve on other planets?" but "what kinds of psychology can produce high technology?"

The core psychology of any species, like its physical form, is largely a product of biological evolution. For example, we can safely predict that land-dwelling animals evolving on a high-gravity planet will tend to be relatively broad and low to the ground and with either very thick and/or numerous legs.[1] Just as gravity is a selection pressure that constrains body shape, the selection pressures that result in intelligence and cultural capacity constrain the psychologies a species can develop.

I am assuming that cultural capacity necessarily goes along with intelligence and that our extraterrestrials will have both. This assumption is a product of a side-step: I am deliberately avoiding the immense topic of "what is intelligence" by defining it operationally. For purposes of this discussion, intelligence is having the ability to visit or otherwise contact us, or receive and decode a message from us (Barkow 2000). Interstellar communication implies an advanced technology, which in turn implies cooperation: it is highly improbable that any single entity could, alone, develop the scientific and engineering knowledge and engage in the multiplicity of tasks such communication requires. Our intelligent extraterrestrials will be a social species organized in terms of cooperative groupings. Their advanced technology implies the accumulation of a vast pool of knowledge, of information that is regularly transmitted both within and across generations. This is a practically a definition of culture, given that the latter can be defined in part as a socially transmitted information pool (Barkow 1989; Barkow, O'Gorman, and Rendell 2012)! So merely by dint of their communicating with us we already know a good deal about our extraterrestrials' psychology—they are intelligent, social, cultural, and capable of cooperation.

---

[1] Before there were xenobiologists there were science fiction writers. See, for example, Hal Clement's 1954 classic, *Mission of Gravity* (Clement 1954).

But will they be altruistic towards us? Or will they be xenophobes, searching the universe for threats? After all, while our own ultimate goal is presumably to engage in reciprocally altruistic communication with extraterrestrials, we also want to minimize the possibility of eliciting their enmity. To learn more about their psychology we need to ask how a social species can evolve intelligence and cooperation. The best place to begin is the literature about how our own species, *Homo sapiens*, may have developed intelligence, altruism, and ethnocentrism. However, lines of reasoning tied to the hominin fossil and archaeological records (e.g., Whiten and Erdal 2012), while valuable for students of our own species, are probably too human-specific to be applied to extraterrestrial intelligence. A broad brush is needed because it is unlikely that the evolution of extraterrestrials would parallel our own in such detail. I will not, therefore, discuss the possible inter-actions between, say, *Homo sapiens* and *Homo neanderthalensis*. Understanding more general evolutionary processes, however, can provide possibilities.

2 Convergent Evolution

My working assumption is that we expect *convergent evolution* to be the usual case everywhere. As Simon Conway Morris (1998, 204) puts it, "again and again we have evidence of biological forms stumbling on the same solution to a problem." After all, whatever common ancestor winged insects, bats and birds have, it most defi-nitely existed long before any of these three lines evolved the capacity for flight. Squid and octopus (cephalopods) have organs of vision, as do arthropods (ticks and spiders, insects, and crustaceans), vertebrates like ourselves, and some jellyfish, not because they are closely related to one another but because similar problems led to similar solutions. Conway Morris (2003) makes a plausible case for the almost inevitability of human-like species evolving on our own planet, which implies that there is at least a real possibility of their evolving on other planets.

Planets appear to be plentiful. The Kepler space observatory was launched in March 2009 to identify terrestrial and larger extrasolar planets, and data continue to be analyzed after observations ended in May 2013. As of August 3, 2013, Kepler had identified 135 confirmed planets (http://kepler.nasa.gov/). Given the likely existence of an immense number of planets orbiting in the habitable zone of their various suns, it is a reasonable assumption that at least some of these bear life. If convergence is indeed the rule, then at least on some of these planets, evolution must have produced intelligent beings. Conway Morris is probably right. Certainly, his reasoning is compatible with the Drake Equation, used by SETI as a way of estimating the number of high technology civilizations in our galaxy (Drake and Sobel 1992).

But are some of these extraterrestrials not only altruistic but also capable of recognizing human altruism? This question obliges us to ask, in turn, how it is that evolution made our own species frequently, if not invariably, altruistic. Why are we ourselves capable of altruism?

3 Why Are Humans Capable of Altruism?

For evolutionists, altruism comes in two flavors.[2] The most familiar is nepotism, kin altruism, favoring one's relatives over others. After all, evolution is about getting as many copies of your genes as possible into the next generation. Kin by definition share copies of your genes. The more closely related you are to them the more altruistic you should be towards them, depending on the cost to you and how likely they are to have additional surviving children due to your intervention. This reasoning implies that individuals should be more likely to help close relatives than distant ones, and those of reproductive age more than those past reproductive age. As with much evolutionary psychology, many find this reasoning makes intuitive sense—danger to one's own children seems more likely to elicit heroism than danger to the children of strangers, and our own culture's old "women and children first" principle when a ship is abandoned is loosely consistent with "help those of higher reproductive value first." Altruism among kin is arguably the most common kind and the most likely to evolve, so it seems reasonable to assume that extraterrestrials will have this kind of altruism. However, if members of a species have no way to differentiate between kin and non-kin because, for example, off-spring disperse widely at birth or hatching and there are no olfactory, visual or other cues to discerning kinship, then kin altruism will not evolve.

The second flavor of altruism is known as "reciprocal altruism." If you and another exchange aid, whether directly or indirectly, immediately or subsequently, you will both probably be ahead. The world over, "you scratch my back and I'll scratch yours" and, "one hand washes the other" are easily understood. Our moral systems, it has been argued, are based on ideas of reciprocity and of social exchange. It seems reasonable to suppose that the capacity for moral indignation and the sense of justice evolved to protect us from those who fail to reciprocate— they function to discourage cheating. These topics continue to be subjects of scholarly discussion. See, for example, Alexander (1987); Cosmides and Tooby (2006); Kurzban, DeScioli, and Fein (2012); or Walsh (2000).

4 Will Extraterrestrials Have Altruism?

Our operational definition of "intelligence" implies the presence of altruism. The development of advanced technology must entail complex, flexible cooperation over long periods of time and this is difficult to imagine without altruism. But which flavor of altruism? It could be that the extraterrestrials developed their technology in the context of cooperating kinship groups, but nepotistic altruism seems more limited than reciprocal altruism, which widens the pool of coopera-tors. It is reasonable to assume that intelligent extraterrestrials will have at least

---

[2] For an introduction to evolutionary psychology, see Buss (2012).

one kind but most likely both kinds of altruism. *If there is reciprocal altruism then mechanisms to generate it would develop, and these mechanisms seem likely to include analogues to our sense of justice and our ideas of morality.* After all, if intelligence implies technology and technology implies cooperation, and if cooperation in turn strongly suggests a reciprocal altruism that is mediated by mechanisms similar to our sense of justice and fairness, then we may find considerable similarity between ourselves and intelligent extraterrestrials.

But altruism is very common among animal species: to conclude that extraterrestrials are likely to be altruistic, at least towards one another, does not account for their intelligence. To do so we need to ask, what are the origins of our own intelligence?

5 Where Did Human Intelligence and Psychology Come From?[3]

Once upon a time, anthropologists believed that we evolved intelligence to help us make and use tools (e.g., Oakley 1952). We now know that tool use is not particularly unusual among animal species, while both our own ancestors and related species were skilful tool-users long before modern human beings and our complex knowledge-accumulating cultures appeared (Tattersall 2012). There must have been other evolutionary processes driving the development of hominin intelligence which do more than further mere survival or the ability to make stone tools. The two apparently sine qua non *intelligence-amplifying processes* are sex and predation. These are very likely to have been involved in the evolution not only of human intelligence but in the development of extraterrestrial intelligence everywhere. Let us begin with predation.

5.1 Predation Theories

A predator like the wolf may, by culling the slow of hoof, cause prey such as deer to evolve fleetness. Offspring resemble their parents, taught Darwin. In an environment that includes hungry wolves, slow deer are a lot less likely to have surviving offspring than are the fast ones: with each generation, the surviving deer are those with the fastest parents, and each generation therefore gets faster (at least, up to the point at which changes needed for additional speed would have negative consequences for survival and reproduction that outweighed the speed advantage).

Of course, evolution works on the wolf, too—the fastest wolves are the ones who catch the most deer, all other factors being equal. So, with each generation,

---

[3] This topic is treated at somewhat greater length in Barkow (1989) and Barkow (2000).

the wolves get faster, because for them, too, offspring resemble their parents. Deer and wolves shape one another in a reciprocal positive feedback, co-evolutionary relationship. What about human beings, were we shaped by wolves so that we became not fast but clever?

In a sense we were, but according to some writers (e.g., Alexander 1971; 1974; 1975; 1979; Barkow 1989, 146–150), the "wolves" that shaped us were other bands of our fellow hominins. It seems very likely that we shaped ourselves. In conflicts both within and between bands, the individuals least likely to survive would have been those worst at making and using weapons and other tools, worst at strategy and tactics, worst at communicating effectively, and worst at cooperating with fellow band members when confronted with danger. At the same time, the bands most likely to survive and grow were those whose cultures and perhaps genetic make-up encouraged in-group cooperation. This scenario has been referred to as "auto-" or "self-predation." The self-predation theory accounts for *ethnocentrism.*

Our ancestors were selected to band together in the face of external threat and to be suspicious of outsiders. We are all aware of ethnocentrism—the tendency to rally around a leader in the face of external threat, for the moment forgetting in-group differences. A definitional component of ethnocentrism is that we favor our fellow in-group members and discriminate against and are mistrustful of outsiders. We are much more likely to be altruistic to members of our in-group than towards out-group members. The origins of ethnocentrism are likely related to kin or nepotistic altruism. In a small band, everyone probably is more closely related to one another than they are to outsiders (Van den Berghe 1981), though today symbolic kinship suffices. In-group members usually feel superior in at least some ways to out-group members (Barkow 1989; LeVine and Campbell 1972; Reynolds, Falger, and Vine 1987), and both external and internal threats (such as a food shortage) may increase in-group loyalty (Cashdan 2001).

Much of human moral history has had to do with redefining the in-group, with people accepting larger and larger groups as their ethnocentric in-group, their unit of cooperation, loyalty and altruism. For a large portion of our population, the in-group (the ethnocentric unit) has moved from the family to the band to members of our religion to fellow citizens of our nation and then perhaps to speakers of our language. For some of us, humanity itself is our in-group. But almost any symbolic similarity can serve to define a group—where one went to school, a love of role-playing games, a fashion sense, vegetarianism, an interest in the search for extraterrestrial intelligence, and so forth. The Internet has made possible the creation of disseminated social networks (Barkow, O'Gorman and Rendell 2012) that at times may function as in-groups. Would-be political leaders try to convince us that they and we are not only members of the same in-group but that that this membership is more important than our other memberships, that this in-group is under great threat, and that he or she is the leader best suited to defend it. The in-group in question, depending on local politics, may have to do with religion, political ideology, citizenship, social class, and so forth.

It is not impossible that during the course of human evolution our ancestors were subject to rivalries not just with bands of members of the same species but with related and nearly equally intelligent species. Each of these two (or more) species would have culled the other(s) of the dull, uncooperative, and uncommunicative. If this process did occur then it is apparent that only one of the species, obviously our own, survived the competition. There is no compelling evidence for such rivalry influencing human evolution but it is worth mentioning because extraterrestrials could owe their intelligence and much of their psychology to this kind of interspecies evolutionary interaction. An evolved psychology produced by prolonged violent conflict with other intelligent species may not predispose members towards interspecies altruism. Indeed, such a psychology could conceivably include obligate xenophobia, or at least predispose members to a readiness to mistrust and hostility towards other intelligent species.

Extraterrestrials whose intelligence was produced (in large measure) by self-predation are likely to resemble us, ethnocentric but capable of alliances and of shifting group boundaries. Extraterrestrials with intelligence produced by intense rivalries with other species may be predisposed to xenophobia. The origins of a species' intelligence have implications for how we should go about eliciting their altruism and avoiding their hostility.

5.2 Sexual Selection Theories

If predation is one possible type of intelligence-amplifying process, the other involves sex.[4] Our having two sexes may have made the amplification of the intelligence and cultural capacity of our early ancestors possible because it led to *sexual selection*. Sexual selection has to do with mate choice—we compete with other members of our sex for the most desirable (best genes) mates; and we compete to be chosen by the other sex. For our own species, in a 37-country survey, David Buss (1989) found that both women and men, when asked to rank 13 traits for desirability in a long-term mate, ranked intelligence second. (Interestingly, "kindness and understanding" ranked first.) If our ancestors were like us, we evolved our intelligence at least in part due to sexual selection, that is, due to mate choice. Men's choices shaped the evolution of women and women's choices the evolution of men, with both preferring (all things being equal) the more rather than less intelligent mates. But not just our intelligence was shaped, so too was our psychology.

Sexual selection means that individuals may go to great lengths to display evidence of their good genes. Zahavi and Zahavi (1997) have even argued for a "handicap principle" in which extravagant and costly morphology and behavior

---

[4] I am assuming that extraterrestrials will have two sexes. Evolutionary biologists have concluded that having two sexes gives plants and animals an adaptive advantage in coevolutionary competition with parasites whose generational period is much shorter and which therefore can evolve more rapidly. For discussion and research, see Brockhurst (2011), Morran et al. (2011), or Ridley (1993).

signal good genes because they honestly convey health and vigor. The classic example of an extravagant and expensive evolved display feature is the peacock's tail (Cronin 1992; Ridley 1993; Zahavi and Zahavi 1997). Miller (2000) argues that, like the peacock's tail, a large vocabulary, a good sense of humor, the ability to dance, sing or play a musical instrument or to paint evolved as sexual display of evidence of good health and genes, leading to the displayer being more frequently chosen as a mate than those who lacked these abilities. Miller (2009) and Saad (2007; 2011), among others, have argued that our conspicuous consumption is a type of sexual display—look at all the goods I can amass and provide for our potential off-spring! Gurven and Hill (2009) and Hawkes and Bird (2002) even argue that men's big-game hunting, in hunter-gathering societies, serves more as a sexual display of prowess (good genes) than as an efficient way to garner high-quality protein. For purposes of communicating with extraterrestrials, however, what is most germane here is the argument that sexual selection is ultimately responsible for the non-utilitarian elaboration of so much that we craft and for the-by-definition expressly non-utilitarian fine arts that we treasure.

Arts and crafts (among other things) are means of displaying good genes and good health and so attracting mates. If we receive messages from extraterrestrials, and if their messages include elaborated or patterned features that have no apparent utilitarian function, then we must entertain the hypothesis that we have been sent art by a species whose evolved psychology is in part a product of sexual selection. The plaques carried by Pioneers 10 and 11, launched by NASA in 1972 and 1973, respectively, were primarily intended to convey information about their originators, on the slim chance that they might eventually be discovered by curious extraterrestrials. The plaques also, however, appear to be beautifully crafted works of art. The golden discs carried by the two Voyagers in 1977 carried both images and music and a variety of sounds. They certainly and deliberately carry art. If the extraterrestrials who find either a plaque or a disk turn out to be evolutionists, their designers may have communicated more than they intended—evolutionist aliens may recognize evidence of a sexually-selected species exhibiting art whose ulti-mate origin was selection for display of good genes and health.

6 Implications for Communication

6.1 Will Extraterrestrials Be Ethnocentric or Even Xenophobic?

If our extraterrestrials got their intelligence even in part through self-predation, as we likely did, then they also should be ethnocentric in the way that we are. This reasoning results in a familiar implication: if we want to communicate altruism to such extraterrestrials, we need to convey that we and they are really members of the same in-group or are available as coalition partners, that we mean them no harm and may be able to do them some good. The challenge is how to convey similarity and altruism, a challenge I will return to. An evolutionary history of self-predation does mean, at least, that a mutually beneficial relationship is a possibility and that enmity would arise only by accident.

What, however, if their intelligence and cultural capacity resulted from the co-predation or interspecies rivalry scenario? To be a bit alarmist about a thought experiment's conclusion, that would mean that our extraterrestrials evolved in company with at least one closely competing species. Their evolved psychology may therefore include an automatic or at least easily triggered xenophobia. We may find ourselves with a xenophobic and even genocidal enemy. Conceivably, their evolved psychology may make it difficult or impossible for them to believe that another intelligent species is friendly. They could be incapable of recognizing our altruism or of showing theirs towards us. Their equivalent of SETI could be a SEHE, a "search for extra-home-planet enemies." Unless we can convince them that we and they are in some crucial way the same species, enmity rather than altruism will result from any encounter. Once again, the implication is that we should strive to at least leave the door open for them to believe that we and they are similar, perhaps even that we are distant members of their own species, or at least closely related. Images of ourselves are out.

Earlier sections of this chapter pointed out that extraterrestrials are very likely to share with us the traits of reciprocal and nepotistic altruism. They are therefore likely to have a sense of justice, or at least a concept of fairness which we can understand. Let us assume that the extraterrestrials are either not xenophobic or else accept us as kin, or that they are ethnocentric but consider us likely coalition partners. How do we trigger their altruism or convey our own? Presumably, we should do so in the same way in which we would elicit altruism from members of our species: Especially in our initial contacts, we should strive to be as generous as possible, while empha-sizing how much we must have in common. Since contact will probably involve only very slow communication, the obvious token of both generosity and similarity we can provide is information. The less obvious one is art.

If the extraterrestrials' intelligence and cultural capacity are in part products of sexual selection then they will have equivalents of art and the potential to rec-ognize our own art. They need not share our sense of aesthetics but merely the ability to appreciate the products of great effort, skill, and elaboration in spheres where these are not necessary for utilitarian purposes. Music is an obvious choice for an art form to be shared with extraterrestrials. After all, on our own planet we seem to be able to appreciate the music of species ranging from whales to song-birds. Perhaps extraterrestrials will have similar breadth.

Art solves two problems for us: How do we convey our similarity to the extraterrestrials, in the hopes of avoiding possible xenophobia and eliciting altruism? By being generous with our art, especially our music. If the logic of two sexes and sexual selection is right then the extraterrestrials will have their own arts and crafts and are likely to at least recognize our efforts (regardless of whether they find them aesthetically pleasing). How do we trigger their altruism? Why, by being generous with information. Scientific information and the details, say, of the construction of Pioneer-style craft, certainly. But, above all, we should be generous with our art.

6.2 Withholding Information from Extraterrestrials

What information should we withhold from extraterrestrials? Given our own evolutionary history and evolved psychology, most of us would probably be inclined to expect or at least hope that extraterrestrials will be friends and allies, or at least will think good thoughts about us in the far future. As with the Pioneer and Voyager communications, we want these others to know about us. Unfortunately, prudence suggests (as has already been suggested) that we include only that information which will at least make it possible for them to believe that we are members of their own species, their close kin, or at least potential coalition partners. This means that we should not (unlike the Pioneer and Voyager spacecraft) initially provide any kind of depiction of ourselves, even one as stylized as the Pioneer spacecraft's figures. If we wish to be extremely prudent, we should not include information about the air we breathe or the foods we eat or how we reproduce, and definitely not what we look like. Basic scientific knowledge is safe and may be provided copiously (with the exception of weapons technology, of course). But art, especially music and visual art that does not include images of living figures, should play a major role in our messaging. Pictures of textiles, tiles, beautiful tools, and abstract art should be safe and (one hopes) effective in perhaps serving as evidence of our essential similarity and as an indication of an altruism worthy of reciprocation. Art permits us to convey the possibility of altruism and kinship while reducing the risk of suggesting to a xenophobic species that we are the hated, dangerous aliens.

References

Alexander, Richard D. 1971. "The Search for an Evolutionary Philosophy of Man." *Proceedings of the Royal Society of Victoria* 84:99–120.

Alexander, Richard D. 1974. "The Evolution of Social Behavior." *Annual Review of Systematics* 5:325–383.

Alexander, Richard D. 1975. "The Search for a General Theory of Behavior." *Behavioral Science* 20:77–100.

Alexander, Richard D. 1979. *Darwinism and Human Affairs*. Seattle: University of Washington Press.

Alexander, Richard D. 1987. *The Biology of Moral Systems*. New York: Aldine de Gruyter.

Barkow, Jerome H. 1989. *Darwin, Sex, and Status: Biological Approaches to Mind and Culture*. Toronto: University of Toronto Press.

Barkow, Jerome H. 2000. "Do Extraterrestrials Have Sex (and Intelligence)?" In *Evolutionary Perspectives on Human Reproductive Behavior*, edited by Dori LeCroy and Peter Moller, 164–181. New York: Annals of the New York Academy of Sciences.

Barkow, Jerome H., Rick O'Gorman, and Luke Rendell. 2012. "Are The New Mass Media Subverting Cultural Transmission?" *Review of General Psychology* 16(2):121–133. doi:10.1037/a0027907.

Brockhurst, Michael A. 2011. "Sex, Death, and the Red Queen." *Science* 333 (6039):166–167. doi:10.1126/science.1209420.

Buss, David M. 1989. "Sex Differences in Human Mate Preferences: Evolutionary Hypotheses Tested in 37 Cultures." *Behavioral and Brain Sciences* 12(1):1–14. doi:10.1017/S0140525X00023992.

Buss, David M. 2012. *Evolutionary Psychology: The New Science of the Mind*. 4th ed. Boston: Pearson Allyn & Bacon.

Cashdan, E. 2001. "Ethnocentrism and Xenophobia: A Cross-Cultural Study." *Current Anthropology* 42 (5):760–765. doi:10.1086/323821.

Clement, Hal. 1954. *Mission of Gravity*. Garden City, NY: Doubleday.

Conway Morris, Simon 1998. *The Crucible of Creation: The Burgess Shale and the Rise of Animals*. New York: Oxford University Press.

Conway Morris, Simon 2003. *Life's Solution : Inevitable Humans in a Lonely Universe*. Cambridge, UK: Cambridge University Press.

Cosmides, Leda, and John Tooby. 2006. "Evolutionary Psychology, Moral Heuristics, and the Law." In *Heuristics and the Law (Dahlem Workshop Reports)*, edited by Gerd Gigerenzer and Christoph Engel, 175–205. Cambridge, MA: MIT Press (in cooperation with Dahlem University Press).

Cronin, Helena. 1992. *The Ant and the Peacock: Altruism and Sexual Selection from Darwin to Today*. Cambridge, UK: Cambridge University Press.

Drake, Frank, and Dava Sobel. 1992. *Is Anyone out There? The Scientific Search for Extraterrestrial Intelligence*. New York: Delacorte Press.

Gurven, M., and K. Hill. 2009. "Why Do Men Hunt?: A Reevaluation of 'Man the Hunter' and the Sexual Division of Labor." *Current Anthropology* 50(1):51–74. doi:10.1086/595620.

Hawkes, K., and R. B. Bird. 2002. "Showing Off, Handicap Signaling, and the Evolution of Men's Work." *Evolutionary Anthropology* 11(2):58–67. doi:10.1002/evan.20005.

Kurzban, Robert, Peter DeScioli, and Daniel Fein. 2012. "Hamilton Vs. Kant: Pitting Adaptations for Altruism Against Adaptations for Moral Judgment." *Evolution and Human Behavior* 33(4):323–333. doi:10.1016/j.evolhumbehav.2011.11.002.

LeVine, Robert A., and Donald T. Campbell. 1972. *Ethnocentrism: Theories of Conflict, Ethnic Attitudes and Group Behavior*. New York: John Wiley and Sons.

Miller, Geoffrey. 2000. *The Mating Mind: How Sexual Choice Shaped Human Nature*. New York: Doubleday.

Miller, Geoffrey. 2009. *Spent: Sex, Evolution, and Consumer Behavior*. New York: Viking.

Morran, Levi T., Olivia G. Schmidt, Ian A. Gelarden, Raymond C. Parrish, and Curtis M. Lively. 2011. "Running with the Red Queen: Host-Parasite Coevolution Selects for Biparental Sex." *Science* 333(6039):216–218. doi:10.1126/science.1206360.

Oakley, Kenneth Page. 1952. *Man The Toolmaker*. 2nd ed. London: British Museum.

Reynolds V., Falger V., and Vine I., eds. 1987. *The Sociobiology of Ethnocentrism. Evolutionary Dimensions of Xenophobia, Discrimination, Racism and Nationalism*. London and Sydney: Croom Helm.

Ridley, Matt. 1993. *The Red Queen: Sex and the Evolution of Human Nature*. New York: Macmillan.

Saad, Gad. 2007. *The Evolutionary Bases of Consumption*. Mahwah, NJ: Lawrence Erlbaum Associates.

Saad, Gad. 2011. *The Consuming Instinct: What Juicy Burgers, Ferraris, Pornography, and Gift Giving Reveal about Human Nature*. Amherst, NY: Prometheus Books.

Tattersall, Ian. 2012. *Masters of the Planet: The Search for Our Human Origins*. New York: Palgrave Macmillan.

Van den Berghe, Pierre. 1981. *The Ethnic Phenomenon*. New York: Elsevier.

Walsh, Anthony. 2000. "Evolutionary Psychology and the Origins of Justice." *Justice Quarterly* 17(4):841–864. doi:10.1080/07418820000094781.

Whiten, Andrew, and David Erdal. 2012. "The Human Socio-Cognitive Niche and Its Evolutionary Origins." *Philosophical Transactions of the Royal Society B: Biological Sciences* 367(1599):2119–2129. doi:10.1098/rstb.2012.0114.

Zahavi, Amotz, and Avishag Zahavi. 1997. *The Handicap Principle: A Missing Piece of Darwin's Puzzle*. Oxford: Oxford University Press.

Predator—Prey Models and Contact Considerations

Douglas Raybeck

Abstract Employing ethological models derived from terrestrial predators and prey, I attempt to evaluate the likelihood that an intelligent alien will be beneficent, neutral or hostile. To this end I review what is known about selective pressures for intelligence generally, and for predators and prey particularly. I also review some of the conditions that would promote or inhibit the development of intelligence. After discussing the contributions to intelligence of tool use and spatial behavior, I—in agreement with the majority of evolutionary biologists, psychologists and anthropologists—settle on social behavior as the most potent contributor to the development of higher intelligence. Predators, although well equipped with fierce dispositions, 'weaponry' and 'armor,' can establish well organized and highly supportive in-groups, such as wolves do. It seems likely that any intelligence that evolves in a social unit will be affected by minimal requirements involved in-group cooperation and cohesion. The result will likely be constraints on agonistic behavior and an ability to engage in cooperative endeavors … within the group. Toward outsiders, the behavior of such organisms may well be far more exploitative. Our own history suggests as much. I conclude with an expression of strong support for the efforts of SETI and others to obtain information about intelligent others. The potential benefits to be gained from this are simply too great to gainsay. However, despite the unlikelihood of actual physical contact, I conclude with a caution about divulging too much information about ourselves.

Keywords Aliens · Intelligence · Extraterrestrial intelligence · Agonistic behavior · Predation · Evolution · Social behavior · Altruism · Aggression

D. Raybeck (✉)
Anthropology, Hamilton College, 500 S. Pleasant, Amherst, MA 01002-2542, USA
e-mail: draybeck@hamilton.edu

D. A. Vakoch (ed.), *Extraterrestrial Altruism*, The Frontiers Collection,   49
DOI: 10.1007/978-3-642-37750-1_4, © Springer-Verlag Berlin Heidelberg 2014

1 Introduction

Within the scientific community, as well as in the popular press and among science fiction writers, there has long been a concern with extraterrestrials and the possibility of communication with them. This concern has led to projects such as the Search for Extraterrestrial Intelligence (SETI) (Morrison, Billingham, and Wolfe 1977) that continues to be a focus of attention for many scientists even though Congress terminated funding in 1993 (Holohan and Garg 2005; Tarter 2007).

Twenty years ago, Harrison (1993) published an intriguing paper concerning extraterrestrial intelligence in one of psychology's major journals. At the same time, scientists have theorized and speculated about the nature of extraterrestrial intelligence and the problems involved in communication between sapient species (Sagan 1973). The consensus has been that the universe is very likely to host other intelligent beings, that some of these will be more technologically advanced than we are today, and that some will be trying to locate other intelligences.

In the science fiction community, images of extraterrestrials have been variegated in form, in intelligence and in intentions. They range from the beneficent aliens of Julian May (1987a; b), who wish only to elevate the lot of humanity and facilitate our participation in an intergalactic "milieu," to the malevolent extraterrestrials of Greg Bear (1987), who travel about the universe locating intelligent life forms and destroying them because they may be potential future competitors. Generally, however, images of aliens in the popular press and among scientists are positive. It is widely believed that if a sapient species can achieve the degree of civilization necessary to support interstellar communication, it is unlikely to be hostile.

In this chapter, I wish to examine this assumption. As an anthropologist, I am aware that there are some markedly different paths to the evolution of intelligence. These differences can provide us with models that can suggest some of the variety we may anticipate among extraterrestrials. I am concerned about the possibility that a technologically-oriented intelligence may as likely be developed by a predatory species as by a non-predatory one. I am particularly concerned with the kinds of stimuli that promote the development of intelligence, and with what sorts of ethical notions might be associated with these varying modes of evolving intelligence. This exercise in modeling should have consequences for how we approach the possibility of extraterrestrial communication.

2 The Case for Intelligence

Among the range of definitions for intelligence, one that is widely accepted is the ability to learn new response patterns (Jerison 1973). Generally, intelligence confers upon an organism greater adaptability and flexibility in dealing with environmental challenges. However, many complex adaptations to the environment do not require the classical concept of intelligence. Scientists have long known that insects are capable of complex adaptations to their environments in a fashion that relies on genetic programming rather than on learning (Wilson 1980). Indeed, Schull (1990) has argued that even the adaptive characteristics of plant and animal species exhibit information processing and that it would be fruitful to view such species as intelligent. Overwhelmingly, however, the scientific community believes that a greater capacity for learning is a superior adaptation to suggested alternatives.

In the evolution of intelligence on earth there has been a consistent trend from relatively closed instinctive patterns toward "open" learning (Hinde 1974; Sluckin 1965). Jastrow (1981) has noted the evolution of intelligence from lower organisms to humanity and to computers. He and others believe that, if one has competing species, the evolution of intelligence is inevitable because of the advantages it confers upon the possessor (Itzkoff 1983; Sagan 1977). However, questions about the rate at which intelligence is developed and the nature of the species that are most likely to possess it are more complex.

Evolutionary theorists and developmental biologists have long been aware that the development of intelligence involves a series of interactions between organisms and their environment (Laughlin and Brady 1978; Laughlin and D'Aquili 1974; Manosevetz et al. 1969; Mazur and Robertson 1972; Tunnell 1973). The environment must contain conditions for which intelligence is an adaptive trait. While "brains" should develop more complexity and flexibility, it is not necessary to anticipate that they become larger (Miller 2007). Beings with greater intelligence then reproduce in increasing numbers, filling their eco-niches and driving out less intelligent competitors. It is important to note, however, that the entities disadvantaged in this scenario are the ones that either compete directly with intelligent others or are directly exploited by them.

Complex environments select for intelligence by creating conditions where more intelligent competitors have an advantage in exploiting limited resources (Evans and Schmidt 1990; Robinson 1990). Animals that proceed by instinct have a limited set of behavioral repertoires with which to respond to changing conditions. They are limited not only by their physiology, but by their ability to perceive the existence of new demands and new resource possibilities. Their coping equipment is genetically based and suited to the environment in which the organism evolved. Should that environment change, the organism may prove unable to adapt to the new circumstances and be seriously disadvantaged in its competition with other species (Daly and Wilson 1978; Dawkins 1976; Smith 1984).

Generally, increasing intelligence gives an organism a better opportunity to model the environment, both natural and behavioral, so that food getting, mating and general survival strategies can be maximized. Intelligence is selected for because it benefits the possessor, not because it is helpful to others.

2.1 Costs and Advantages of Intelligence

An increase in intelligence has meant a corresponding rise in brain size. As Jerison (1973, 8) has noted, "The mass of neural tissue controlling a particular function is appropriate to the amount of information processing involved in performing the function." This has been true in organic evolution and in the evolution of artificial intelligence as well (Gardner 1985; Goldstein and Papert 1977; Jastrow 1981; Llinas 1990; Nelson and Bower 1990; Schank and Childers 1984). It seems likely that, however information is processed, it would also be true for extraterrestrials.

Intelligence is not without certain physical costs. Particularly in the case of higher mammals, intelligence has been found to be expensive in terms of the body's resources. Brain tissue requires large supplies of glucose and oxygen (Milton 1988), but these are justified by the advantages that intelligence confers. Indeed, the costs of intelligence are evidence of its importance and success as an environmental adaptation.

There are also social consequences that accompany the development of significant intelligence. An increasing reliance on a learned repertoire implies an increased period of dependency on the part of the young. The need for learning plus the problems of rearing learning-based offspring involve a very high cost from an evolutionary perspective. Such organisms have few offspring and this means that, unlike lower organisms that reproduce in greater numbers, the survival of each offspring is important. This longer maturation period and the need for security create a trend toward social living, because the infant and its mother need the support of others (Laughlin and D'Aquili 1974). This model is true not only for humans but also apes, cetaceans, elephants, and most other mammals with appreciable intelligence. Further, as we shall see, the exigencies of social life can prove to be as strong a stimulus for the evolution of increased intelligence as any other factor. This creates a positive feedback loop in which intelligence promotes social living, which, once established, makes increased intelligence highly adaptive.

Even among lower animals, greater intelligence means more flexibility in dealing with environmental conditions. For predators, this implies a greater ability to locate and consume prey, while, for prey, greater intelligence increases the likelihood of avoiding such a fate (Byrne and Whiten 1988).

As intelligence increases, other emergent properties appear which reflect the expanded complexity of the system, and which confer still greater advantages on the possessor. At some point, increasing intelligence should lead to self-awareness (Itzkoff 1985; Jastrow 1981; Laughlin and D'Aquili 1974). An organism equipped with self-awareness can model not only the external environment, but also include itself as an element of attention. It has a self-concept separable from the environment and capable of conscious examination and reflection (Tunnell 1973). Concurrent with such a development is an increase in the organism's ability to construct an internal environment that can not only represent the external world, but also make possible the construction of symbols which are, by definition, arbitrarily related to their referents (Gazzaniga 1992; Laughlin and D'Aquili 1974; Laughlin et al. 1990).

The capacity for symbolism represents an enormous evolutionary advantage for any intelligent species. Prior to its appearance, communications are limited by environmental stimuli in what is termed a "closed" system (Hockett 1973). In such circumstances, an organism emits a signal that is automatically called forth by an external stimulus. There is no displacement in time or space, and such calls are generally mutually exclusive. The information-carrying capacity of the system is thus limited to the number of calls hard-wired into the organism. With symbolism, organisms gain the ability to displace their messages and to combine them in ever more complex and novel assemblages. Further, they can assign meanings in complex ways influenced, but not dictated, by biology. This opens up the realm of culture, a learned set of patterns for behavior that are far more malleable than the biological substrate that made them possible.

While symbolism involves greatly increased freedom from the constraints of the organism's biological limitations, this freedom is not absolute. For humans, the structure of our brain imposes limits both on the amount of information we can process at any given time (Greenspan 2004; Miller 1951; 1956), and on the kinds of information we can process (Ardila and Ostrosky-Solis 1989; Jerison 1990; Lenneberg 1967; Thompson and Green 1982). There is reason to believe that similar limitations and perceptual dispositions would attend any evolving sentience (Gazzaniga 1992; Sauer and MacNair 1983; Stokoe 1989; Wasserman 1989). Given such an expectation, it seems likely that sentients who have evolved from a predator background would differ markedly from sentients whose gustatory preference run to plants.

3 Predator Intelligence Models

There are a variety of relations that obtain between predator and prey. Some predators, such as the anteater, specialize in a single prey; others, like the wolf, ingest a wide range of prey, but most probably fall in the mid-range (Evans and Schmidt 1990). All predators need strategies to locate, obtain and consume prey, but the nature of these strategies can range from the genetically programmed activities of spiders, to the complex hunting practices of the !Kung bushmen of the Kalahari Desert (Lee 1979; 1984; Marshall 1976). In the latter case, intelligence not only makes it more likely that prey will be obtained, but also promotes an optimal distribution of calories and even saving to meet future needs.

In assessing whether or not predators are as likely as others to develop high intelligence, the answer is unequivocal—they are not less, but more likely than others to evolve a high intelligence. This somewhat surprising conclusion results from an examination of ethological research, as well as contemplation of the models purporting to describe factors that promote intelligence.

Recall that intelligence is selected for when it enables an organism to exploit resources that would otherwise elude it. This argument holds for both predators and prey, but, for reasons I will discuss below, its selective pressure is greater for predators. Recall also, that complex environments select for intelligence by creating conditions where more intelligent competitors have an advantage in exploiting limited resources (Evans and Schmidt 1990; Robinson 1990). Predators have a more difficult set of problems to solve and these involve environmental conditions that are more complex for the predator than for its prey. Said another way, predators are more environmentally challenged than prey and this increases the selective advantage of increased intelligence.

Prey need to locate resources which, in the case of herbivores, are nicely stationary. Further, they need to survive the depredations of predators, *but* it is not necessary that all individuals need to endure to insure the perpetuation of the prey species. Indeed, many prey adapt to the competition with predators by becoming more fecund rather than more elusive.

In contrast, predators must actively solve their problems, including locating prey. As Malthus would suggest, there are always more prey than predators, but such prey may prove difficult to find. To survive, predators must prove more capable than their prey. The complexity of a predator's environment includes not only those elements also encountered by prey, but also the behavior of the prey itself. It might be argued that the prey could benefit from being able to better model the behavior of predators but, given their higher birth rate and the costs of intelligence, the selective advantage of intelligence is actually less for prey than for predators.

An intelligent predator is likely to view other entities in an extremely utilitarian, probably gustatory, fashion. There would likely be constraints on exploitative behaviors, since no intelligent predator would wish to extirpate a source of calories, but there is no reason to anticipate much in the way of altruism between sapient species. Indeed, should extraterrestrial visitors prove to be evolved from a consistent predator base, it seems likely that their interest in us would, at least from our perspective, be quite malevolent.

One can argue that the assumption of uniform hostility on the part of extraterrestrials descended from predator stock is too simplistic since it does not incorporate the meliorating influence of adaptation to social life over a prolonged period of evolution. My image of predators also obfuscates the possible role of culture in reducing an "us–them" view of the universe. In fairness, then, we should examine a wider range of possibilities in which intelligence can be promoted by a variety of circumstances in addition to predation.

4 Evolutionary Sources of Intelligence

4.1 Tool Use

Since the middle of this century, one of the classic arguments in anthropology concerning a probable stimulus for intelligence focused on early tool use (Oakley

1959). Tool use and, especially, tool manufacture place a premium on eye-hand coordination, the ability to visualize a future result, and other capacities associated with intelligence (Washburn 1960; Wynn 1988). To the extent that tool use and tool making represent an adaptive advantage in a competitive environment, the qualities on which they depend will be selected for. It is argued that our *australopithecine* forbears, who first used tools, and *Homo habilis,* who first constructed tools, set in motion a positive feedback loop, an ineluctable chain of events that culminated in *Homo sapiens sapiens.* The selection for better eye-hand coordination and greater intelligence resulted in organisms that could construct more effective tools. These tools conferred an even greater adaptive advantage which, in turn, increased the selective pressure for better eye-hand coordination, greater intelligence, and so forth.

Although it is now regarded as unlikely that this model best accounts for the evolution of human intellectual capacities (Wynn 1988), it does seem probable that constructing tools helped to further human intelligence. It also seems possible that the development of a tool tradition would have a similar influence on extraterrestrial life forms.

Interestingly, while the role of tool reliance is relevant to the development of intelligence, it seems to tell us nothing about the ethical implications of that intelligence. Tools can be used for a variety of purposes, both malignant and benign. The purpose for which tools are used will depend on considerations that are essentially independent of tool manufacture. Tool use means greater efficiency, but it does not suggest toward what end.

4.2 Spatial Behavior

Most evolutionary scenarios for our hominid past include a prolonged period of foraging. Except for carnivores, it seems likely that a lengthy interval of gathering would characterize many organisms as they evolved toward higher intelligence. Several anthropologists have argued that the demands of foraging behavior make increased intelligence highly adaptive. Foraging puts a premium on memory and on the ability to locate and exploit ephemeral resources. Further, foraging through a defined domain rewards the ability to estimate the location and reoccurrence of seasonal resources. One authority on primate foraging behaviors has argued that primates with larger brains also have larger ranges and more varied diets, suggesting a causal relationship (Milton 1988).

Whatever the role of foraging in selecting for intelligence, it seems likely that it would be only one factor among many. Some authorities have suggested that the evolution of the nervous system was partly due to the memory requirements described above and partly due to a more general need for problem solving skills. It is thought that there were selective pressures calling for the mind to make ever-finer discriminations (Iran-Nejad et al. 1992).

The ability to develop accurate cognitive maps of an organism's territory would confer a variety of advantages ranging from more reliable resource exploitation to fewer encounters with dangerous competitors. However, again, this adaptation would seem to provide little indication of the ethical implications of an intelligence derived from such stimuli. To encounter matters of ethical moment we must, almost by definition, look to the social realm where organisms interact with one another.

4.3 Social Behavior

In my opinion, the best argument for the importance of the social environment in creating pressure for increased intelligence was advanced by Alison Jolly (1985), a noted primatologist. Jolly's study of lemurs revealed that there were significant, complex, social problems that needed to be solved for an organism to mate, cooperate with others, and maintain a viable group status. She argued that the need to adapt to complex social circumstances selected for intelligence in both males and females. Further, the slow maturation of young lemurs created a situation in which learned social skills had an early impact on dominance relations and, later, on mating opportunities. This reproductive concern is not limited to males, because it has been shown that dominant females tend to have more opportunities for mating and a greater likelihood of raising dominant males.

Several studies have supported Jolly's contribution and elaborated some of the mechanisms involved (Lewin 1988; Paoli and Borgognini 2006; Stanford 1998). Cheney, working with vervets, found that their adaptive social behaviors and social learning were significantly more complex than behaviors related to other tasks such as foraging (Cheney and Seyfarth 1988). There is currently general agreement that demands of social participation are perhaps the most powerful stimuli for the development of higher intelligence. Authorities assert that socially-skilled organisms have significant advantages over others, including a better ability to foresee the behavior of competitors (Smith 1984, 69), and greater skill in constructing and maintaining profitable alliances (Harcourt 1988).

Portions of this scenario seem foreordained by the nature of intelligence itself. As we noted earlier, greater intelligence means a prolonged period of dependency, a greater need for a learned behavioral repertoire, and a general trend for social living to support the first two. The complexities of social life, the differential access to resources, and mating opportunities that accompany high levels of social skill all place considerable selective pressure on increased intelligence and, to some extent sociability. Ethological studies indicate that any organism whose behavior puts the group at risk suffers exclusion, injury and/or a loss of mating opportunities.

This model would seem to have some utility for conjecturing about the nature of extraterrestrial intelligence and attitudes. It seems likely that any intelligence that evolves in a social unit will be affected by the minimal functional requirements involved in-group cooperation and cohesion. The result will likely be an organism that has serious constraints on agonistic behavior and an ability to engage in cooperative endeavors. This scenario is markedly more hopeful than the one suggested above for intelligent predators, but it would still be wise to consider the probable nature of social behavior, for there are often marked differences between in-group behavior and that directed toward outsiders. All one need do to realize the significance of this distinction is to reflect on human history.

5 Machiavellian Social Behavior

If this material can be used to project extraterrestrial intentions, an examination of group behavior among monkeys, apes and humans reveals some rather disquieting social trends. Indeed, according to recent authorities, the altruism and cooperation that characterize social life appears to have roots in a rather ominous social calculus. Smith has argued that the exigencies of social life provide a powerful stimulus for increased intelligence, the capacity for symbolism and the ability to abstract patterns: "…an animal would have to think of others as having motivations similar to its own, so that it could foresee their future behavior, and it would have to communicate symbolically" (Smith 1984, 69).

However, the question remains as to what end these abilities are directed, and a collection of essays suggests that Machiavellianism is evolutionarily adaptive:

> …in most cases where uses of social expertise are apparent, they are precisely what Machiavelli would have advised! Cooperation is a notable feature of primate society, but its usual function is to out-compete rivals for personal gain. [However,]…it seems likely that the later course of human evolution has been characterized by a much greater emphasis on altruistic uses of intelligence. (Byrne and Whiten 1988, vi)

Unfortunately, the authors also note that the weight of evolutionary evidence supports the idea that our intelligence evolved principally from "a need for social manipulation" (Byrne and Whiten 1988, vi). Basically, it seems that it is in the individual's interest to take advantage of others, as long as doing so does not jeopardize social standing, mating possibilities, and access to resources.

If the nature of in-group dynamics seems a somewhat unpromising suggestion of what extraterrestrial contact might hold, the character of out-group relations is even less encouraging. Nobel laureate Konrad Lorenz (1963) has argued that inter-group relations among many species are characterized by aggression and that this agonistic behavior has a positive function. He suggests that intra-group aggression serves as a spacing mechanism to promote a dispersal of populations throughout the environment, thereby facilitating a more efficient utilization of resources. He notes that such behavior is particularly true for members of the same species and for others that exploit the same resources.

In instances of confrontations between carnivores, Lorenz believes that there are instinctive inhibitions on the use of deadly force. He suggests that these have evolved because carnivores are very well equipped to damage each other. Thus, the result of an aggressive encounter would probably mean the death or maiming of both parties. Instead, intra-carnivore contests, rather than extending to deadly action, are limited to displays of ferocity. However, herbivores and omnivores are less well equipped to seriously injure one another and, as a consequence, are presumed to lack instinctive checks on the display of intra-species aggression. Indeed, since both parties can survive the encounter, intra-species aggression among non-carnivores may help select for increased intelligence, because more intelligent organisms avoid contests they are apt to lose but initiate ones that they are likely to win (Cheney and Seyfarth 1988; Harcourt 1988). This would increase mating opportunities and inclusive fitness.

According to Borgia (1980), who has examined human aggression as a biological adaptation, individuals will participate in aggression when it improves their fitness relative to other behaviors in which they could engage. Thus, an accurate assessment of complex social circumstances where aggression may be directed toward others or toward oneself is a highly adaptive skill, and one that also places an emphasis on and selects for intelligence.

Intra-species behavior ranges from Machiavellian to agonistic according to whether the principles are members of the same or of different groups, among other relevant social variables. However, inter-species behavior displays a far narrow set of behaviors. Simply put, with the exception of some symbiotes, the record of inter-species behavior is clearly one of competition and aggression (Byrne and Whiten 1988; Hinde 1974; Lorenz 1963). It seems that the only consideration that tempers inter-species aggression is self-interest. Thus, some predators limit their kills and increase their territories in order to preserve the availability of prey (Lorenz 1963).

Thus, whether a species derives its intelligence from tool use, territorial exploration, an adaptation to complex social life, or some combination of the three, there seems to be no reason to anticipate the evolution of an intelligence characterized by beneficence. On the contrary, it seems that one of the functions of intelligence is to promote a more efficient exploitation of the environment, an environment that contains other organisms, including members of one's own group.

6 Conclusion

I confess to having begun the research for this chapter in a mood of optimism, anticipating that extraterrestrial intelligences would be at least as likely to display benevolence as malevolence since they would have mastered a complex technology, survived their own evolutionary challenges, and learned sufficient cooperation to make high civilization possible. My research has led to a reevaluation of my original expectations and, to the extent that these models are applicable to future encounters with extraterrestrials, a much more somber conclusion.

Obviously models such as these, which are grounded in the particular nature of terrestrial organisms, especially mammals, cannot presume to anticipate all possibilities. It is possible, though not probable, that an extraterrestrial intelligence would be telepathic, hive oriented or significantly different in a variety of ways (Hanlon and Brown 1989; Wasserman 1989). In such circumstances, models such as those proposed here may have limited, or even no, utility. However, several authorities believe there are good reasons to anticipate a sentience significantly different from our own but sharing sufficient characteristics to enable communication (Raybeck 1992; Sagan 1973; 1977).

I have not argued that a species must be a carnivore to be a predator. Indeed, some omnivores, such as ourselves, are truly formidable predators. Neither have I argued that a species must be exclusively a predator to be influenced by selective pressures appropriate for a predatory evolutionary scenario. However, if predation is a major means of environmental adaptation, then the presumed result is a simplistic world view representing a consistent "us–them" dichotomy in which "us" are fine... but "them" are dinner.

The assessments of non-predator forms of intelligence, while more complex and somewhat more encouraging than the models suggested by a presumed intelligent predator, still imply a rather unpromising set of circumstances. As noted earlier, intra-group behavior among non-predators seems best characterized by Machiavellianism rather than by disinterested altruism. As for inter-group relations, the likelihood of violence seems greatly increased. Still worse is the prognostication for inter-species violence, which would seem to approximate that suggested by the models for predator behavior.

If these scenarios seem too pessimistic, we should recall our own recent history and current state of affairs. As an omnivore with a rather predatory past, our treatment of our own species has not generally been characterized by an enlightened altruism. Slavery, colonialism, religious wars, and inter-ethnic violence have marked our history and continue to mar our present. This is not a necessary state of affairs, as there are societies, such as the Semai, where war and even interpersonal violence are effectively unknown (Dentan 1968; Knauft 1987). However, when humans compete for limited resources, inter-group violence is a common, and often predictable, response (Ferguson and Farragher 1988; Harrison 1973; Montagu 1968). Indeed, competition within groups can, in several social settings, also readily yield agonistic behavior (Chagnon 1983; Meggitt 1977). Thus, it would seem naive to anticipate better behavior from extraterrestrials than we manifest ourselves.

While the speed-of-light limitations on space travel make it unlikely that any extraterrestrial could readily visit us, such things are within the realm of possibility. The best analogy might be with early European exploitation of Southeast Asia. The distance was impressive, communications haphazard, and the risks great. Nonetheless, a small European power, Portugal, managed to enslave populations, devastate property and destroy small states. It also led to Portuguese control of the spice trade, and to Portuguese ascendancy back in Europe (Hall 1955; Harrison 1968; Swearer 1984).

Despite the rather negative conclusions of this study, I would not counsel the abandonment of SETI or any reduction of the current efforts to listen in on intelligent extraterrestrial life forms. On the contrary, I think we would be well advised to be as informed as we can concerning the possibility of other sentients. This is particularly the case as evidence continues to point to the increasing likelihood of extraterrestrial intelligence (Shostak 2009; Tarter 2007). Indeed, in light of the behavioral significance of differing gustatory patterns, I would particularly like to know what they had for dinner. I would feel much more comfortable entering into discussions with a salad-eater than with an entity that derives its nourishment from higher on the food chain. Nonetheless, as I have suggested, it is just these latter entities that we are most apt to encounter. What then?

The potential benefits to be gained from interstellar communication are too great to be ignored or avoided. Certainly the listening should continue but, as I have suggested, the potential danger of attracting the attention of an extraterrestrial sentient is also too great to be ignored. I recommend carefully assessing the location of any future extraterrestrial communicants, and gathering whatever information about them might be possible, prior to contemplating an active exchange of messages. Finally, if we do find reason to send forth a message, I recommend we break with the model established by Pioneers 10 and 11, which included a detailed representation of our solar system and some hints on how to get here. At the minimum, we should try to avoid including a return address.

References

Ardila, Alfredo, and Feggy Ostrosky-Solis, eds. 1989. *Brain Organization of Languages and Cognitive Processes*. New York: Plenum Press.

Bear, Greg. 1987. *The Forge of God*. New York: TOR.

Borgia, Gerald. 1980. "Human Aggression as a Biological Adaptation." In *The Evolution of Human Social Behavior*, edited by Joan S. Lockard, 165–191. New York: Elsevier.

Byrne, Richard W., and Andrew Whiten, eds. 1988. *Machiavellian Intelligence*. New York: Clarendon Press.

Chagnon, Napoleon. 1983. *Yanamamo: The Fierce People*. New York: Holt, Rinehart, and Winston.

Cheney, Dorothy L., and Robert M. Seyfarth. 1988. "Social and Non-Social Knowledge in Vervet Monkeys." In *Machiavellian Intelligence*, edited by Richard W. Byrne, and Andrew Whiten, 255–270. New York: Clarendon Press.

Daly, Martin, and Margo Wilson. 1978. *Sex, Evolution, and Behavior*. North Scituate, MA: Duxbury Press.

Dawkins, Richard. 1976. *The Selfish Gene*. New York: Oxford University Press.

Dentan, Robert Knox. 1968. *The Semai: A Nonviolent People of Malaya*. New York: Holt, Rinehart and Winston.

Evans, David L., and Justin O. Schmidt, eds. 1990. *Insect Defenses: Adaptive Mechanisms and Strategies of Prey and Predators*. Albany, NY: State University of New York Press.

Ferguson, R. Brian, and Leslie E. Farragher. 1988. *The Anthropology of War: A Bibliography*. New York: Harry Frank Guggenheim Foundation.

Gardner, Howard. 1985. *The Mind's New Science: A History of the Cognitive Revolution*. New York: Basic Books.

Gazzaniga, Michael S. 1992. *Nature's Mind: The Biological Roots of Thinking, Emotions, Sexuality, Language and Intelligence*. New York: Basic Books.

Goldstein, Ira, and Seymour Papert. 1977. "Artificial Intelligence, Language, and the Study of Knowledge." *Cognitive Science* 1:84–123.

Greenspan, Stanley I., and Stuart G. Shanker. 2004. *The First Idea: How Symbols, Language, and Intelligence Evolved from Our Early Primate Ancestors to Modern Humans*. Cambridge, MA: Da Capo Press.

Hall, D. G. E. 1955. *A History of South-East Asia*. London: Macmillan & Company.

Hanlon, Robert E., and Jason W. Brown. 1989. "Microgenesis: Historical Review and Current Studies." In *Brain Organization of Languages and Cognitive Processes*, edited by Alfredo Ardila and Feggy Ostrosky-Solis, 3–15. New York: Plenum Press.

Harcourt, Alexander H. 1988. "Alliances in Contests and Social Intelligence." In *Machiavellian Intelligence*, edited by Richard W. Byrne and Andrew Whiten, 132–152. New York: Clarendon Press.

Harrison, Albert A. 1993. "Thinking Intelligently about Extraterrestrial Intelligence: An Application of Living Systems Theory." *Behavioral Science* 38:189–217.

Harrison, Albert A., and Alan C. Elms. 1990. "Psychology and the Search for Extraterrestrial Intelligence." *Behavioral Science* 35:207–218.

Harrison, Brian. 1968. *South-East Asia: A Short History*. London: Macmillan.

Harrison, Robert. 1973. *Warfare*. Minneapolis: Burgess Publishing Company.

Hinde, Robert A. 1974. *Biological Bases of Human Social Behaviour*. New York: McGraw-Hill Book Company.

Hockett, C. F. 1973. *Man's Place in Nature*. New York: McGraw-Hill Publishers.

Holohan, Anne, and Anurag Garg. 2005. "Collaboration Online: The Example of Distributed Computing." *Journal of Computer-Mediated Communication* 10(4), article 16. Accessed December 28, 2012. http://jcmc.indiana.edu/vol10/issue4/holohan.html.

Iran-Nejad, Asghar, George E. Marsh, and Andrea C. Clements. 1992. "The Figure and the Ground of Constructive Brain Functioning: Beyond Explicit Memory Processes." *Educational Psychologist* 27:473–492.

Itzkoff, Seymour W. 1983. *The Form of Man: The Evolutionary Origins of Human Intelligence*. Ashfield, MA: Paideia Publishers.

Itzkoff, Seymour W. 1985. *Triumph of the Intelligent: the Creation of Homo sapiens sapiens*. Ashfield, MA: Paideia.

Jastrow, Robert. 1981. *The Enchanted Loom: Mind in the Universe*. New York: Simon and Schuster.

Jerison, Harry, J. 1990. "Paleoneurology and the Evolution of Mind." In *The Workings of the Brain: Development, Memory and Perception*, edited by Rudolfo R. Llinas, 3–16. New York: W. H. Freeman and Company.

Jerison, Harry J. 1973. *Evolution of the Brain and Intelligence*. New York: Academic Press.

Jolly, Alison. 1985. *The Evolution of Primate Behavior*. New York: Macmillan Publishing Company.

Knauft, Bruce M. 1987. "Reconsidering Violence in Simple Human Societies: Homicide among the Gebusi of New Guinea." *Current Anthropology* 28:457–499.

Laughlin, Charles D., Jr., and Eugene G. D'Aquili. 1974. *Biogenetic Structuralism*. New York: Columbia University Press.

Laughlin, Charles D., Jr., and Ivan A. Brady, eds. 1978. *Extinction and Survival in Human Populations*. New York: Columbia University Press.

Laughlin, Charles D., Jr., John McManus, and Eugene G. D'Aquili. 1990. *Brain, Symbol & Experience*. Boston: New Science Library.

Lee, Richard B. 1979. *The !Kung San: Men, Women, and Work in a Foraging Society*. New York: Cambridge University Press.

Lee, Richard B. 1984. *The Dobe !Kung*. New York: Holt Rinehart and Winston.

Lenneberg, Eric H. 1967. *Biological Foundations of Language*. New York: Wiley.

Lewin, Roger. 1988. *In the Age of Mankind*. Washington, DC: Smithsonian Books.

Llinas, Rodolfo R., ed. 1990. *The Workings of the Brain: Development, Memory, and Perception*. New York: W.H. Freeman and Company.

Lorenz, Konrad. 1963. *On Aggression*. Translated by Marjorie Kerr Wilson. New York: Harcourt, Brace & World, Inc.

Manosevetz, Martin, Gardner Lindzey, and Delbert D. Thiessen, eds. 1969. *Behavioral Genetics: Method and Research*. New York: Appleton-Century-Crofts.

Marshall, Lorna. 1976. *The !Kung of Nyae Nyae*. Cambridge, MA: Harvard University Press.

May, Julian. 1987a. *The Metaconcert*. New York: Ballantine Books.

May, Julian. 1987b. *Surveillance*. New York: Ballantine Books.

Mazur, Allan, and Leon S. Robertson. 1972. *Biology and Social Behavior*. New York: The Free Press.

Meggitt, Mervyn. 1977. *Blood is Their Argument: Warfare among the Mae Enga Tribesmen of the New Guinea Highlands*. Palo Alto, CA: Mayfield Publishing Company.

Miller, Geoffrey F., and Lars Penke. 2007. "The Evolution of Human Intelligence and the Coefficient of Additive Genetic Variance in Human Brain Size." *Intelligence* 35(2):97–114.

Miller, George A. 1951. *Language and Communication*. New York: McGraw-Hill Book Company, Inc.

Miller, George A. 1956. "The Magical Number Seven, Plus or Minus Two: Some Limits on our Capacity for Processing Information." *Psychological Review* 63:81–97.

Milton, Katherine. 1988. "Foraging Behavior and the Origin of Primate Intelligence." In *Machiavellian Intelligence*, edited by Richard W. Byrne and Andrew Whiten, 285–305. New York: Clarendon Press.

Montagu, M. F. Ashley, ed. 1968. *Man and Aggression*. London: Oxford University Press.

Morrison, Philip, John Billingham, and John Wolfe, eds. 1977. *The Search for Extraterrestrial Intelligence: SETI*. Washington, DC: National Aeronautics and Space Administration.

Nelson, Mark E., and James M. Bower. 1990. "Brain Maps and Parallel Computers." *Trends in Neurosciences* 13:403–408.

Oakley, Kenneth. 1959. *Man the Tool-Maker*. Chicago: University of Chicago Press.

Paoli, T., E. Palagi, and S. M. Borgognini Tarli. 2006. "Reevaluation of Dominance Hierarchy in Bonobos (*Pan paniscus*)." *American Journal of Physical Anthropology* 130(1):116–122.

Raybeck, Douglas. 1992. "Problems in Extraterrestrial Communication." In *Proceedings*. Ninth CONTACT Conference. Palo Alto, CA: CONTACT.

Robinson, Michael H. 1990. "Predator-Prey Interactions, Informational Complexity, and the Origins of Intelligence." In *Insect Defenses: Adaptive Mechanisms and Strategies of Prey and Predators*, edited by David L. Evans, and Justin O. Schmidt, 129–149. Albany, NY: State University of New York Press.

Sagan, Carl, ed. 1973. *Communication with Extraterrestrial Intelligence (CETI)*. Cambridge, MA: The MIT Press.

Sagan, Carl. 1977. *The Dragons of Eden: Speculations on the Evolution of Human Intelligence*. New York: Ballantine Books.

Sauer, Charles H., and Edward A. MacNair. 1983. *Simulation of Computer Communication Systems*. Englewood Cliffs, NJ: Prentice-Hall, Inc.

Schank, Roger C., and Peter G. Childers. 1984. *The Cognitive Computer: On Language, Learning and Artificial Intelligence*. Reading, MA: Addison-Wesley Publishing Co., Inc.

Schull, Jonathan. 1990. "Are Species Intelligent?" *Behavioral and Brain Sciences* 13:63–109.

Shostak, Seth. 2009. *Confessions of an Alien Hunter: A Scientist's Search for Extraterrestrial Intelligence*. Washington, DC: National Geographic.

Sluckin, W. 1965. *Imprinting and Early Learning*. Chicago: Aldine Publishing Company.

Smith, John Maynard. 1984. "The Evolution of Animal Intelligence." In *Minds, Machines and Evolution*, edited by Christopher Hookway, 63–71. New York: Cambridge University Press.

Stanford, Craig B. 1998. "The Social Behavior of Chimpanzees and Bonobos." *Current Anthropology* 39(4):399–420.

Stokoe, William C. 1989. "Language: From Hard-Wiring or Culture?" *Sign Language Studies* 63:163–180.

Swearer, Donald K. 1984. *Southeast Asia*. Guilford: The Dushkin Publishing Group, Inc.

Tarter, Jill C. 2007. "The Evolution of Life in the Universe: Are We Alone?" *Highlights of Astronomy* 14:14–29.

Thompson, Richard A., and John R. Green, eds. 1982. *New Perspectives in Cerebral Localization*. New York: Raven Press.

Tunnell, Gary G. 1973. *Culture and Biology: Becoming Human*. Minneapolis: Burgess Publishing Company.

Washburn, S. L. 1960. "Tools and Human Evolution." *Scientific American* 203:62–75.

Wasserman, Philip D. 1989. *Neural Computing: Theory and Practice*. New York: Van Nostrand Reinhold.

Wilson, Edward O. 1980. *Sociobiology*. Cambridge, MA: The Belknap Press.

Wynn, Thomas. 1988. "Tools and the Evolution of Human Intelligence." In *Machiavellian Intelligence*, edited by Richard W. Byrne and Andrew Whiten, 271–284. New York: Clarendon Press.

Harmful ETI Hypothesis Denied: Visiting ETIs Likely Altruists

Harold A. Geller

Abstract In the spring of 2010, after completing a new series for television, Stephen Hawking expressed his concern about any attempts to make contact with ETIs (extraterrestrial intelligences). He hypothesized that direct ETI contact would result in harm to our species. In this chapter we will examine the harmful ETI hypothesis of Hawking. We will demonstrate that this hypothesis is not logically consistent with what we know about the physical universe and the evolution of life on Earth. If anything, visiting ETIs will turn out to be altruists. We will demonstrate that the harmful ETI hypothesis necessarily leads to logical inconsistencies with the physical laws of the known universe. The efforts required to travel interstellar distances necessarily force ETIs to be altruists or remain in their corner of the galaxy.

Keywords Extraterrestrial intelligence · Stephen Hawking · Christopher Columbus · Fermi Paradox · Interstellar travel · Engineering · Reliability · Space colonization · Belligerence · Altruism

1 Introduction

After completing the filming of a series for television, Stephen Hawking was interviewed by a London reporter for the *Sunday Times*. Hawking said:

> To my mathematical brain, the numbers alone make thinking about aliens perfectly rational. The real challenge is to work out what aliens might actually be like.... [They] would be only limited by how much power they could harness and control, and that could be far more than we might first imagine.... Such advanced aliens would perhaps become

H. A. Geller (✉)
School of Physics, Astronomy and Computational Science, George Mason University,
4400 University Drive, Fairfax, VA 22030-4444, USA
e-mail: hgeller@gmu.edu

D. A. Vakoch (ed.), *Extraterrestrial Altruism*, The Frontiers Collection,
DOI: 10.1007/978-3-642-37750-1_5, © Springer-Verlag Berlin Heidelberg 2014

nomads, looking to conquer and colonize whatever planets they can reach.... I imagine
they might exist in massive ships, having used up all the resources from their home
planet.... If aliens ever visit us, I think the outcome would be much as when Christopher
Columbus first landed in America, which didn't turn out very well for the Native
Americans (Leake 2010).

We will first summarize the harmful ETI (extraterrestrial intelligence)
hypothesis as presented by Stephen Hawking in the above mentioned interview.
Hawking based his harmful ETI hypothesis on several premises: there are a large
number of possible sources of ETI in the galaxy; the biochemistry of ETIs is
comprehensible and similar to the human species; ETIs are capable of harnessing
and controlling much energy; ETIs would build spacecraft of enormous dimensions to colonize other worlds; and ETIs would use up the resources of their own
stellar system. From these premises Hawking concludes that if ETIs visit Earth, the
interaction would be similar to the situation experienced by the North American
natives at the time of Columbus' journey, thus we should not even attempt to
contact any ETIs in our galaxy.

2 The Large Numbers Argument

Let us examine individually each of the premises posited by Hawking. His first
premise is that "the numbers alone make thinking about aliens perfectly rational."
It would be self-defeating to disagree with Professor Hawking regarding the
rationality of thinking about ETIs within the confines of this volume on extraterrestrial intelligences, but let us examine his point explicitly. It appears Hawking
is referring to the very large numbers of stars in our Milky Way galaxy. Actually,
we assume Hawking is limiting his scope to our own galaxy. Perhaps he meant to
consider the universe at large. However, applying the Drake Equation (Drake
1965) to this analysis of the harmful ETI hypothesis limits us to our own galaxy,
which bounds our analysis appropriately. ETI space travel across intergalactic
distances and time can be considered a trivial extension to the harmful ETI
hypothesis.

If we are considering the entire universe, as opposed to the Drake Equation,
which addresses the number of civilizations in our galaxy, then we first need to
estimate the number of galaxies in the universe. The Hubble Space Telescope
provides deep space images which allow us to estimate about 150 billion galaxies
in the observable universe, which we will round off to 10^{11} galaxies (Beckwith
et al. 2006).

Let us next examine the number of stars in the Milky Way galaxy. Current
estimates of the number of stars in the Milky Way are on the order of 400 billion
stars (van Dokkum and Conroy 2010). However, again for ease of calculations, let
us assume that the number of stars in the Milky Way is 10^{11}.

We next consider the number of planets orbiting the stars in the Milky Way.
Best estimates for this are based upon the percentage of stars examined by the

Kepler mission which have discovered to have at least one planet. The results from the Kepler mission have us conclude that the assumption of the number of planets in the galaxy is ten times the number of stars in the galaxy (Basri et al. 2005).

For ease of calculations, let us assume that there is just one planet available in the habitable zone for each of the 10^{22} stars in the universe. Applying the law of large numbers and binomial statistics, it can be shown that the probability of having one planet with intelligent life is approximately e^{-1} or roughly 37 % (Stewart 2011). In fact, using these statistics, the probability of there being no planets with intelligent life is also 37 %. The probability that there are two or more planets with intelligent life in the universe—the only alternative to there being either one planet or no planets with intelligent life—is therefore $1 - e^{-1} - e^{-1}$ or roughly 26 % (Stewart 2011).

Thus, assuming this is the line of thought of Hawking, we have reasonable probabilities for intelligent life on planets in the universe. If this line of thought is continued, we could arrive at an estimate for the number of planets with intelligent life in the Milky Way as being on the order of 10^4 (Stewart 2011). That may put the average distance between planets with intelligent civilizations at about 1,000 light years. So, any intelligent civilization wishing to reach out to other civilizations in the physical way, sometimes known as contact of the third kind, would expect to travel 1,000 light years (Stewart 2011).

This kind of reasoning with the law of large numbers and binomial statistics is probably what led Enrico Fermi to the statement we know today as the Fermi Paradox. That is, if intelligent life is as widespread, even at 1,000 light years apart, then why has there been no contact with any extraterrestrial intelligent at this point? There are of course many alternative approaches to solving this paradox.

3 The Fermi Paradox Versus the Large Numbers

In 2002, physicist Stephen Webb compiled a group of 50 so-called solutions to the Fermi Paradox (Webb 2002). The title of the book is one way that Fermi put his question to his lunchtime friends. That is, *If the Universe Is Teeming with Aliens ... Where Is Everybody?* As part of his own 50th solution to the Fermi Paradox, Webb takes an alternate approach to the formulation of the Drake Equation. Webb proposes the development of an analogue to the Sieve of Eratosthenes, which was fundamentally a clever algorithmic methodology for uncovering prime numbers up to a given value first developed circa 240 BCE. In this case, the Sieve of Fermi would be used to examine all the planets in the galaxy and leave the most likely to have an intelligent civilization present, in lieu of prime numbers. The eight-step algorithm applied to the Sieve of Fermi leads Webb to conclude that the human species is the only intelligent species in the galaxy today. Of course, if this is the case, then there is no reason to worry about any attempts to communicate with another civilization, because we are the only ones in this galaxy at this moment (Webb 2002).

Now back to Hawking's harmful ETI hypothesis and its axioms. The next premise within the Hawking hypothesis deals with the chemical makeup of the ETI with which we make contact. In fact, the type of contact is also under consideration here by Hawking. The biochemistry of *Homo sapiens* has developed over 4.56 billion years. There is evidence that all life on Earth is derived from a common biochemistry found in some primordial living cell. The evidence for this includes a number of facts which individually would not be prohibitive statistically, but taken together are very improbable. For example, all sugars in every living creature on this planet use sugars of the dextro form (Zubay 2000). On the other hand (pun intended) all amino acids utilized in every cell on this planet is of the levo isomer form (Zubay 2000). Also, while there are 62 known amino acids, only 20 amino acids are utilized by every living cell on planet Earth (Zubay 2000).

It is difficult to understand how so many humans currently believe that we have not only been visited by ETIs but individual members of ETIs have had sexual relations with humans. While any single chemical similarity between ETIs and us humans may not seem like small odds, the combined probability of the chemical uniqueness of the biochemicals of our cells is another matter. Assuming the probability of ETIs having the same isomers of sugars is $\frac{1}{2}$ and the probability of having the same isomers of amino acids is also $\frac{1}{2}$, and the added probability of the same amino acids is 1/3, leading to a probability of all occurring as being 1/12. Now consider the probability of the same interpretation of the amino acid codons. The likelihood of the same nucleotide sequence is based upon the number of nucleotides. While every cell on the planet Earth may only make use of 4 nucleotides, there are 8 known nucleotides. That leads to a probability of getting the correct nucleotide as being $(1/8)^3$ or 1/512. So the probability of getting the correct nucleotide sequence, assuming the codons are interpreted the same in both species is 1/6144. Again, that is a probability without considering the fact that the codon interpretation by the transfer RNA must be identically complementary, which is a calculably small number of about 1/37748736.

Let us back up and consider some of the precursors to the chemistry of life. Hawking's second premise has deep underpinnings. Hawking assumes that the basic chemical elements that comprise our human bodies are indeed the same as the chemical elements that are used by the corporeal bodies of the ETIs. This is not a far-fetched premise, but its underpinnings are based upon the chemistry of Population I stars. Our Sun is fortunate to have been born in a nebula whose composition was enriched by previous stars of Population II and Population III. These generations of stars allowed our solar nebula to contain a large amount of the heavier chemical elements. After all, the universe is dominated by hydrogen and helium. However, if our solar nebula was comprised only of these two elements, then obviously *Homo sapiens* could not exist.

Now on to the next premise of Hawking's in his harmful ETI hypothesis. That is, the ETI must be capable of harnessing and controlling a great deal of energy. How much energy are we talking about? To say "a lot" doesn't even breach the subject.

4 The ETI Energy and Engineering Issues

In 1998, Edward Zampino of NASA's Lewis Research Center produced a review entitled *Critical Problems for Interstellar Propulsion Systems* (Zampino 1998). Although it is a dozen years old now, the critical problems addressed in this report still hold true today. In fact, they will hold true for the future because they are not just engineering issues of the present, but they address "the fundamental physics from which technology emerges." Let us examine the four physics and engineering problems that underpin these critical problems. The first is the so-called propellant mass problem (Zampino 1998).

For years, engineers have known that chemical rocket propulsion is simply not feasible for the journey to other stars. Even allowing for a lengthy trip of say a thousand years, engineers can calculate that the amount of chemical propellant required is on the order of 10^{119} kg (Millis 1997). Now it looks so simple an estimate, but let us examine that amount within the context of the total chemical production of the USA in 2011, which we will round off to 10^{12} kg (American Chemistry Council 2012). So turning the entire chemical industrial output of the USA towards rocket fuel would require 10^{107} years. This is an incredible 10^{97} times the age of the universe! Even adding the entire world's chemical production output would not get this time requirement anywhere near the age of the universe. So, for interstellar space travel you can safely forget about chemical propellants.

The next major issue that needs to be addressed Zampino refers to as the "round trip problem" (Zampino 1998). Here Zampino is referring to the minimum time it must take for a round-trip ticket to say the nearest stars in the Alpha Centauri system. Zampino forgoes any consideration of a fusion reactor, concluding that this too would not be sufficient energy generation for a round-trip interstellar journey. So Zampino jumps to what he calls the perfect rocket —that is, one that utilizes anti-matter. The energy conversion for Zampino's perfect rocket is assumed to be 50 %. While it doesn't appear to be shooting for the stars, Zampino provides an appendix which spells out the necessary time and propellant (in the form of anti-matter) that is required for such a journey.

For a journey that would appear to take the inhabitants of the spaceship some ten years, Zampino demonstrates that the mass ratio of initial vehicle mass to payload mass would still be on the order of 4, and it would be possible to envision a velocity of greater than 80 % the speed of light in vacuum. So for a payload of 10^6 kg, you would need about 360 times that much fuel in the form of antimatter (Zampino 1998). The problem is, how do you confine the antimatter? Don't forget that it will experience annihilation, or the formation of gamma-ray radiation, when contacting any ordinary matter. The engineering of this is unimaginable, in spite of the many science fiction stories that achieve this so easily.

The third major issue that needs addressing before interstellar travel can be achieved is related to the consequences of Einstein's Special Theory of Relativity. There are many aspects to this application of the relativity theory. Two of the most relevant consequences of relativity to ETIs traveling vast distances to get to Earth include time dilation and the relativistic increase of mass.

Many science pundits are quick to point out that the vast distances of space can be traversed within a human lifetime. In fact, due to relativistic time dilation, if humans were to travel to the nearest stars in the Alpha Centauri system, you could conceivably traverse the distance in ship time of just 1.745 years at 98 % the speed of light in vacuum, while here on Earth of course 8.77 years would have passed (Zampino 1998). But that is just the nearest of stars. Let us examine the case of space travel to something much further. Assume you are able to obtain a velocity very close to the speed of light, specifically 99.9999999999995 % that of c, the speed of light in a vacuum (Zampino 1998). You could conceivably reach the Andromeda galaxy in about 56 years, still within a human lifetime. Of course the time lapsed on Earth would be over 4 million years. So you would not have the same humans to return to, not even the same species. However, it still is theoretically possible to traverse galactic distances in a lifetime within the confines of the theory of relativity. However, please note that we have not said anything about the amount of energy needed to traverse this distance.

Relativity allows one to calculate the energy required to make such trips as noted above. With respect to the trip to the Andromeda galaxy, it would require about 10^{15} kWhr (Zampino 1998). So what does that number represent? It is on the same order as all of the electrical energy generated on planet Earth in a single year! In this case it is the electrical energy output for the year 2009. Would the Spanish government have agreed to sponsor Columbus' voyage if it required all of the energy of the known world at the time?

5 The ETI Reliability Issues

The fourth and last major problem addressed by Zampino is the reliability issue. Even a refrigerator does not last forever—in fact, we can estimate how long an appliance may last, and then when it needs maintenance or replacement.

A major parameter used by the engineering community in determining the reliability of a piece of equipment is known as the mean time between failures, or MTBF. Let us return to the very simple voyage to Alpha Centauri at a modest 0.98 the speed of light in vacuum. This would have the crew experiencing 1.745 years on this voyage, while the stay-at-home family of the crew would experience a passage of 8.77 years (Zampino 1998). Utilizing a standard reliability function R(t) for any device, one needs to solve the equation:

$$R(t) = e^{-\int h(x)dx}$$

Zampino applies the often utilized simplification by assuming the failure-rate function is a constant, leading to:

$$R(t) = e^{-\lambda t}$$

Using the above reliability function and assuming a terminal velocity of 0.98 c and a 30-day stay at the destination leads one to conclude that a total of 1.75 years will pass for the crew and 8.77 years of time will pass for the stay-at-home family members. It can be shown that the mean time between failures must be 1,774 years for a reliability of 0.999 (Zampino 1998, 5). Remember, you would be relying on a 24-h, 7-day-a-week, continuous-duty cycle. Can you imagine a refrigerator which can last for 1,774 years? The engineering of this situation is far from trivial! Zampino (1998, 5) himself concludes that any "space vehicle reliability will have to be 'ultra-high' and must 'transcend' the reliability of any space vehicle hardware, which we design and build today."

6 The Clash of Civilizations and Species

Let us now recall that Hawking's statement specifically provides an analogue encounter using Christopher Columbus. With respect to a visit from ETIs, Hawking stated that "I think the outcome would be much as when Christopher Columbus first landed in America, which didn't turn out very well for the Native Americans." The first thing we need to do is clarify the details of Columbus' story. While many young students in school are taught a myth and legend about the voyages of Columbus, historians have pieced together a rather different story. First of all, it needs to be pointed out that Columbus and the American natives were of the same species. This is a fundamental flaw in Hawking's analogy. It has already been established herein that there is an infinitesimally small probability that any ETI will be able to have a sexual encounter with a human. It will just not be genetically feasible.

So why did Columbus choose to voyage to the Americas? Actually, before we address this question, it should be pointed out that Columbus was not the first to encounter the natives of the Americas. In his book *Lies My Teacher Told Me: Everything Your American History Textbook Got Wrong*, James W. Loewen provides a summary of no less than 15 previous encounters in history with natives in the Americas. Now, it is admitted that not all of these have been historically certified, so Loewen provides his readers with a relative measure of the quality of the evidence in support of these encounters. Nonetheless, Columbus was not the first human from another continent to encounter the natives of the Americas, who themselves were not native to the Americas if you go back far enough in time, say 10,000–50,000 years ago (Loewen 2007).

Now, can the reason for Columbus' journey really be analogous to ETIs' reasons for journeying to our planet Earth? First, we must clarify why Columbus chose to journey to the Americas, which he himself may not have been certain even existed. In spite of what some are taught in grade school, Columbus did not set out to prove the world round. This was known for centuries, if not millennia prior to Columbus' journey. While there are those who argue as to whether or not Columbus had the correct circumference of the Earth, no scholars seriously argue that Columbus was the only one of a handful of people of his era who actually thought the world was spherical, not flat. Nonetheless, the myth continues to this day.

One thing all historians appear to agree upon is the fact that Columbus did not seek out a new continent with any altruistic underpinnings (Loewen 2007). However, please note that Columbus did indeed know that he had discovered a new continent, as his own journal alludes to this fact. Nonetheless, Columbus and his backers, even the Queen of Spain, sought benefits for themselves in a number of ways. Whether gold or spices or some combination thereof, plundering all newly discovered landforms and their inhabitants was certainly on the list (Loewen 2007).

There is evidence that whenever the Spaniards landed on an unknown land they would read what was called "the Requirement." This included a statement that I must include here, if just to demonstrate their selfish ends, with whatever means necessary, to subjugate the natives:

> I implore you to recognize the Church as a lady and in the name of the Pope take the King as lord of this land and obey his mandates. If you do not do it, I tell you that with the help of God I will enter powerfully against you all. I will make war everywhere and every way that I can. I will subject you to the yoke and obedience to the Church and to his majesty. I will take your women and children and make them slaves (Loewen 2007, 36).

The above statement may be what Hawking had in mind, but while it applies to Columbus and his contemporaries, it will be demonstrated that this attitude cannot be the attitude of ETIs who actually reach the Earth.

Even Columbus' own son, Ferdinand Columbus described a tribute system which his father had set up:

> [The Indians] all promised to pay tribute to the Catholic Sovereigns every three months, as follows: In the Cibao, where the gold mines were, every person of 14 years of age or upward was to pay a large hawk's bell of gold dust; all others were each to pay 25 pounds of cotton. Whenever an Indian delivered his tribute, he was to receive a brass or copper token which he must wear about his neck as proof that he had made his payment. Any Indian found without such a token was to be punished (Loewen 2007, 56).

Loewen summarizes the situation succinctly as "Spaniards hunted American Indians for sport and murdered them for dog food" (Loewen 2007, 56). Certainly, such a summary may lead Hawking to request that our species not alert any ETIs to our existence. But again, the analogy is not valid. ETIs will not be able to treat humans this way if our biochemistries are so different. ETIs could not even use our species for "dog food" as their equivalent dogs would not be able to metabolize our biochemicals.

Even the native females, when impregnated by their overlords, chose to "abort and have aborted" their pregnancies rather than "leave them in such oppressive slavery." But any ETIs would be hard pressed to copulate with our species, due to the near-certainty that our DNA or generational information transference would not be alike.

It should also be pointed out that even Columbus had a very different view of the meeting of another civilization on his initial voyage. His first words used to describe the natives included "well built" and "of quick intelligence" (Loewen 2007, 62). In his later descriptions from his later voyages, his description transmutes into calling the natives as "cruel," "stupid" and "warlike and numerous, whose customs and religion are very different from ours." The change in attitude towards the natives was largely due to Columbus' failure to discover much gold, and turn the ventures into profitable ones. Large profits could be reaped only by the slave trade, first of Indians to Europeans and then from Africa. Psychologists refer to this as a cognitive dissonance.

We should note that there is evidence that there was a slave trade even prior to Columbus. We have already noted that there was much evidence for exposure to Western Civilization many years prior to Columbus, and some of their reasoning was based upon the slave trade. It is possible that even Phoenician sailors traveled to the so-called New World, and that they brought their own slaves with them.

It is commonly-held misconception that the natives easily succumbed to the superior technology and strengths of the European invaders. However, evidence-based history teaches us a different story. The natives were in many cases very able warriors and had superior numbers. The real gift relayed by the European explorers which allowed the conquering of native populations was "the diseases Europeans brought with them that aided their conquest" (Loewen 2007, 73). As has been discussed already, the diseases that any ETIs bring to our planet Earth will not be evolutionarily compatible with our own body chemistry.

Now to the question of the likelihood that if a species does spread through the galaxy it would be a harmful ETI. Do we have any examples of this? We have seen that the examples used by Hawking do not apply as they inevitably are based upon a single species. As Columbus and all following European invaders were of the same species as the natives discovered on the islands or mainland of North and South America, they cannot be used in any comparison. So is there any available to us?

Since it has been demonstrated that ETIs would not likely be able to make use of human biochemistry, as it would necessarily be quite different from ETI biochemistry, what about the interaction examples with different species? It is likely that if Hawking would accept the lack of an analogue with Columbus or any other human prospective conqueror, then he might put forward examples here on Earth between interactions of humans and non-human species. One example might be the dodo bird. Recall that the dodo bird was a flightless bird endemic to the island of Mauritius, an island east of Madagascar in the Indian Ocean. On the island, the dodo birds had no native predators (Hume 2006). Unfortunately for them, human European explorers discovered their existence as early as 1598. By 1662 the dodo bird was extinct. The island paradise of the dodo bird had been overwhelmed by humans, without a fight, as the humans used the dodo bird as a source of food (Hume 2006). This certainly is an example of two different species encountering each other, and as noted, it didn't end well for the native biota. However, please note that dodo bird biochemistry was evolved from the same cellular ancestry as

Homo sapiens and thus allowed for the bird to act as a food source for members of the invading species. Might this happen if we were visited by an ETI? The only reasonable answer is "no"! You see, while the dodo bird was derived from a common ancestor to *Homo sapiens*, even though we shall never know how much was truly common without any DNA to examine, any visiting ETI would not have evolved from the same common ancestor. Thus, the visiting ETI would not be afforded any beneficial biochemical nourishment in consuming the human species. In fact, such a difference in cellular biochemistry may cause the consumption of humans to be detrimental to the health of the visiting ETI.

The dodo bird is just one example of what is nominally considered to be an encounter between an intelligent species, i.e., *Homo sapiens*, and a classic "dumb" species, the dodo bird. So what about encounters between intelligent species?

Here on Earth those species considered most intelligent outside of the genus *Homo* include the dolphin and chimpanzee. Encounters between these species have not gone too well. Dolphin is considered a delicacy in some areas of the world. Even when our own species honors the ecosystem of an intelligent species like dolphin, the results can be disastrous for the species. One only has to realize that dolphins were killed by the thousands when they were caught in the nets of those trolling for fish, such as tuna. So again, a result for dolphin that is less than ideal. And finally, what about the fate of that species on Earth considered most intelligent next to humans, the chimpanzee? One only has to read of the fate of chimpanzees as pets, although not always successful ones, and the fate of chimpanzees as experimental guinea pigs for human studies of disease and other conditions that lead to death, to exemplify such a species interaction.

Once again, while the meeting of the only known intelligent species today never ended up with stories of mutual satisfaction, the meetings of species has always been on this planet, between different species which nevertheless have the same common ancestor and identical underpinning biochemistry.

Thus, at this point, there is no reason for ETIs to have a desire to conquer the dominant species on Earth, which is, by the way, insects. Assuming that ETI noticed the modulation of electromagnetic radiation and sought to discover the source of modulation, the species it would be led to is *Homo sapiens*. And what about the travelers in the craft utilized by the ETI to reach our planet? What would they be like? We are arguing that they would not be belligerent. They would have no cause to be belligerent because they would not require any of our natural resources. They don't need our resources because they have mastered interstellar travel, which implies that they have an energy source which they can utilize. With such an energy source, they could manufacture any foodstuffs, which is actually a chemical energy source for life, and thus don't require humans as food. Such a concept, of using humans like cattle being led to the slaughter, was a plot in a 1962 episode of *The Twilight Zone*. It just does not make any logical sense to travel across interstellar distances just for food, when all you need is an energy source for your life whose value must exceed the energy source utilized to travel the interstellar distances.

Now consider the ETI interstellar traveler. Since belligerence would not be a useful characteristic of any ETI on an interstellar journey, what type of being must such a traveler be? It seems reasonable to conclude that the interstellar traveler would be altruistic in nature.

This begs the question as what altruism can be. There is a fascinating article by Benjamin Kerr, Peter Godfrey-Smith, and Marcus W. Feldman simply titled "What Is Altruism?" that appeared in *Trends in Ecology and Evolution* in 2004 (Kerr et al. 2004). Their viewpoint is from the science of evolutionary biology and their abstract provides a simple definition of altruism: "Altruism is generally understood to be behavior that benefits others at a personal cost to the behaving individual" (Kerr et al. 2004, 135). As the authors are quick to point out, "other authors have interpreted the concept of altruism differently, leading to dissimilar predictions about the evolution of altruistic behavior" (Kerr et al. 2004, 135).

In 1977, G. Harry Stine wrote a popular level book *Living in Space*. One of his chapters necessarily examined the social aspects of living in space. Stine utilized data developed by NASA regarding the volume of a spacecraft needed to support a person for a given duration of time. NASA examined three types of habitats, classified as civilian habitats with recycling and agriculture, civilian habitats with limited recycling, and military-type habitats. These data were extrapolated for missions on the order of 70 years (Stine 1997). NASA concluded that military-type habitats were the most efficient ones for both short voyages and long term voyages. NASA noted that the most efficient long-term habitats in the military were nuclear submarines, which have about ten cubic meters per person. For long-term space habitats approaching 70 years, NASA concluded that about 1,000 cubic meters per person was required for civilian models (Stine 1997). About one-tenth of that value was considered as being necessary for military-type habitation models. Nonetheless, all models had at their underpinning an understanding that the "good of the many, outweighs the good of the individual" (Helweg 1996). This is, at its core, an altruistic characteristic. After all, on a long journey such as an interstellar one must be, there cannot be any room for selfishness. There is just no benefit to a selfish individual in the confines of a habitat that must endure for years, even greater in length than a single lifetime.

Now do we have any examples of species that have ventured off of their own planet? We only know of one species, *Homo sapiens*, which has ventured to another celestial body, that is the Moon. But the history of the exploration of the Moon, which began in the late 1950s, demonstrates that the reason for sending people to the Moon was anything but altruistic. There is the evidence of the plaque deposited on the surface of the Moon by the first visitors of the *Homo sapiens* species which states that "We came in peace for all mankind" (Dick 2008). However, historians such as Steven Dick have long known of the complexities of the gargantuan effort which ultimately succeeded in placing that plaque on the surface of the Moon, and the many different reasons that numerous leaders of our society really had in mind in accomplishing such a feat. This included every human driving force from altruism to greed.

7 Conclusion

There was a conference held in Washington, DC celebrating the anniversary of the launch of the Sputnik satellite. The scholars who attended were pretty much in agreement that the reason for the motivation for the exploration of the Moon was not altruistic. The United States of America was in competition with its chief opponent on the planet Earth at that time, the Soviet Union. In fact, once it was apparent that there was no longer a competition with an enemy, the USA never returned to the Moon, and in fact, is not even able to do so now with a manned operation, approaching 50 years after that initial foray into space.

Thus, as seen by the manned exploration effort buoyed by the Cold War, selfishness does not result in a continuing effort to explore another celestial body. While there is talk about returning to the Moon by our species, the only motivation given these days is the exploitation of the minerals and natural resources by for-profit corporations. Due to the effort required and the amount of funds that need to be invested, these commercial ventures are not likely to succeed. This is something that the author was a part of studying when under contract with NASA Head-quarters Office of Commercial Programs via Science Applications International Corporation (SAIC).

The question of announcing our existence to rest of the universe is already moot. Modulated radio waves have already left this planet announcing that some kind of intelligent life form exists on here. We cannot halt these waves.

The questions raised by the harmful ETI hypothesis have been asked before, and will be asked again in the future. In 1985, within the volume *Interstellar Migration and the Human Experience*, Ben Finney stated that "the prospect of colonizing already inhabited worlds would be as illogical as it would be morally repugnant" (Finney 1985, 197). In the same book, Michael Hart addresses ETI belligerence. He says "interstellar wars will be extremely rare, much rarer than warfare has been on Earth. This will be a consequence of the enormous distances between the stars and the large travel times between civilizations" (Hart 1985, 290). Alfred Crosby, in the same volume, states categorically that "the most dangerous enemy of humans beyond Earth's atmosphere and beyond the Moon's orbit in the long term will be the implacably hostile environment" (Crosby 1985, 210) of space itself. Perhaps even Crosby has missed the mark, and perhaps Walt Kelly was right when he said that "We have met the enemy and he is us" (Kelly et al. 1982). The enemy is not any ETI.

References

American Chemistry Council. 2012. *Guide to the Business of Chemistry—2012*. Washington, DC: American Chemistry Council.
Basri, Gibor, William J. Borucki, and David Koch. 2005. "The Kepler Mission: A Wide-field Transit Search for Terrestrial Planets." *New Astronomy Reviews* 49(7–9):478–485.

Beckwith, S.V.W., M. Stiavelli, A. M. Koekemoer, J. A. R. Caldwell, H. C. Ferguson, R. Hook, R. A. Lucas, L. E. Bergeron, M. Corbin, S. Jogee, N. Panagia, M. Robberto, P. Royle, R. S. Somerville, and M. Sosey. 2006. "The Hubble Ultra Deep Field." *The Astronomical Journal* 132:1729–1755.

Crosby, Alfred W. 1985. "Life (with All Its Problems) in Space." In *Interstellar Migration and the Human Experience*, edited by Ben R. Finney and Eric M. Jones, 210–219. Los Angeles: University of California Press.

Dick, Steven J. 2008. *Remembering the Space Age*. Washington, DC: National Aeronautics and Space Administration.

Drake, Francis. 1965. "The Radio Search for Intelligent Extraterrestrial Life." In *Current Aspects of Exobiology*, edited by G. Mamikunian and M.H. Briggs, 323–345. New York: Pergamon.

Finney, Ben R. 1985. "The Prince and the Eunuch." In *Interstellar Migration and the Human Experience*, edited by Ben R. Finney and Eric M. Jones, 196–208. Los Angeles: University of California Press.

Hart, Michael H. 1985. "Interstellar Migration, the Biological Revolution and the Future of the Galaxy." In *Interstellar Migration and the Human Experience*, edited by Ben R. Finney and Eric M. Jones, 278–291. Los Angeles: University of California Press.

Helweg, Otto J. 1996. "Teaching Ethics: Are We Emphasizing the Right Thing?" Paper presented at the Conference on Values in Higher Education, Ethics and The College Curriculum: Teaching and Moral Responsibility, Knoxville, Tennessee, April 11–13.

Hume, Julian P. 2006. "The History of the Dodo *Raphus cucullatus* and the Penguin of Mauritius." *Historical Biology* 18:2, 69–93.

Kelly, S., W. Crouch, and G. Trudeau. 1982. *The Best of Pogo: An Exuberant Collection of Walt Kelly's Immortal Cartoons Plus Photos, Articles and Other Pogo Memorabilia from the Pages of The Okefenokee Star*. New York: Simon and Schuster.

Kerr, Benjamin, Peter Godfrey-Smith, and Marcus W. Feldman. 2004. "What Is Altruism?" *Trends in Ecology and Evolution* 19(3):135–140.

Leake, Jonathan. 2010. "Don't Talk to Aliens, Warns Stephen Hawking." *The Sunday Times* (London), April 25. Accessed December 28, 2012. http://www.thesundaytimes.co.uk/sto/news/uk_news/Science/article272392.ece.

Loewen, James W. 2007. *Lies My Teacher Told Me: Everything Your American History Textbook Got Wrong*. New York: Simon and Schuster.

Millis, M.G. 1997. "Breaking Through the Stars." *Ad Astra* 9(1):36–40.

Stewart, Ian. 2011. *The Mathematics of Life*. New York: Basic Books.

Stine, G. Harry. 1997. *Living in Space: A Handbook for Work and Exploration Beyond the Earth's Atmosphere*. New York: M. Evans and Company, Inc.

van Dokkum, P. G., and C. Conroy. 2010. "A Substantial Population of Low-mass Stars in Luminous Elliptical Galaxies." *Nature* 468(7326):940–942.

Webb, Stephen. 2002. *If the Universe Is Teeming with Aliens…Where Is Everybody?: Fifty Solutions to the Fermi Paradox and the Problem of Extraterrestrial Life*. London: Copernicus Books and Praxis Publishing.

Zampino, Edward. 1998. *Critical Problems for Interstellar Propulsion Systems*. Internal Report. Cleveland, OH: NASA Breakthrough Propulsion Physics.

Zubay, G. 2000. *Origins of Life on the Earth and in the Cosmos*. New York: Harcourt Academic Press.

Altruism Toward Non-Humans: Lessons for Interstellar Communication

Abhik Gupta

Abstract Discussions to date about communicating concepts of altruism to extraterrestrial intelligence (ETI) have focused on examples of altruism between members of the same species. However, because ETIs belong to species different from humans, analysis of yet another dimension of human altruism, the one which encompasses not only the non-human living world, but also non-living components of nature such as rivers, forests, mountains, and the like, is essential in light of the concepts and theories of evolutionary sociobiology. Because these beings have evolved independently, notions of kin selection would not guide their relations with humans. Similarly, given the vast distances of interstellar space and the long durations of round-trip exchanges of information, notions of reciprocity might also be irrelevant—at least at the level of biological individuals. We might, however, gain some clues to possible motivations for interstellar dialogues by considering altruism as revealed by human biophilia, or the affinity for other forms of life. Humans exhibit altruism towards others, including non-humans, because they assign some value to the receiver who may be a relative, a non-relative, a non-human plant or animal, or even a non-living river, mountain, or forest. Such value could be of two basic types, viz., extrinsic or instrumental value, which derives from the use value of a given subject, and intrinsic or inherent value, which exists regardless of utility. Altruism or cooperation appears to be a characteristic feature of mature and advanced societies, and is expected to be possessed by highly-advanced ETI societies. I argue in this chapter that the phenomenon of biophilia, which is perhaps a remnant of the nature-religions practiced by hunter-gatherer societies and is still practiced by indigenous ethnic groups all over the world, provides the basis for our ability to extend cooperation to ETI. If nature-religions

A. Gupta (✉)
Department of Ecology & Environmental Science, Assam University,
Silchar 788011 Assam, India
e-mail: abhik.eco@gmail.com

A. Gupta
Home: 1st Floor, Kuntila Apartment, Opp. Adharchand H.S. School, Ambicapatty,
Silchar 788004 Assam, India

D. A. Vakoch (ed.), *Extraterrestrial Altruism*, The Frontiers Collection,
DOI: 10.1007/978-3-642-37750-1_6, © Springer-Verlag Berlin Heidelberg 2014

are perceived as a meme-complex, the rites and rituals accompanying nature worship could form just one such meme in this meme-complex. The other memes in this complex may include an ability to appreciate nature's beauty and bounty, recognition of an intrinsic value in nature, and a general feeling of biophilia. As we contemplate the possibility of communication with extraterrestrial intelligence, the meme-complex of biophilia may provide insights into a plausible prerequisite for interstellar discourse: altruism that extends far beyond the care of conspecifics.

Keywords Altruism · Extraterrestrial intelligence · Extrinsic value · Intrinsic value · Non-human organisms · Nature religions · Memes

1 Introduction

One of the most mysterious and exciting questions that still eludes human knowledge is whether life, especially intelligent life, exists only on planet Earth or elsewhere in the universe as well. If we accept the notion that there is a high probability of the existence of extraterrestrial intelligent life (ETI), then what do we communicate to it to give it a fair idea about the human species? Besides our knowledge of science, technology, art, literature and philosophy, it may also be essential to enlighten ETIs about the evolutionary traits of the human mind, so that they have prior knowledge of the nature of human response to their presence. It may be important for them to know whether humans would receive them (either physically or the mere knowledge of their existence through radio signals) with hostility and mistrust or with love and joy. Ironically, I feel that humans need to pose this question to themselves as well, and in this sense, the composition of interstellar messages becomes a soul-searching endeavor for the human species. It has been said that just looking for ETI has expanded humanity's perception of its "self-image" (Tough 1998, 745), and in the poignant words of Carl Sagan, "In the deepest sense, the search for extraterrestrial intelligence is a search for who we are" (Carl Sagan quoted in Tough 1998, 746). It is in this context that altruism becomes a very important evolutionary trait and a human value that needs to be communicated to extraterrestrial intelligence.

Altruism, which literally means unselfishness, an act of benevolence or welfare to others, or in a broader sense, cooperative behavior, has stirred the imaginations of philosophers, political scientists, economists, psychologists and evolutionary biologists. Does altruism come naturally to living beings in which it is encountered, or is it exceptional behavior? The latter is more likely not the case, as altruism is encountered in a wide array of organisms ranging from the lowly protozoa, to social insects, to relatively highly evolved vertebrates like fishes, birds, mammals, and among the latter, in humans, who are believed to be the most highly evolved form of life on Earth. Extending the same logic to the ET, one would expect altruism to be a characteristic feature of many ET species, including

ETI, as evolution of life anywhere in the Universe is likely to proceed along Darwinian pathways (Dawkins 1976). Furthermore, convergence is also of very widespread occurrence in the evolution of life on Earth, and is likely to occur on other life-bearing planets.

Altruism baffled Darwin so much that he called it an apparently "insuperable problem," a "special difficulty" and consequently "fatal" to his theory of natural selection, and he had to come up with an answer, which of course he did. However, the genetic basis of altruism through kin selection was first postulated by Hamilton (1964a; b). Subsequently, the question of altruism among unrelated individuals was explained through the theory of reciprocity and the strategy of Tit for Tat (TFT) (Trivers 1971; Axelrod and Hamilton 1981) as well as through trait-group selection (Wilson 1975; Wilson and Sober 1994) and by-product mutualism that involves indirect reciprocity (Connor 1986). Nevertheless, it has also been shown that cooperative and defecting strategies could co-exist at equilibrium (Dugatkin and Wilson 1991; Wilson et al. 1998). In fact, human altruism exhibits so much heterogeneity that mere gene-based evolutionary theories are inadequate to explain all the patterns. The role of gene-culture coevolution and transmission via cultural traits or memes/culturgens needs to be taken into account when explaining the evolution of altruism in human societies (Dawkins 1976; Boyd and Richerson 1982; Delius 1991; Fehr and Fischbacher 2003; Gintis 2011). Culture has been defined as, "all of the information that individuals acquire from others by a variety of social learning processes including teaching and imitation" (Boyd and Richerson 1985, cited in Richerson et al. 2010, 8985). It has also been proposed that altruistic norms that are not individually fitness-enhancing can nevertheless "hitchhike" on individually fitness-enhancing internal norms and thereby ensure their transmission and continued survival (Gintis 2003). Reciprocity in repeated interactions, "reputation-based cooperation" and "strong reciprocity," the latter being a combination of altruistic rewarding and punishment, irrespective of reputation or other gains to the strong reciprocators, are believed to be powerful determinants of human altruistic behavior. Despite these expositions about the nature of human altruism, several questions remain, such as the exact nature of the relationship between altruism and cultural and economic institutions. Another grey area is the role of altruistic rewarding for cooperation in large groups (Fehr and Fischbacher 2003). It has also been shown that both cooperative behavior and punishment of defectors led to activation of neural circuits associated with rewards and a negative dopamine response in the event of the partner defecting. This indicates that evolution has equipped humans with proximate mechanisms that make altruistic behavior rewarding (Fehr and Rockenbach 2004). It also leads to the supposition that cooperation serves as a positive trait in the survival and success of a prosocial species like *Homo sapiens*. A totally new face of human altruism is now unfolding, and that is toward ETI. Here we ought to analyze a kind of "two-way traffic," where we try to predict possible ETI responses to humans while delving into the expected human responses to the presence of ETI. In order to fully understand the constructs of this aspect of the human (and the extraterrestrial) mind, it is perhaps not enough to analyze only the evolutionary sociobiological basis of altruism among unrelated individuals and strangers. This is primarily due to the fact that ETIs belong to distinctly different biological species. Hence, in order to understand human altruism toward ETI and vice versa, we may have to probe into yet another dimension of human altruism that takes into its fold non-human organisms, including both plants and animals. And if we accept that convergence could not only be widespread in genetic but also in cultural evolution, we might venture to conjecture that ETI could also have experienced similar transitions during its evolution. However, before we go to the intricacies of human altruism toward non-humans, it may be worthwhile to look into the linkages between evolutionary sociobiology and normative ethics with relation to the emergence of this type of altruism.

2 Human Altruism: A Function of Values

Why do humans exhibit altruism towards others, including non-humans? It occurs basically because we assign some value to the receiver, who may be a relative, a nonrelative, a non-human plant or animal, even a non-living river, mountain or forest. Given this fact, we must now consider if we would also probably be altruistic toward ETs. Such values could be of two basic types, viz., extrinsic or instrumental value, and intrinsic or inherent value. Extrinsic value derives from the objective properties of something or somebody that has use for others (in this case to the doer of an altruistic act) by virtue of its functions (Martell 1994). Thus humans can exhibit altruism to another living or non-living entity simply because they derive some utility or expect some reciprocity, direct or indirect, in the present or the future, from the latter. For instance, when we protect an animal or a plant, or an ecosystem, or the entire biosphere, guided by the principles of sustainable development for building a safe and sustainable future for ourselves and/ or our progeny, we exhibit an act of altruism which emanates from the fact that we assign such extrinsic value to these entities. This is because we expect some reciprocal benefits, direct or indirect and for present and/or future generations, to accrue to us from these entities. These altruistic acts to non-humans are, therefore, essentially anthropocentric. In contrast to such extrinsic values, an intrinsic value is a value in itself regardless of its utility for us. For example, we (or at least most of us) tend to believe that other human beings have a value in themselves as conscious, intelligent beings regardless of their use or value for others (Martell 1994). When human beings do an altruistic act to total strangers even in purely one-shot Prisoner's Dilemma Games, they do so because they recognize some intrinsic value in that other being and not because of the latter's relationship with the former, or in expectation of some benefit or reward from the latter. Such intrinsic value may be nurtured in association with some internal norm(s), which may be defined as a behavior pattern enforced by internal sanctions such as shame, guilt, sense of honor, empathy for others, and loss of self-esteem, etc., and transmitted both vertically and horizontally. This is opposed to external norms, which are driven by external sanctions like rewards and punishments (Gintis 2003). External norms and extrinsic values, therefore, may be explained by the concept of both direct and indirect reciprocal altruism. Are we going to be altruistic toward ETIs because we expect some benefits or rewards—directly or indirectly—from them? However, the scale of time and space over which signals or messages are likely to be exchanged is too vast to rule out any immediate benefits, except perhaps the social and political ones that might result from the decision to transmit and act of doing so (Langston 2013). If reciprocity is ruled out, are we then showing altruism to ETI because we recognize some intrinsic value in these creatures that belong to another intelligent species? If this is the case, the ability of humans to assign intrinsic values to non-human organisms may be worth examining if we are to properly understand the different facets of human altruism to ETI.

3 The Basis of Moral Concerns for Other Species

Wilson (1993, 31) has termed the "innately emotional affiliation of human beings to other living organisms" as biophilia. He has further said that, "Innate means hereditary and hence part of ultimate human nature." This trait has evolved in humans because of their existence as members of hunter-gatherer bands for more than 99 % of their history, during which they had close association and interactions with other organisms as well as with inanimate elements of nature such as rivers, lakes, rocks and mountains. Because of this long and deep association, human affinity for nature does not stem merely from the need to extract material benefits from it, but also because of the latter's contributions to our cognitive, aesthetic, emotional and spiritual development (Kellert 1993). In an experiment, kindergarteners were allowed to experience common animals like bats and then asked to write stories about what would happen if a person met a bat. The stories were mostly marked by human-animal friendship and kinship. In one story, a bat met a girl at night and they learned from each other what they ate and became friends. In another story, the bat was thought to be angel-like and, while flying, lost its balance, dropped to the ground, and visited the boy's home, where it was offered food, which in the case of the bat was bugs. In the end, the bat got a friend and the boy got a pet (Fawcett 2002). In these stories, animals were treated as moral agents and assigned intrinsic value by very young children whose innate properties were not yet modified by cultural factors. On the contrary, most of the older children's views on bats consisted of fear and aversion. Another experiment (Bering 2006, as cited in Wilson and Green 2007), which investigated whether belief in afterlife was a social construct in children or a part of their innate make-up, found that very young children believed that a mouse retained its feelings even after death, indicating that belief in afterlife was an innate property. It also showed that young children spontaneously recognized intrinsic value and hence gave moral standing to other species without hesitation.

4 Speculating About ETI

There are two distinct streams of thought regarding the possible attitude of ETI towards humans. One view advocates extreme caution about announcing our presence in the galaxy to any strange civilization because it might be aggressive and violent, and consequently attack and colonize us and plunder our resources. Therefore, announcing our presence into space before ascertaining the nature and intent of ETI is tantamount to children shouting their presence inside a jungle where unknown dangers may lurk. However, if ETI did not attack Earth during years when humans did not have the technology to fight back and the planet had a much larger storehouse of natural resources, why would it do so at a time when we have a greater capability to fight back from a resource-depleted planet (Brin 2002)? It has also been argued that since ETI societies are likely to be at least 10,000 years older than ours, they are expected to have given up strife and violence and be based on altruism and cooperation, or they would have annihilated themselves before reaching a higher level of scientific and technological attainments (Tough 1986). In fact, altruism could be a characteristic feature of all mature civilizations (Brin 2002). It is known that selection pressure in ecosystems shifts from a preponderance of "r-selection" characterized by rapid growth to "k-selection" marked by increased dependence on feedback control. At the same time, internal symbiosis and stability increase with a concomitant reduction in entropy. Similarly, human societies, despite fluctuating and often turbulent deviations, can also be said to have moved away from, or are at least striving to move away from, high birth rates, rapid growth, obsessive emphasis on economic profit, unchecked exploitation of resources, oppression of weaker societies, and violation of human and animal rights and to the realm of human and animal rights and welfare, birth control, education, culture, peace and cooperation (Odum 1969). Cooperation, therefore, is a hallmark of mature ecosystems and advanced civilizations, and there is insufficient reason to think that the evolution of ETI would be drastically different. However, it is also likely that although compassionate, ETs may be ready to use their highly advanced weapon systems by adopting a "tit-for-tat" strategy, if faced with a "hawkish" attitude from our side. Hence, a more pragmatic human approach could be a "dove"-like or better a "bourgeois" strategy, where we forcefully assert our ownership over our territory while welcoming friendly interactions with ETI. However, should it happen that we come across the ETI in outer space or in their world(s), we should instead wave a flag of peace and cooperation in a dove strategy.

5 Nature Can Be Our Kin, and So Can ETI

Hamilton (1964a; b) first introduced the concept of inclusive fitness to explain the evolution of altruism among kin. Simply stated, it means that individuals are likely to perform altruistic acts towards others with whom they share the maximum number of their genes. For example, an individual shares on an average 50 % of its genes with siblings and offspring, 25 % with grandchildren and nephews/nieces, and a mere 12.5 % with first cousins. Hence an individual is more likely to be altruistic towards its sibling/offspring than to a nephew/niece than to a first cousin and so on. If we try to fit in the concept of assigning intrinsic value into the theory of kin selection, we may say that an individual tends to assign more intrinsic value to offspring and siblings than to a nephew or niece and so on. However, in order to understand the origin and perpetuation of nature worship and biophilia in the light of the theory of kin selection, we would have to move away from the genetic basis of kinship to subjective and cultural ideas of kinship, as is often understood by social anthropologists. In other words, we have to invoke not only the gene, but the meme as well (Dawkins 1976). Early hunter-gatherer societies (and even later-day tribal societies) operated in a very limited geographical area and were almost totally dependent on resources extracted from it. Such communities have been termed as the "ecosystem people" (Dasmann 1988, 277). Nature worship and offering protection to nature through the creation and maintenance of sacred groves, taboos on hunting, minimizing resource use overlap, and other mechanisms of prudent resource use are the characteristic features of such communities (Gadgil and Guha 1992; Gadgil and Vartak 1994; Gadgil 1995). In such situations, where people lived and moved in a small area, kin recognition was governed by the simple rule that anybody encountered every day was kin (Dawkins 1976). Thus non-kin could also be regarded as "kin" provided they were members of the same group. In small areas, people not only came across people, but also the same forest, river, mountain, trees or even particular animals in their daily wanderings for food and other resources. It is likely that they began to perceive these entities as their "kin" as well. That the tribal mind is capable of such a non-dualistic pattern of behavior in which they can recognize non-humans as kin, is also revealed by their myths and folklore which do not make any distinction among God, man, plants, animals and even inanimate natural entities. Many tribal groups also acknowledge the contribution of non-humans towards the creation of the world and the imparting of knowledge to the humans (Saraswati 1993). Thus it appears that in the hunter-gatherer worldview, humans could consider non-humans as kin and accept gifts of knowledge from them. Interestingly, one of the oldest known folktales describes the altruism of an Egyptian prince who spared the life of a dog and a crocodile in spite of the threat to his own. Further, in most folktales, more animals were altruistic donors or recipients than were humans (Arkhipova and Kozmin 2013). Therefore, on the basis of our hunter-gatherer tribal past, we could be altruistic not only to ETI, but to other forms of ET as well. Functional magnetic resonance imaging (fMRI) scans of brains of people offered fair and unfair bargains revealed that unfair offers activated the bilateral insula, which is known to be associated with negative emotional factors (Sanfey et al. 2003, cited in Fehr and Rockenbach 2004). On the contrary, mutual cooperation between human partners stimulated the brain's reward circuit, which includes the mesolimbic dopamine system in the striatum and the orbitofrontal cortex, while defection by one partner led to a negative response in the dopamine system (Rilling et al. 2002, cited in Fehr and Rockenbach 2004). Thus altruism is shown to have a concrete, neural basis too.

6 Use Versus Inherent Value of Non-Humans

Though it is generally believed that altruistic acts towards non-humans in many indigenous cultures are linked to the use-value of the latter, evidence to the contrary is also plentiful. Deb and Malhotra (2001) have shown that the tribes of West Bengal, India, hold as sacred a tree (*Adina cordifolia*) and a shrub (*Euphorbia neriifolia*) that have no direct-use values. Similarly, the sparrow, the jackal, the tiger, several species of songbirds, and various species of snakes are held sacred in many indigenous cultures of Assam, Manipur and Tripura (Gupta and Guha 2002). Soulé (1993, 443) has suggested that biophilia comprises several "bioresponsive behavioral systems," including affinity for a particular scenery or habitat, domestic as well as wild animals, and aversion to snakes, spiders, etc. However, in many oriental cultures, fear or aversion to snakes had led to worship of the Snake God or Goddess and then to empathy for snakes, which leads people to avoid harming snakes, unless they must as a last resort in self-defense. Wilson (1993, 34) has provided an "ophidian version of the biophilia hypothesis" where he elucidates the role of serpents as "prominent agents in mythology and religion in a majority of cultures." Many cultures, exemplified by the Meiteis of Manipur and Assam, North East India, go beyond biophilia to "ecophilia" or "cosmophilia" through the practice of *Chingoiron*—the worship of hills—and *Nungoiron*—the worship of rocks (Singh et al. 2003). In Korean shamanism, deities could exist, among other places, in trees, the ground, rocks, springs, rivers, and the sea (Rhi 1993).

   In contrast to these altruistic overtures, living in small groups also resulted in the development of strong ethnocentrism and nepotistic altruism, leading to a situation where any other human being (of a different tribe from a different area) who the members of a given tribe did not encounter regularly was not considered as kin and hence subjected to conflict and violence, often resulting in ruthless intertribal conflicts accompanied by practices like headhunting. Thus the same tribe members, who could be altruistic towards non-human entities and worship them as or along with their ancestors, could fail to exhibit this behavior when in contact with conspecifics from another tribe. Thus ethnocentrism could be a two-edged sword as it would give rise to altruism towards in-group and defection towards out-group members. The question is whether the ETI and we would treat each other as in-group or out-group members.

7 From Anthropophilia to Ecophilia

Does the human propensity to be altruistic towards non-humans provide a basis for human altruism towards ETI? Adherents of nature religion, nature worship and totemism are now encountered only in small, isolated communities scattered throughout the world. Being essentially a replicator, the survival value of a meme, just like that of a gene, depends on its longevity, fecundity and copying-fidelity. Because nature-religions are practiced by a very small fraction of the world's population, their fecundity is said to be very low. Does it then imply that nature-religion or nature worship as a meme is virtually removed from the meme pool of the human species, at least in mainstream societies? On the contrary, the picture changes if we think of nature-religion as a meme-complex. It has been suggested that many bioresponsive systems are likely to exist and not just one biophilia (Soulé 1993). Wilson (1993, 34) refers to the evolution of a complex of biophilic responses governed by different gene assemblages and neural structures and mechanisms. The rites and rituals accompanying nature worship are just one such meme in this meme-complex. Other memes include an ability to appreciate nature's beauty and bounty, recognition of an intrinsic value in nature, and a general affinity and compassion for other forms of life and even entire ecosystems or landscapes. These latter memes are not lost from the meme pool, in spite of the man-nature dualism meme holding sway in western societies for a long time. The Dutch philosopher Baruch Spinoza, for example, developed a pantheistic nonanthropocentric philosophical system in which God was identified with Nature. Unlike Descartes, Spinoza found mental attributes throughout nature. His teachings influenced many leading figures of the eighteenth century European Romantic Movement such as Coleridge, Wordsworth, Shelley and Goethe, who sang the praises of nature and found a soul in nature and its various components. In the philosophical front, nonanthropocentric ideas were put forward by Mill, Thoreau, Marsh, Muir, Santayana and others in the nineteenth and early twentieth centuries, and by the likes of Leopold, Brower, Carson, Udall, White, Nash, Ehrlich, Naess and their numerous adherents in the middle and late twentieth century (Sessions 1995). Needless to say, this trend continues to spread and is poised to remain as a significant worldview during the present century and beyond.

Besides Western poets and philosophers, the image of nature in the writings of many Eastern writers also deserves mention. The "religion of soul" perceived in natural entities like flowers, clouds, seas and mountains found expression in the songs, poems and philosophical musings of Rabindranath Tagore, a poet and Nobel Laureate from India. In Japanese, the oldest word for beautiful is *kuashi*, which means a dense growth of leaves. In mediaeval Japanese poetry, *hana* (flower) is a symbolic term for rhetorical beauty, *tane* (seed) stands for the creative moment of poetic activity, and *taketakashi* (tall tree) stands for sublime. Basho, a well-known poet of the haiku (short poem), is reputed to have told his disciples to learn the spirit of the pine tree from the pine tree itself (Imamichi 1998, 286).

The meme of ecophilia survives not only among poets, philosophers and eco-centric thinkers, but in the minds of the common people as well. Trees, animals, lakes, rivers, mountains and the sea, or even a man-made city park, form an integral part of childhood associations we fondly remember, often mixed with a sense of nostalgia, just like that of old friends and relations. Even the most hardcore pro-development maniac is likely to find pleasure in watching a flower in full bloom or fish in an aquarium. The flourishing tourist trade centered around wilderness areas is another example of the meme of ecophilia running strong in the minds of countless people all over the world. In India, many professional car drivers—who are barely educated—consciously avoid running over frogs and toads while journeying down the road from one pond to another in driving rain. Some of these people are not vegetarians and have no religious taboo on killing animals. We may not be very far from the truth to say that the nature-religions of our remote ancestors form the memetic core of such properties of the human mind. These examples are by no means exhaustive, but merely illustrate the meme of "nature appreciation" or biophilia or ecophilia, or a feeling of kinship with nature, which continues to be copied and perpetuated in almost all societies the world over, but perhaps running stronger in small indigenous cultures.

8 Can We (and They) Do It?

The human attitude towards nature in general and non-human organisms in particular is probably coming full circle. We began as ecocentric worshippers of nature, became conquerors and exploiters, and then started recognizing the rights of the "others" as our co-inhabitants of this planet. At this point in our history, therefore, we are better oriented to recognize the rights of ETs in the universe, and perceive them as our kin or in-group members. This memetic transformation from unabashed anthropophilia to biophilia and ecophilia may not be true or complete for all humans, but it definitely is a major worldview today. It is possible that the evolution of intelligent species throughout the cosmos follows a similar pathway, which is expected to be similar in ETI.

Barkow (2013) believes it is vitally important to convince ETI that we regard them as in-group. In order to do that, we have to transcend the species barrier. This would become more difficult if they or we learn that we look very different because of our inter-species differences. Here our biophilia meme is likely to help us accept a morphologically different being as our in-group. In his film *ET*, director Steven Spielberg introduced us to Elliott, a young boy, who accepts a very different-looking ET as a friend. Accepting non-humans as friends and even kin may be compared to the worldview of children, who live in the realm of imagination where species barriers do not exist. Hence the biophilia or ecophilia meme could provide us the ethical framework for accepting ET as our kin and for behaving altruistically towards them.

The other question is this: even if we are able to be altruistic towards ET, would they be able to reciprocate? It has been argued that since both genetic and memetic convergence is likely to be a rule of thumb not only in on Earth but also in the whole universe, there is reason to be assured that ETI would reciprocate our altruistic gesture. However, we may confront a different situation if ETI (or at least some ETI species) are "ecological refugees" who have migrated to another planet after exhausting the resources and exterminating the biodiversity of their "mother" planet. Gadgil (1995, 107–108) has suggested that ecological refugees, when displaced from their original territory and forced to colonize new areas, mostly fail to exhibit their original altruistic attitude towards nature because of their lack of attachment to their new home. However, this is probably a sweeping generalization, because certain central Indian tribes that were forced to work in the tea gardens of North East India continued their ecocentric practices in their new home through the establishment of small sacred groves where total protection was offered to plants and animals (Gupta and Guha 2002). Therefore, we may still remain optimistic that ETI would be able to reciprocate with altruistic overtures to our altruism.

References

Arkhipova, A., and A. Kozmin. 2013. "Do We Really Like the Kind Girls and Animals?: Cross-cultural Analysis of Altruism in Folktales." In *Altruism in Cross-Cultural Perspective*, edited by Douglas A. Vakoch, 57–70. New York: Springer.

Axelrod, R., and W. D. Hamilton. 1981. "The Evolution of Cooperation." *Science* 211:1390–1396.

Barkow, J. H. 2013. "Eliciting Altruism While Avoiding Xenophobia: A Thought Experiment." In *Extraterrestrial Altruism: Evolution and Ethics in the Cosmos*, edited by Douglas A. Vakoch, 37–48. Heidelberg: Springer.

Boyd, R., and P. Richerson. 1982. "Cultural Transmission and the Evolution of Cooperative Behavior." *Human Ecology* 10:325–351.

Brin, D. 2002. "A Contrarian Perspective on Altruism: The Dangers of First Contact." Accessed October 30, 2012. http://www.setileague.org/iaaseti/brin.pdf.

Connor, R. C. 1986. "Pseudoreciprocity: Investing in Mutualism." *Animal Behavior* 34:1652–1654.

Dasmann, R. F. 1988. "Towards a Biosphere Consciousness." In *The Ends of the Earth: Perspective on Modern Environmental History*, edited by D. Worster, 277–288. Cambridge, UK: Cambridge University Press.

Dawkins, R. 1976. *The Selfish Gene*. Oxford: Oxford University Press.

Deb, D., and K. C. Malhotra. 2001. "Conservation Ethos in Local Traditions: The West Bengal Heritage." *Society and Natural Resources* 14:711–724.

Delius, J. D. 1991. "The Nature of Culture." In *The Tinbergen Legacy*, edited by M. S. Dawkins, T. R. Halliday, and R. Dawkins, 75–99. London: Chapman and Hall.

Dugatkin, L. A., and D. S. Wilson. 1991. "ROVER: A Strategy for Exploiting Cooperators in a Patchy Environment." *American Naturalist* 138(3):687–701.

Fawcett, L. 2002. "Childrens Wild Animal Stories: Questioning Inter-species Bonds." *Canadian Journal of Environmental Education* 7(2):125–139.

Fehr, E., and U. Fischbacher. 2003. "The Nature of Human Altruism." *Nature* 425:785–791.

Fehr, E., and B. Rockenbach. 2004. "Human Altruism: Economic, Neural, and Evolutionary Perspectives." *Current Opinion in Neurobiology* 14:784–790.

Gadgil, M., and R. Guha. 1992. *This Fissured Land.* Delhi: Oxford University Press.

Gadgil, M., and V. D. Vartak. 1994. "The Sacred Uses of Nature." In *Social Ecology*, edited by R. Guha, 82–89. Delhi: Oxford University Press.

Gadgil, M. 1995. "Prudence and Profligacy: A Human Ecological Perspective." In *The Economics and Ecology of Biodiversity Decline: The Forces Driving Global Change*, edited by T. M. Swanson, 99–110. Cambridge, UK: Cambridge University Press.

Gintis, H. 2003. "The Hitchhiker's Guide to Altruism: Gene-culture Coevolution, and the Internalization of Norms." *Journal of Theoretical Biology* 220:407–418.

Gintis, H. 2011. "Gene-culture Coevolution and the Nature of Human Sociality." *Philosophical Transactions of the Royal Society* B 366:878–888.

Gupta, A., and K. Guha. 2002. "Tradition and Conservation in Northeastern India: An Ethical Analysis."*Eubios Journal of Asian and International Bioethics* 12:15–18.

Hamilton, W. D. 1964a. "The Genetical Evolution of Social Behaviour. I." *Journal of Theoretical Biology* 7:1–16.

Hamilton, W. D. 1964b. "The Genetical Evolution of Social Behaviour. II." *Journal of Theoretical Biology* 7:17–52.

Imamichi, T. 1998. "The Character of Japanese Thought." In *The Humanization of Technology and Chinese Culture: Chinese Philosophical Studies*, XI, edited by T. Imamichi, M. Wang and F. Liu, 279–296. *Cultural Heritage and Contemporary Change*, Series III, Asia, Volume 11, General editor G. F. McLean. Washington, DC: The Council for Research in Values and Philosophy.

Kellert, S. R. 1993. "The Biological Basis for Human Values in Nature." In *The Biophilia Hypothesis*, edited by S. R. Kellert and E. O. Wilson, 412–472. Washington, DC: Island Press.

Langston, M. C. 2013. "The Accidental Altruist: Inferring Altruism from an Extraterrestrial Signal." In *Extraterrestrial Altruism: Evolution and Ethics in the Cosmos*, edited by Douglas A. Vakoch, 131–140. Heidelberg: Springer.

Martell, L. 1994. *Ecology and Society: An Introduction.* Cambridge: Polity Press.

Odum, E. P. 1969. "The Strategy of Ecosystem Development." *Science* 164:262–270.

Rhi, B. Y. 1993. "The Phenomenology and Psychology of Korean Shamanism." In *Contemporary Philosophy: A New Survey,* Vol. 7, *Asian Philosophy*, edited by G. Fløistad, 253–268. Dordrecht: Kluwer Academic Publishers.

Richerson, P. J., Boyd, R. T., and J. Henrich. 2010. "Gene-culture Coevolution in the Age of Genomics." *Proceedings of the National Academy of Sciences USA* 107:8985–8992.

Saraswati, B. 1993. "The Implicit Philosophy and Worldview of Indian Tribes." In *Contemporary Philosophy: A New Survey,* Vol. 7, *Asian Philosophy*, edited by G. Fløistad, 121–136. Dordrecht: Kluwer Academic Publishers.

Sessions, G. 1995. "Ecocentrism and the Anthropocentric Detour." In *Deep Ecology for the 21st Century*, edited by G. Sessions, 156–183. Boston: Shambhala.

Singh, L. J., N. B. Singh, and A. Gupta. 2003. "Environmental Ethics in the Culture of Meeteis from North East India." In *Bioethics in Asia in the 21st Century*, edited by S. Y. Song, Y. M. Koo, and D. R. J. Macer, 320–326. Tsukuba: Eubios Ethics Institute.

Soulé, M. E. 1993. "Biophilia: Unanswered Questions." In *The Biophilia Hypothesis*, edited by S. R. Kellert and E. O. Wilson, 441–455. Washington, DC: Island Press.

Tough, A. 1986. "What Role Will Extraterrestrials Play in Humanity's Future?" *Journal of the British Interplanetary Society* 39:491–498.

Tough, A. 1998. "Positive Consequences of SETI before Detection." *Acta Astronautica* 42(10–12):745–748.

Trivers, R. 1971. "The Evolution of Reciprocal Altruism." *The Quarterly Review of Biology* 46:35–57.

Wilson, D. S. 1975. "A Theory of Group Selection." *Proceedings of the National Academy of Sciences USA* 72:143–146.

Wilson, D. S. and E. Sober. 1994. "Re-introducing Group Selection to the Human Behavioral Sciences." *Behavioral and Brain Science* 17:585–654.

Wilson, D. S., D. C. Near, and R. R. Miller. 1998. "Individual Differences in Machiavellianism As a Mix of Cooperative and Exploitative Strategies." *Evolution and Human Behaviour* 19:203–212.

Wilson, D. S., and W. S. Green. 2007. "Evolutionary Religious Studies: A Beginner's Guide." Evolutionary Studies Program, Binghamton University, Binghamton, NY. Accessed on October 30, 2012. http://evolution.binghamton.edu/religion/wp-content/uploads/2009/09/BeginnersGuide.pdf.

Wilson, E. O. 1993. "Biophilia and the Conservation Ethic." In *The Biophilia Hypothesis*, edited by S. R. Kellert and E. O. Wilson, 31–41. Washington, DC: Island Press.

Caring Capacity and Cosmocultural Evolution: Potential Mechanisms for Advanced Altruism

Mark L. Lupisella

Abstract This chapter proposes a model for "advanced altruism," building from what are fairly well understood biological-genetic dynamics ("biological altruism") leading to more complex social and cultural forms of altruism ("biocultural altruism"), which can act as a springboard for "advanced altruism" that sufficiently transcends biological self-interest. Two mechanisms are emphasized: *increasing caring capacity* and *cosmocultural evolution*, both of which can lead to advanced forms of altruism. Increasing caring capacity involves making cost-benefit ratios of altruistic acts increasingly favorable to all actors—especially by reducing the cost of caring. Cosmocultural evolution emphasizes the coevolution of culture and cosmos and suggests the universe can be a common objective framework. Cosmocultural evolution appeals to the more "subjective" power and value of cultural evolution, which can result in a broad-based respect for the universe and all beings in it. The model may have general applicability to beings that have evolved via natural selection, including possibly single "collective" biological or machine intelligences. However, the model appears to be only marginally applicable to a single being that never evolved through phases of social evolution, requiring additional speculation, some of which is provided to address, for example, how a notion of self and others could develop in such an extreme example.

Keywords Altruism · Biological altruism · Biocultural altruism · Advanced altruism · Extraterrestrial intelligence · SETI · Caring capacity · Cosmocultural evolution · Collective intelligence

M. L. Lupisella (✉)
NASA Goddard Space Flight Center, Greenbelt, USA
e-mail: mark.l.lupisella@nasa.gov

M. L. Lupisella
20945 Winola Terrace, Ashburn, VA 20147, USA

D. A. Vakoch (ed.), *Extraterrestrial Altruism*, The Frontiers Collection,
DOI: 10.1007/978-3-642-37750-1_7, © Springer-Verlag Berlin Heidelberg 2014

1 Introduction

The possibilities regarding extraterrestrial altruism seem almost endless, provoking us to take on uncomfortable and sometimes daunting levels of extrapolation and speculation that often accompany questions about extraterrestrial life and intelligence. There is, however, the proverbial lamppost under which we may look, as practitioners of astrobiology and the search for extraterrestrial intelligence (SETI) have been doing for decades by basing searches and analyses on what we know of humanity and life on Earth.

Such an approach may also be applicable to the psychological and social sciences that study phenomena such as altruism. This challenging extrapolation may in fact pay great dividends not only because contemplating extraterrestrial altruism may be a way to help better understand human altruism and its potential, but also to be prepared for if and when we communicate with extraterrestrial intelligence (ETI). The latter includes the practical policy challenges we face today regarding sending messages to ETI without having first received any ("Active SETI"). Altruism may play a key role in these considerations (Vakoch 2001; 2006).

So how do we increase our chances of getting it right? How do we begin to have a sense for what we might be dealing with in communicating with ETI so that we can at least partially address hopes and fears (Michaud 2007; Tough 2000; Vakoch 2011; Lupisella 2011)? To what extent can we count on mechanisms and implications of altruism that we are familiar with today being applicable to ETI? Do we even understand altruism well enough to justify an extrapolation to putative ETI?

This chapter is based on the premise that even for the ambitious purposes of addressing extraterrestrial altruism, we not only have a fairly good understanding of non-human and human altruism, but also that many of the mechanisms could in fact be applicable to other intelligent species. The degree of confidence we can have regarding this potential applicability is admittedly uncertain, and may well be impossible to assess in the near-term, but is potentially more knowable in the longer term (Denning 2009; Rolston 2013 in this volume). Nevertheless, this chapter will try to show that the basic potential for broad applicability is defensible. There may however be a potential limit of sorts, not only as a kind of epistemological horizon beyond which our lamppost may shine dimly, if at all, but perhaps an altruistic threshold that cannot easily, or perhaps ever, be exceeded by any species—namely, a kind of "selfishness trap."

This chapter will first articulate a high-level framework for the evolution of altruism and then apply that model to help assess its potential implications regarding ETI, focusing on two examples of a "single" intelligent entity. The framework will use three basic categories of evolution and altruistic behavior, (1) biological, (2) biocultural, and (3) "neocultural" or "advanced," all of which will include brief commentaries regarding motive and circumstance—two key factors of altruistic behavior.

A minimalist view of altruism suggests it can fairly easily evolve via a number of different but reasonably well-understood and related mechanisms that might be broadly categorized as biological and cultural. Cultural evolution could include (a) "biocultural" dynamics, i.e., cultural phenomena that can be directly or indirectly explained by selfish biological interests, and (b) "advanced," "neocultural" or "new" cultural phenomena that somehow do not sufficiently depend on biological explanations. This latter set of advanced neocultural phenomena might include neocultural altruism or, more generally, "advanced altruism," where the altruistic acts are not explainable via biological or biocultural mechanisms.

Advanced altruism arguably does not yet exist to any great extent on Earth, but it may be emerging. We can ask the theoretical question: what, if anything, would it take for such advanced altruism to emerge? Advanced altruism is not the only kind of altruism of interest when considering ETI, but it is arguably the most compelling scenario since it may help address a fundamental challenge regarding altruistic behavior that may be of particular relevance regarding ETI, namely "the out-group problem."

Within the three broad categories of altruism noted above (biological, biocultural, and advanced), two mechanisms are emphasized: (a) *increasing caring capacity* via technological and cultural evolution (within the category of biocultural evolution) and (b) *Cosmocultural evolution* (within the category of advanced altruism).

Caring capacity can be seen as an indicator of the capacity that an individual or group has for caring for others and is analogous to the concept of ecological "carrying capacity," which indicates how many individuals an ecological system can sustain. This caring capacity might be assessed in many ways, but the central point for the purposes of this chapter will be that as the relative cost of caring is reduced via a number of mechanisms ranging from technological to cultural, biological instincts for caring can be more easily realized and all group members can ultimately be well-cared for as a result. But what about out-group members? Increasing caring capacity may or may not be extended to out-group members depending on many circumstances, including philosophical principals and worldviews.

A worldview used in this chapter is "cosmocultural evolution," which suggests that cultural evolution and physical-cosmic evolution co-evolve, and perhaps are already co-evolving, with the potential for cultural evolution to play an important, if not critical, role in the overall evolution of the universe (Lupisella 2009b). Such a worldview could imply that a default position toward other cultural beings is to respect them. This respect could be motivated by number of interpretations of cosmocultural evolution: (a) cultural beings are a product of cosmic evolution and may be seen to have some degree of "cosmic value," (or perhaps "intrinsic value" if intrinsic value is defined as being grounded in the universe), (b) cultural beings have a potential for playing an important role in cosmic evolution, (c) diverse cultural beings may see themselves in some sense as part of the same group (as compared to non-cultural or "unintelligent" beings), which could lead to (d) a dramatically expanded notion of selfhood, where cosmocultural beings sharing the same universe may have a radically-expanded notion of selfhood, perhaps ultimately equating selfhood with the universe itself.

The applicability of the model varies substantially between two different scenarios of a single intelligence: one in which the intelligence evolved via natural selection on individuals, and the other in which it did not. For a single entity that did not first evolve through a process of individual natural selection, the framework is only marginally applicable. The framework is more helpful in assessing the potential for levels of altruism for beings that evolved via natural selection, particularly through phases of social evolution.

2 Mechanisms for Altruism

This section will touch on three broad categories of altruism: (1) biological, (2) biocultural, and (3) advanced altruism, with an emphasis on the last of these since that may be the most promising way to develop "out-group altruism." These three categories overlap but may provide useful distinctions that form the basis of a framework in which to understand the evolution of altruism and its application to potential extraterrestrial altruism.

2.1 Biological Altruism

Biological altruism can be thought of as altruism easily tied to fairly well-understood biological dynamics such as kin selection/inclusive fitness, cooperation and reciprocal altruism, and certain broader group/social dynamics (Hamilton 1964; Price 1970; 1972; Trivers 1971; Axelrod and Hamilton 1981; Dawkins 1989; Wright 1997; Wilson and Wilson 2007; Churchland 2011). Psychological altruism, which is usually meant to suggest an internal psychological state that reflects an intention or "ultimate desire" to further the well-being of others (Sober and Wilson 1998) would be considered a subset of biological altruism in the sense that natural selection would presumably select for the relevant internal states, such as empathy, which are often needed for psychological altruism (Batson 2011).

The motive in biological altruism is narrow and "primitive" in the sense that it is directly motivated by genetically-selected predispositions that increase the probability of propagating genetic copies into the next generation. Circumstance is important to the extent that it determines which actions with kin or potential reciprocators can vary and affect outcomes.

2.2 Biocultural Altruism

Some form of basic biological altruism is important for a certain amount of individual cooperation and for broader cooperation within groups, but it is probably not sufficient for larger-group stability. Biological altruism can provide a foundation for transitioning to biocultural constructs and norms (e.g., art, effective social rules of behavior, certain "formal" ethics, etc.) to help reflect and enforce genetically-driven altruistic tendencies, leading to group stability and longevity. Such a transition is similar to the notion of major transitions in evolution (Maynard Smith and Szathmáry 1995) and may correlate roughly with the major transition from primate groups to human societies. In biocultural altruism, the primary mechanisms are largely social and cultural, which can co-evolve with genes (Richerson and Boyd 2005), but are directly tied to the innate genetic mechanisms of biological altruism noted previously. These cultural norms can improve cooperation and social stability, and can often allow some in a group to benefit more than others.

Biocultural altruism suggests that much of the behavioral motive is socially influenced, but still ultimately rooted in selfish genetics of biological altruism. Circumstance is important in part because conflicts between groups often arise from specific details such as environment and ideology (e.g., commonly held worldviews which can provide effective group cohesiveness), which may drive important group-selection effects (Sober and Wilson 1998; Wilson and Wilson 2007; Haidt 2007). Biocultural altruism moves us from informal, tenuous social stability to more formalized socio-cultural norms that formally and explicitly codify behavior and worldviews, but that still rely on a noteworthy reward of some sort (e.g., many forms of religion).

So biocultural altruism can be seen as arising out of at least three related areas that reflect an overall progression: (a) personal and social expression, e.g., communication of personal sentiments, entertainment, art, music, writing, social symbols (Deacon 1997), worldviews, etc.], (b) social norms and organization (e.g., "laws," etc.), and (c) technological advancements. A more detailed way of saying this is that biocultural altruism results from (a) ways of identifying and empathizing with others, particularly their inner worlds (Hunt 2007), (b) social norms of behavior, including norms based partly on some forms of reason (Pinker 2011), as well as (c) means for increasing efficiencies in resource acquisition and time management (e.g., technology). These processes could also be seen as stability-seeking (or at least stability-creating) mechanisms for a group.

An extreme example that may reflect biocultural altruism is the United Nations Declaration of Human Rights, which was formed primarily in the wake of the deeply destabilizing effects of World War II. Unfortunately, this declaration has not been adequately realized by many, if not most, standards. Nevertheless, human violence appears to have decreased over long periods (Pinker 2011).

The combination of personal and social expression, explicit social rule-making, and technological advancement have a key interplay and feedback loop with another stability-seeking and altruistic mechanism, namely the increase of what might be thought of as the "caring capacity" of a group or society.

2.2.1 Caring Capacity

Darwinian evolution suggests that altruistic behaviors can occur when the genetic cost-benefit ratio for the actor is favorable in some way. If the cost is low enough, the benefit can also be low for the behavioral trait to persist. If the cost is low, many in a group may act more readily on their caring instincts. Caring capacity can be loosely defined as the ability of an individual or group to care for others. The ecological concept of "carrying capacity," which measures how many individuals an ecological system can sustain, is somewhat analogous.

We might consider something like a "caring-capacity principle," which would suggest that when the cost-benefit ratio becomes increasingly favorable for enough actors, it is easier to trigger biologically-altruistic acts and to form explicit bio-cultural altruistic social norms. Many factors, such as social norms and technology, can reduce the cost of caring or increase the cost of not caring (e.g., by increasing motive and by decreasing the time, effort, resources and complexity required to care for others), therefore creating favorable cost-benefit ratios for altruistic acts. This can allow the overall "caring capacity" of a group or society to increase—without necessarily any explicit intent to do so (see Wright 2000 for a related treatment).

But technology also has the effect of facilitating more selfish acts, in part because our dependence on others is decreased. Technological advancement, then, can be seen as a multiplier of both our innate selfish and altruistic tendencies. Personal and social expression, as well as social norms, can help control the selfish implications of technology.

An increased caring capacity can be applied to future generations as well, contributing to what might be called "multigenerational altruism" (Lupisella 2001). Presumably, societies that had their caring capacity increased through social and technological advances would be more likely to be long-lived, assuming that increased caring capacities were applied to future distant generations.

2.2.2 The Out-Group Problem

While group selection may create in-group altruism, it is unclear how altruism toward out-group members would be achieved via a strict individual or group-selection processes. Perhaps a kind of "out-group common-threat altruism" can develop in order to increase sensitivity to other groups so that separate groups can use each to address common threats as they arise. But it may be more plausible that in-group altruism might evolve simultaneously with out-group hostility in order to increase group survival fitness relative to other groups (Choi and Bowles 2007). When it comes to extraterrestrial intelligence, this may be one of the most important questions since extraterrestrial civilizations could easily see each other as ultimate out-groups.

This out-group problem may not be a practical one if extraterrestrial civilizations are far from each other and circumstances are favorable (e.g., if it's difficult or impossible to physically affect or compete with each other)—we would simply coexist and possibly minimally cooperate at low cost and low risk. However, the more challenging questions on the extremes arise when ETI could either be hostile or altruistic—and importantly, the same question applies to humanity.

If we have "intelligence" in common, the out-group distinction may not be very dramatic since ETI civilizations might see themselves as part of a broader group of intelligent beings. This seems theoretically plausible (Smith 2009) and will be touched on more in the next section, but if we consider how hostile human beings have been to intelligent members of our own species, this might be an unwarranted assumption. Even if we have very basic core values in common, such as reciprocity (see Harrison 2013 in this volume), the Golden Rule, etc, many details could be different enough (Ruse 1985) to set up strong out-group distinctions, as is prevalent with human beings. And if biocultural norms (e.g., details of ethics) are essentially created or "invented" by intelligent beings, and hence not objective in the strict philosophical sense (Mackie 1977), then there may be more reason to expect strong out-group distinctions based on those ideological details. But intelligence can lead to broader intellectual worldviews that might ultimately be sufficiently inclusive to help reduce out-group hostility and perhaps even engender out-group altruism.

There appear to be examples of out-group altruism in humans, but they are apparently limited. Also, in a broader sense, altruism toward members of other human groups can be seen as altruism toward members of the larger human group as a whole, hence it can be seen as an extended form of in-group altruism. Nevertheless, substantial out-group altruism today appears to be relatively minimal compared to in-group altruism.

Biocultural altruism is a form of optimizing the biological cost-benefit for individuals as well as groups and can be seen as a kind of bridge—with slowly increasing caring capacity as a key example—between biological altruism and advanced altruism.

2.3 Advanced Altruism

Advanced altruism moves beyond the motives of biologically driven self-interest and most of the social biocultural norms that result. Many ideas that appear to get close to advanced altruism exist (e.g., Kant's Categorical Imperative, Utilitarianism, certain tenets of both Eastern and Western religions, certain forms of environmental ethics, etc.), and have culminated, at least intellectually, in the United Nations Declaration of Human Rights, or perhaps even recent developments in Bolivia that have codified the rights of nature in law as the Law of the Rights of Mother Earth. But these articulations still appear to be based on some combination of (a) biological and biocultural altruism (e.g., providing group stability which is ultimately in the self-interest of many individuals), (b) seeking reward (e.g., an afterlife, etc.) and have (c) been only been minimally realized when we consider typical human behavior. Human behavior has not met ideological aspirations.

Advanced altruism would call for at least two things: (1) altruistic motivations that ultimately are not driven by some form of self-interest, and (2) widespread behavior directly consistent with those motivations. Advanced altruism should also probably (a) address the out-group problem, perhaps leading in some sense to a kind of "universal altruism," which might then (b) relate to and/or directly inform features of Metalaw (Haley 1956; 1963; Fasan 1970) and (c) perhaps, in stronger forms, be considered "proactive advanced altruism" in the sense that beings would be compelled to take the initiative to actively help others at some cost to themselves for reasons that are not tied to self-interest. Another way to think about advanced altruism is that it might correlate with high peaks of Sam Harris' Moral Landscape (Harris 2010) as extended to a truly universal scale that would include other ETI, and for which there could be multiple, and perhaps quite different, high peaks.

For advanced altruism, circumstance is less important because a principled approach would tend to dominate in almost all circumstances. Motives would be truly idealistic, based on rational, intellectual, and philosophical principals (Batson, Ahmed, and Stocks 2011) and worldviews that do not depend on self-interest.

2.3.1 General Mechanisms for Advanced Altruism: Escaping the Selfishness Trap?

An interesting possibility is that evolution via natural selection locks beings into a kind of "selfishness trap" (Lupisella 2001). That is, cultural methods for becoming far less selfish (e.g., enforced norms or physical modifications), or possibly even completely unselfish, could be envisioned and perhaps even implemented by many individuals. However, such methods may not be adequately realized because the selfishness of many individuals may prevent them from giving up their selfishness—in part because it provides advantages, especially when many others might forego their selfishness. Escaping the selfishness trap might be an all-or-nothing problem.

So what might help beings address the biological depth of selfishness? Would it be possible to escape the selfishness trap? This is very similar to asking how advanced altruism can be realized. This raises a related question: can and should advanced altruism require the starkness of absolute equity as Andrew Haley (1956) first suggested? It is one thing to have dramatically reduced selfishness, but can and should selfishness be "eradicated" and/or require complete equity of completely different intelligent beings so that when there is a conflict, or difficult choices to make, no inequity exists at all?

Advanced altruism may not have to rise quite to the level of perfect equity for all rational beings at all times under all circumstances (what might be thought of "strong advanced altruism")—"provisional morality" may be good enough (Shermer 2004). But advanced altruism should probably at least provide a framework for avoiding harm, seeing other rational beings as equal enough to be entitled to respect and rights, and preferably provide guidance for why, and perhaps how, we should actively enhance the existence of other rational beings.

One possible direct mechanism to realize advanced altruism might be to physically create it. For example, humanity might be faced with the possibility to moderate selfishness directly through brain chemical control (as is partially done today for a variety of mental conditions) or via genetic manipulation. However, if something like this were feasible, it would still require significant motivation to implement the change. Certain philosophical worldviews might be an effective way to motivate some level of selfishness control (as has been done throughout human history), including forms of "memetic engineering" that might protect humans from our future technological descendents (Gardner 2009) or other advanced forms of technological beings.

Religions have, in many cases, addressed advanced altruism, ranging from the Golden Rule, to "turn the other cheek," to the extreme non-violence of Jainism. A belief in an objective transcendent source can be an effective motivator for advanced altruism (e.g., God or spirit). However, religion seems to have fallen well short of realizing advanced altruism as defined here because (a) much religion is based on a reward system (hence self-interest is a key motivator) and group stability (biocultural altruism), (b) the truth value of a transcendent objective source of value has been problematic as it is unconvincing to many, and (c) most human behavior is not consistent with many religious altruistic aspirations.

So too have theoretical principle-based philosophical approaches, based largely on reason, such as the Categorical Imperative (similar to the Golden Rule of many religions), Utilitarianism, modern animal rights (Singer 1981) and environmental ethics (Hargrove 1989) not given rise to truly advanced altruism. Such approaches may be biocultural anyway in the sense that they may be (a) motivated by societal stability and/or (b) a sense of personal and collective aesthetics and/or (c) in some sense marginally affordable because of the reduced cost of caring afforded by much of modern human life (e.g., affording rights to animals and caring for the environment).

Self-interest or "egoism" could conceivably be leveraged to realize advanced altruism by dramatically expanding the notion of self to include much that is beyond the traditional self, potentially to include the whole of the universe. Expanded selfhood has a certain logical appeal, and by some accounts, the universe could qualify as having a sense of selfhood (Mathews 1991). It isn't clear that individual selfish beings could adopt this kind of broad sense of identity, but if it were, a compelling philosophical foundation/framework would be important to motivate it.

It is also possible that other kinds of worldviews might evolve and spread to help encourage deep and widespread advanced altruism beyond groups, extending to all intelligent beings and perhaps even "non-rational" beings and possibly beyond to include non-living entities and the universe itself (Singer 1981; Rolston 1990; Swimme and Berry 1992; Swimme 1995; Lupisella and Logsdon 1997; Dick 2000; Gupta 2013 in this volume). Such worldviews would presumably need to have something sufficiently in common with putative ETI in order to help provide a common, preferably "objective," frame of reference for perceptions of value and morality. The universe itself may provide that common objective frame of reference.

2.3.2 Cosmocultural Evolution

Cosmocultural evolution suggests that the cosmos and culture tightly co-evolve, with cultural evolution being important for cosmic evolution. Culture may be critically important as a dominant factor in the evolution of the universe, including the potential for creating a "moral universe" (Lupisella 2009b; for related recent treatments see Davies 2009; Smart 2009; Gardner 2009). Cosmocultural evolution suggests that the emergence of intelligent cultural beings (which are presumably valuing agents of some sort) "bootstraps" the universe into the realm of value, meaning, and purpose, and hence morality as a derivative—where none may have existed prior. This, combined with the potentially infinite physical and conceptual power of cultural evolution, can give rise to a "morally creative cultural cosmos" (Lupisella 2009a; b). This kind of cosmocultural worldview appeals to both the objective reality of the universe as well as the more "subjective" realities of cultural beings and forms a kind of fused worldview that can leverage the best of both—the common objectivity of the universe and the intentional creative moral power of cultural beings.

Such a view can provide an objective frame of reference (the universe) that we presumably have in common with ETI, and possibly engender a broader sense of selfhood as well as an appreciation for pluralism and diversity of beings—all of which contributes to cosmocultural evolution. Cosmic significance for cultural beings may provide a compelling (and possibly even an objective) justification of their value, since the universe significantly transcends cultural beings and has a knowable physical reality.

Cosmocultural evolution does not prescribe a particular cultural ideology except to suggest generally that cultural evolution and intelligent cultural beings, as the creators of culture, have significance for the evolution of the universe. In some sense, cosmocultural evolution suggests that some degree of cultural pluralism is favored since the more diverse cultural evolution there is, the more the cosmic significance of cultural evolution can be realized, which would obviously have to be balanced against unhealthy destructive tendencies of cultures.

2.3.3 A Synthesis for Advanced Altruism

This treatment suggests that powerful forms of biocultural altruism (ultimately rooted in genetic self-interest) can result from a number of mechanisms, including the increase of caring capacities by making the cost-benefit ratios of caring for others increasingly favorable. Cost-benefit ratios for altruism can be made favorable by slowly reducing costs of caring over time due to efficiency gains (e.g., technology), as well as increasing benefits of caring (or increasing costs of *not* caring) via social norms. Increasing the caring capacity of a group or society can produce a robust version of biocultural altruism (e.g., laws, social norms, declarations of universal rights, etc.) that can act as a sustainable foundation and jumping-off point for advanced altruism that can be extended to out-groups.

When caring is less costly (requiring less time and people, hence allowing enough individuals to develop philosophical cultural norms), and when caring is motivated by social norms, the philosophical choice of true altruism toward all potential beings can more easily follow. As cost-benefit ratios of altruistic acts become more favorable, the overall caring capacities of groups and societies can increase. Increasing caring capacity helps enable the philosophical choice of advanced altruism. Increasing caring capacity may also make the cultural choice for advanced altruism much easier, but it is still a choice that has to be made in the face of deep selfish pressures that will still persist.

But broader perspectives such as cosmocultural evolution can help provide an intellectual, and perhaps to some extent emotional, context to help motivate advanced altruism. Indeed, humanity appears to be slightly transcending our biological heritage of Darwinian evolution. Among our many unique characteristics, it is the modern extension of our unusual biological capacity to care for others that suggests humanity may be breaking the shackles of our evolutionary chains. We may perhaps be entering a more truly advanced altruistic phase of human development for which certain kinds of worldviews such as cosmoculturalism may play a role. We are seeing only glimpses of such advancement in *Homo sapiens sapiens* today, but it may serve as pointer, a data point for the advanced evolution that may be occurring—or perhaps has long ago occurred in other intelligent species.

3 Implications for Extraterrestrial Altruism and a "Single" Intelligence

As with most philosophizing and moralizing, the real challenge is the devil in the details with the complexities of actual real world behavior. Philosophical principles are easy, clear, and simple (hence part of their appeal), but behavior is often complex, messy and unpredictable. Principles are often blind to real world details. Principles are easy, behavior is not.

The model suggested here probably applies, at least in a general sense, to intelligent species that have evolved via natural selection into broader social cultural groups. That is, some amount of biocultural altruism seems reasonable to expect if individuals evolve sociality, which requires some degree of stability and reciprocity. And some degree of advanced altruism seems plausible as well, especially if ETI appeals to the universe as a whole for an objective framework for itself and other potential intelligent beings (Vakoch 2009; Lupisella 2009b).

But a very different case, perhaps a stressing case of sorts, is a "single" or "unitary" intelligent being. How would this model of altruism apply to such an entity? There are at least two general possibilities for such a being: (1) it may have evolved via natural selection of many separate individuals leading to a single collective intelligence, or (2) it may have come into being by some other non-Darwinian mechanism.

3.1 A Darwinian "Collective Mind"

In the case of natural selection of individuals over time leading to a single collective intelligence, the model suggested here appears to have applicability since the entity could have evolved through the three phases of altruistic development noted previously that would eventually become a single collective intelligence or "collective mind."

However, it also seems possible that separate individuals could become a highly coupled, or highly integrated, collective intelligence without going through the more complex social phases of biocultural evolution and advanced altruism if there were pressures to drive rapidly toward a collective intelligence. This kind of rapid collective intelligence might be akin to the social insects we see today. For this case, one option is that advanced levels of altruism (or intelligence more generally) may not evolve because a strong sense of individualism, when combined with the delicate balance with other individuals needed to drive stable forms biocultural and advanced forms of altruism in social beings, would likely be lacking.

Nevertheless, given enough time or the right circumstances (e.g., reduced competition with other species), a social insect-like intelligence could give rise to an emergent phenomenon that has the essential operational features of advanced altruism, assuming the right kind of collective awareness could emerge (e.g., see Sharkey 2006 for the possible emergence of a "group mind"). For example, an awareness of a very large and diverse environment as noted below (see Oudeyer and Kaplan 2004 for how the right levels of curiosity can give rise to collective machine self-development) could give rise to a surprisingly deep form of altruism.

But if ETI did evolve more slowly via natural selection on many individuals leading to a single collective intelligence (either biological or machine, the latter being more likely than we think, e.g., Dick 2003), it is possible that very advanced forms of altruism could develop and persist. Such an intelligence would have to preserve its sense of "the other" after having evolved to a state where the notion of the other would presumably be unimportant, and perhaps even resisted. But if such an entity were able to preserve and maintain its sense of the other, and especially if it were able to maintain and develop advanced forms of altruism either from its past or as part of an increasing theoretical and/or experiential understanding the universe, then there's a reasonable chance that such a being would have altruistic inclinations toward other extraterrestrial beings.

3.2 Non-Darwinian Intelligence

But what if a single entity were to come into being without having gone through the process of natural selection as we understand it? What if a being had never developed a sense of "the other" because "the other" was never in its environment? The model suggested here is based on individuals, social dynamics, multi-level selection and "advanced" intellectual worldviews that can arise out of socio-cultural constructs. But what if an entity never experienced a social reality at all?

This may seem like a misguided thought-experiment given what is presumably a low probability that such a being could even arise (but see alternative definitions of life from Lupisella 2004), let alone with sufficient intelligence to be highly communicative, especially over long distances. Nevertheless, in addressing extraterrestrial altruism, it seems reasonable to consider such a possibility given the powerful and wide-ranging implications associated with potential extraterrestrial beings.

A single, non-social being that didn't evolve via natural selection poses an interesting challenge in the sense that the model presented here is only marginally relevant because it is unclear how altruism could arise if there are no others in the environment for which altruism would be relevant. This case suggests a different model, or at least additional speculation, would be needed to address such a possibility. That model could vary widely, ranging from (a) alternative natural mechanisms for the emergence of altruism to (b) other extreme mechanisms akin to intervention from another being (or beings) that created or instilled a sense of "the other," or a sense of altruism directly, in the being in question. In theory, the latter could include "divine" intervention, although that will not be the focus here.

A single being of this kind suggests an interesting dichotomy. On the one hand, such a being, while not altruistic in the sense explored here (where behavior stems from socially selected internal sentiments and/or explicit worldviews), may be benign toward other new beings simply because it never developed an adversarial or hostile posture. In such a case, this being might be seen as altruistically neutral, harmless, and unable to respond to an intrusion and/or defend itself. However, if such a being needed to consume or modify its environment in some way to survive (a virtual certainty?), then those abilities could be aggressively applied to any unusual environmental change (or "threat") if it were treated as an environmental factor that had to be dealt with for continued survival. That is, its lack of sense of "otherness" could be totally benign and harmless, or completely "mindlessly" aggressive and destructive.

But what if such a being were able to gain an understanding of itself and its broader environment? If some level of self-awareness were possible, if it were possible that a level of awareness of a broader environment in which its sense of self was contextualized, then it seems plausible that such a being could develop an understanding of the possibilities of others like itself in its wider world. Similarly, it seems possible that awareness of a tight integration or dependency on its wider environment may drive the emergence of a kind of deep "dependency awareness" where a model of itself would include seamless "give and take" interactions with its wider world. This could give rise to a deep knowledge or even appreciation or wisdom for this interchange between being and environment.

But how broad would its environment, its "world," be? This is similar to asking how broad would its "worldview" be? How broad *could* it be? I would like to suggest the possibility that a being able to "observe" or contemplate its larger environment, and/or perhaps move in space (e.g., a being that seeks out the energy of stars) may very well develop a conceptual understanding of the broader cosmos and the potential for other beings in it, which, when coupled with an awareness of environmental dependency, could give rise to a form of broadly applicable altruism—i.e., applicable to everything in its environment.

It is certainly possible that such a being could also be completely unaware in any way that would allow for an altruistic orientation of any kind. But as suggested above, if there is some contemplation of environment, if there is perception of a large space and variation the environment of that large space that requires responsiveness to survive, then that being may ultimately achieve a broader understanding of a large, diverse environment. This could lead to a form of appreciation or even reverence for its environment and anything in it since its survival and well-being would depend on its environment. This could be seen as, or at least become, a weak form of cosmoculturalism in the sense that its awareness, dependency, and appreciation could be extended broadly to all that is (effectively, the cosmos) and to any other beings within the universe, with whom it may ultimately participate in the coevolution of culture and cosmos.

4 Conclusion

The model of altruism suggested here builds from what are fairly well understood biological dynamics based primarily on genetic fitness (biological altruism) which can lead to more complex "biocultural altruism." Biocultural altruism is still ultimately rooted in biological altruism but it can act as a springboard to "advanced altruism" which effectively transcends biological self-interest.

Two mechanisms were emphasized within the categories of biocultural and advanced altruism: *increasing caring capacity* (within the category of biocultural altruism) and *cosmocultural evolution* (within the category of advanced altruism), both of which could help give rise to advanced forms of altruism that might be applicable to out-groups. Increasing caring capacities can happen without much, if any intention, but when it does, advanced forms of altruism can more easily emerge. Cosmocultural evolution, which emphasizes the coevolution of culture and cosmos, invokes both, a potentially common or "universal" objective framework (the cosmos), as well as the more subjective diversity and potentially infinite power of cultural evolution, both of which many beings may be able to identify with and value, perhaps resulting in a broad-based respect for the universe and all beings within it.

The model seems to be at least generally applicable to individual social beings that have evolved via natural selection, especially over a long enough time period, including single "collective" biological or machine intelligences that may emerge. However, the model is only marginally relevant to a single being that never experienced "the other" and requires additional speculation such as a means for developing of a notion of self and others. A notion of self or other could be triggered by a deep and broad environmental awareness and dependency. This might lead to fundamental respect for environment and all within it, including on very large scales (possibly the universe), and including other beings that it may ultimately interact with as part of a broader cosmocultural dynamic.

References

Axelrod, Robert, and William D. Hamilton. 1981. "The Evolution of Cooperation." *Science* 211(4489):1390–1396.

Batson, Charles D. 2011. *Altruism in Humans*. New York: Oxford University Press.

Batson, Charles, N. Ahmad, and E. L. Stocks. 2011. "Four Forms of Prosocial Motivation: Egoism, Altruism, Collectivism, and Principlism." In *Social Motivation*, edited by David Dunning, 103–126. New York: Psychology Press.

Choi, Jung-Kyoo, and Samuel Bowles. 2007. "The Coevolution of Parochial Altruism and War." *Science* 318(5850):636–640. doi:10.1126/science.1144237.

Churchland, Patricia. 2011. *What Neuroscience Tells Us about Morality*. Princeton, NJ: Princeton University Press.

Davies, Paul C. W. 2009. "Life, Mind, and Culture as Fundamental Properties of the Universe." In *Cosmos and Culture: Cultural Evolution in a Cosmic Context*, edited by Steven J. Dick and Mark L. Lupisella, 383–397. NASA SP-2009-4802. Accessed December 28, 2012. http://history.nasa.gov/SP-4802.pdf.

Dawkins, Richard. 1989. *The Selfish Gene*. Oxford: Oxford University Press.

Deacon, Terrence. 1997. *The Symbolic Species: The Co–evolution of Language and the Brain*. New York: W. W. Norton.

Denning, Kathryn. 2009. "Social Evolution: State of the Field." In *Cosmos and Culture: Cultural Evolution in a Cosmic Context*, edited by Steven J. Dick and Mark L. Lupisella, 63–124. NASA SP-2009-4802. Accessed December 28, 2012. http://history.nasa.gov/SP-4802.pdf.

Dick, Steven. J. 2000. "Cosmotheology: Theological Implications of the New Universe." In *Many Worlds: The New Universe, Extraterrestrial Life, and the Theological Implications*, edited by Steven J. Dick, 191–210. Philadelphia: Templeton Foundation Press.

Dick, Steven. J. 2003. "Cultural Evolution, the Postbiological Universe and SETI." *International Journal of Astrobiology* 2(1):65–74.

Fasan, Ernst. 1970. *Relations with Alien Intelligences*. Berlin: Berlin Verlag.

Gardner, James. 2009. "The Intelligent Universe." In *Cosmos and Culture: Cultural Evolution in a Cosmic Context*, edited by Steven J. Dick and Mark L. Lupisella, 361–382. NASA SP-2009-4802. Accessed December 28, 2012. http://history.nasa.gov/SP-4802.pdf.

Gupta, Abhik. 2013. "Altruism Toward Non-Humans: Lessons for Interstellar Communication." In *Extraterrestrial Altruism: Evolution and Ethics in the Cosmos*, edited by Douglas A. Vakoch, 79–91. Heidelberg: Springer.

Haidt, Jonathan. 2007. "The New Synthesis in Moral Psychology." *Science* 316:998–1002.

Haley, Andrew G. 1956. "Space Law and Metalaw—A Synoptic View." *Harvard Law Record* 23.

Haley, Andrew G. 1963. *Space Law and Government*. New York: Appleton-Century-Crofts.

Hamilton, William. 1964. "The Genetical Evolution of Social Behaviour."*Journal of Theoretical Biology* 7(1):1–52.

Hargrove, Eugene C. 1989. *Foundations of Environmental Ethics*. Englewood Cliffs: Prentice-Hall.

Harris, Sam. 2010. *The Moral Landscape: How Science Can Determine Human Values*. New York: Free Press.

Harrison, Albert A. 2013. "Cosmic Evolution, Reciprocity, and Interstellar Tit for Tat." In *Extraterrestrial Altruism: Evolution and Ethics in the Cosmos*, edited by Douglas A. Vakoch, 3–22. Heidelberg: Springer.

Hunt, Lynn. 2007. *Inventing Human Rights: A History*. New York: W. W. Norton and Company, Inc.

Lupisella, Mark. 2001. Participant statement in *Humanity 3000 Seminar No. 3 Proceedings*, 37. Bellevue, WA: Foundation For the Future. See also pages 251 and 330 for additional references to "selfishness trap." Accessed December 28, 2012. http://www.futurefoundation.org/documents/hum_pro_sem3.pdf.

Lupisella, Mark. 2004. "Using Artificial Life to Assess the Typicality of Terrestrial Life."*Advances in Space Research* 33:1318–1324.

Lupisella, Mark. 2009a. "The Search for Extraterrestrial Life: Epistemology, Ethics, and Worldviews." In *Exploring the Origin, Extent, and Future of Life*: *Philosophical, Ethical and Theological Perspectives,* edited by Connie Bertka, 186–204. Cambridge: Cambridge University Press.

Lupisella, Mark. 2009b. "Cosmocultural Evolution: The Coevolution of Cosmos and Culture and the Creation of Cosmic Value." In *Cosmos and Culture: Cultural Evolution in a Cosmic Context*, edited by Steven J. Dick and Mark L. Lupisella, 321–359. NASA SP-2009-4802. Accessed December 28, 2012. http://history.nasa.gov/SP-4802.pdf.

Lupisella, Mark. 2011. "Pragmatism, Cosmocentrism, and Proportional Consultation for Communication with Extraterrestrial Intelligence." In *Communication with Extraterrestrial Intelligence (CETI)*, edited by Douglas A. Vakoch, 319–331. Albany: State University of New York Press.

Lupisella, Mark, and John Logsdon. 1997. "Do We Need a Cosmocentric Ethic?" Paper IAA-97-IAA.9.2.09, presented at the 48th International Astronautical Congress, Turin, Italy, October 6–10.

Mackie, J. L. 1977. *Ethics: Inventing Right and Wrong*. London: Penguin.

Mathews, Freya. 1991. *The Ecological Self*. London: Routledge.

Maynard Smith, John, and Eörs Szathmáry. 1995. *The Major Transitions in Evolution*. Oxford: Oxford University Press.

Michaud, Michael A. G. 2007. *Contact with Alien Civilizations: Our Hopes and Fears about Encountering Extraterrestrials*. New York: Springer.

Oudeyer, Pierre-Yves, and Frederic Kaplan. 2004. "Intelligent Adaptive Curiosity: A Source of Self-development." In *Proceedings of the 4th International Workshop on Epigenetic Robotics*, edited by L. Berthouze, H. Kozima, C. G. Prince, G. Sandini, G. Stojanov, G. Metta, and C. Balkenius, 117. Lund: Lund University Cognitive Studies.

Pinker, Steven. 2011. *The Better Angels of Our Nature: Why Violence Has Declined*. New York: Viking.

Price George R. 1970. "Selection and Covariance." *Nature* 227:520–521.

Price George R. 1972. "Extension of Covariance Selection Mathematics." *Annals of Human Genetics* 35(4):485–490.

Richerson, Peter J., and Robert T. Boyd. 2005. *Not by Genes Alone: How Culture Transformed Human Evolution*. Chicago: University of Chicago Press.

Rolston, Holmes, III. 1990. "The Preservation of Natural Value in the Solar System," In *Beyond Spaceship Earth: Environmental Ethics and the Solar System,* edited by Eugene C. Hargrove. San Francisco: Sierra Club Books.

Rolston, Holmes, III. 2013. "Terrestrial and Extraterrestrial Altruism." In *Extraterrestrial Altruism: Evolution and Ethics in the Cosmos*, edited by Douglas A. Vakoch, 211–222. Heidelberg: Springer.

Ruse, Michael. 1985. "Is Rape Wrong on Andromeda?: An Introduction to Extraterrestrial Evolution, Science, and Morality." In *Extraterrestrials: Science and Alien Intelligence*, edited by Edward Regis, 43–78. Cambridge: Cambridge University Press.

Sharkey, Amanda J. C. 2006. "Robots, Insects and Swarm Intelligence." *Artificial Intelligence Review* 26(4):255–268.

Shermer, Michael. 2004. *The Science of Good and Evil*. New York: Times Books/Henry Holt & Company.

Singer, Peter. 1981. *The Expanding Circle: Ethics and Sociobiology*. Oxford: Oxford University Press.

Smart, John M. 2009. "Evo Devo Universe? A Framework for Speculations on Cosmic Culture." In *Cosmos and Culture: Cultural Evolution in a Cosmic Context*, edited by Steven J. Dick and Mark L. Lupisella, 201–295. NASA SP-2009-4802. Accessed December 28, 2012. http://history.nasa.gov/SP-4802.pdf.

Smith, Kelly C. 2009. "The Trouble with Intrinsic Value: A Primer for Astrobiology." In *Exploring the Origin, Extent, and Future of Life: Philosophical, Ethical and Theological Perspectives*, edited by Connie Bertka, 261–280. Cambridge: Cambridge University Press.

Sober, Elliott, and David Sloan Wilson. 1998. *Unto Others: The Evolution and Psychology of Unselfish Behavior*. Cambridge, MA: Harvard University Press.

Swimme, Brian. 1995. *The Hidden Heart of the Cosmos: Humanity and the New Story*. New York: Orbis Books.

Swimme, Brian, and Thomas Berry, 1992. *The Universe Story*. New York: Harper Collins Publishers.

Tough, Allen, ed. 2000. *When SETI Succeeds: The Impact of High Information Contact*. Humanity 3000 Workshop Proceedings. Bellevue, WA: Foundation For the Future. Accessed December 28, 2012. http://www.futurefoundation.org/documents/hum_pro_wrk1.pdf.

Trivers, Robert L. 1971. "The Evolution of Reciprocal Altruism." *Quarterly Review of Biology* 46(1):35–57.

Vakoch, Douglas A. 2001. "Altruism as the Key to Interstellar Communication." *Research News & Opportunities in Science and Theology* 2(3):2, 16.

Vakoch, Douglas A. 2006. "Evolution, Culture, and Extraterrestrial Altruism: The Importance of Active SETI in a Selfish Universe." Paper presented at the Symposium on Culture, Anthropology, and the Search for Extraterrestrial Intelligence (SETI), annual conference of the American Anthropological Association, San Jose, CA.

Vakoch, Douglas A. 2009. "Encoding Our Origins: Communicating the Evolutionary Epic in Interstellar Messages." In *Cosmos and Culture: Cultural Evolution in a Cosmic Context*, edited by Steven J. Dick and Mark L. Lupisella, 415–439. NASA SP-2009-48022. Accessed December 28, 2012. http://history.nasa.gov/SP-4802.pdf.

Vakoch, Douglas A., ed. 2011. *Communication with Extraterrestrial Intelligence (CETI)*. Albany: State University of New York Press.

Wilson, David Sloan, and Edward O. Wilson. 2007. "Rethinking the Theoretical Foundation of Sociobiology." *The Quarterly Review of Biology* 82(4):327–348.

Wright, Robert. 1997. *The Moral Animal: Evolutionary Psychology and Everyday Life*. New York: Vintage.

Wright, Robert. 2000. *Non-Zero: The Logic of Human Destiny*. New York: Vintage.

The Precautionary Principle: Egoism, Altruism, and the Active SETI Debate

Adam Korbitz

Abstract At its crux, the debate whether Active SETI is a dangerous endeavor focuses on differences of opinion as to whether ETI would engage in ethical egoism (selfishness) toward humanity or ethical altruism (benevolence). Many critics of Active SETI employ a line of reasoning similar to the Precautionary Principle often utilized in the regulation of health and the environment. Several aspects of human behavior actually render the Precautionary Principle and related risk-balancing methods of dubious utility in sorting out competing risks. These same flaws apply equally to the Active SETI debate, in particular to the question of ETI egoism versus altruism. The Precautionary Principle encourages those engaged in the Active SETI debate to focus solely on one risk (egoistic and dangerous ETI) while ignoring other risks that are at least as likely if not more likely (such as lost "opportunity benefits" from contact with altruistic and benevolent ETI). Active SETI critics also ignore the very real political risks to science in general and SETI in particular that are created by possibly unfounded assertions of danger.

Keywords Active SETI · Altruism · Egoism (ethics) · Extraterrestrial intelligence · METI · Precautionary Principle · Risk analysis · Risk communication · Risk perception · SETI

1 Introduction: Egoism, Altruism, and the Active SETI Debate

Active SETI (the deliberate transmission of messages into space in the hope that they will be detected someday by an extraterrestrial civilization) has generated debate in the SETI community for years (Denning 2010a; Vakoch 2011). Critics of

A. Korbitz (✉)
CeleJure, 410 Midland Lane, Monona, WI 53716-3827, USA
e-mail: a.korbitz@att.net

D. A. Vakoch (ed.), *Extraterrestrial Altruism*, The Frontiers Collection,
DOI: 10.1007/978-3-642-37750-1_8, © Springer-Verlag Berlin Heidelberg 2014

Active SETI have suggested that its proponents are at best naïve and at worst engaged in a practice that might endanger all of humanity and its future on Earth. Proponents of Active SETI have responded by suggesting that its critics are, at the very least, needlessly paranoid if not silly (Denning 2010a). The debate, which previously had rarely drawn much public attention, received widespread coverage in the popular media when Stephen Hawking opined in 2010 that communicating with or alerting an advanced extraterrestrial civilization to our existence may risk colonization or conquest by that advanced civilization (Khan 2010).

Dire warnings regarding the potential dangers of Active SETI often apply a line of reasoning analogous to the Precautionary Principle developed several decades ago in the context of policy debates regarding possible threats to the environment and public health. However, several aspects of human behavior render the Precautionary Principle of dubious utility when deciding which risks to regulate and which to ignore. When applied to Active SETI, a Precautionary Principle-type analysis may forestall a line of human scientific and technological activity and intellectual inquiry without adequate or even plausible evidence of risk.

Standing as a backdrop to this debate is the stark reality that we do not even know if extraterrestrial intelligence (ETI) exists anywhere in the universe, let alone in our own Milky Way galaxy. Therefore, we really know nothing about the ethics, laws, motivation or capabilities of ETI, if ETI exists at all contemporaneously with our own civilization. In particular, we have no way of knowing if ETI's disposition toward humans or other extant civilizations would be one of egoism (selfishness or even predation), ethical altruism, or something in between resembling a recognition of equality between races or civilizations. Given this vacuum of knowledge, we do not currently have reason to believe that Active SETI is inherently risky, presenting a clear and present danger to the human race (akin to the proverbial should of "Fire!" in a crowded theatre), one that must be regulated or forestalled in some manner.

Explicit and implicit appeals to the Precautionary Principle as an argument against engaging in Active SETI—or even for imposing a moratorium on such activity—are easy to find in the literature (Billingham and Benford 2011; Michaud 2001; 2003; Shuch and Almár 2007). Also easy to find in the literature are other arguments both for and against engaging in Active SETI, such as who "speaks for Earth" or whether Active SETI is actually science or if it should categorized as some other type of activity, such as art or diplomacy. A discussion of those arguments is beyond the scope of this chapter, and other authors have summarized them quite ably (Musso 2012). The intent of this chapter is to hopefully lay to rest but one argument against engaging in Active SETI, which is that—with an eye to the Precautionary Principle— it is necessarily somehow wise, or cautious, or precautionary, to refrain from or even enjoin (via the force of law) such activities based on unsupported assumptions about the ethical nature of ETI, i.e., egoism versus altruism.

2 The Precautionary Principle

2.1 Definition

This chapter's description and general critique of the Precautionary Principle is drawn heavily from the work of Cass N. Sunstein (2005) (apart from the principle's application to both extraterrestrial altruism and the Active SETI debate, topics which Sunstein has not addressed). Until 2012, Sunstein served under President Barack Obama as the administrator of the White House's Office of Information and Regulatory Affairs. Sunstein (2005) has provided both a concise definition and a withering analysis of the Precautionary Principle, which has become a common way of analysing and regulating risks related to health, safety and the environment, both in Europe and the United States (albeit usually in reaction to different perceived risks).

To briefly summarize Sunstein's (2005, 4) argument, the Precautionary Principle in its many forms is animated by the idea that regulators should take steps to guard against potential harms, even when causal relationships are unclear and even if it is not known that those harms will come to fruition. The Precautionary Principle enjoys widespread international support as a basis for regulatory action (Sunstein 2005, 15), and some have argued that the principle has status as binding and customary international law (McIntyre and Mosedale 1997). The first known use of the Precautionary Principle was in Swedish environmental law in 1969, and appeals to it have grown common since then, especially (but not exclusively) in European law; it was written into both the 1992 Maastricht Treaty on the European Union and the now-defunct draft Constitution for the European Union (Sunstein 2005, 16–18).

The Precautionary Principle appears in various forms in many national and international legal documents, but can be easily categorized into both a weak and a strong version. In its simplest form, the Precautionary Principle advocates the avoidance of any measures that will create a risk of harm. As Sunstein (2005, 13) describes it, the weak version requires caution until safety is established and does not require unambiguous evidence or harm prior to regulation: "In a catchphrase: Better safe than sorry," Sunstein states.

As Sunstein (2005, 18) points out, the weak version of the principle is hardly objectionable: no reasonable person would argue that a lack of decisive harm is grounds for refusing to regulate a potential hazard. As an example of this sensible approach to the Precautionary Principle, Sunstein (2005, 18) quotes the 1992 Rio Declaration regarding global sustainable development: "Where there are threats of serious or irreversible damage, lack of full scientific certainty shall not be used as a reason for postponing cost-effective measures to prevent environmental degradation."

However, Sunstein (2005, 19) identifies much stronger versions of the Precautionary Principle, quoting as an example the following 2002 Congressional testimony from a representative of an environmental organization: "… the Precautionary Principle mandates that when there is a risk of significant health or

environmental damage to others or to future generations, and when there is scientific uncertainty as to the nature of that damage or the likelihood of the risk, then
decisions should be made so as to prevent such activities from being conducted
unless and until scientific evidence shows that the damage will not occur."

A concise and relevant example of the strong version of the Precautionary
Principle is also quoted by Sunstein (2005, 20) from the Final Declaration of the
First European "Seas at Risk" Conference in 1994: if "the 'worst case scenario'
for a certain activity is serious enough then even a small amount of doubt as to the
safety of that activity is sufficient to stop it taking place." This appears to be
closest to the version of the Precautionary Principle that is most often implicit in
arguments against Active SETI.

2.2 The Precautionary Principle in Practice: Different Cultures, Different Risks

As Sunstein (2005, 21) points out, every culture is precautionary about some risks
but not others. European nations have adopted a precautionary stance toward
hormones in beef, whereas the United States has been more concerned with mad
cow disease in beef and blood donations; Europe has essentially banned genetically modified foods, whereas U.S. regulators have been more aggressive
regarding carcinogenic additives in food; American occupational safety law is far
more precautionary than that of Sweden (Sunstein 2005, 20). Americans tend to be
highly precautionary regarding risks from terrorism and abandoned toxic waste
dumps, but not regarding risks from global warming, indoor air pollution, poverty,
poor diet, and obesity (Sunstein 2005, 34).

While Sunstein (2005, 21) is correct that every culture chooses the risks it
decides to fear, he also points out that the Precautionary Principle appears to be
more formally enshrined in European law than in American law. European legal
authorities have not yet settled upon a weak or strong version of the principle, but
the stronger versions appear to prevail (Sunstein 2005, 21).

3 Criticism of the Precautionary Principle

3.1 Failure to Provide Guidance Regarding When to Regulate

The fact that humans from different cultures are selective, in different ways, about
the risks they choose to fear points to the central weakness of the Precautionary
Principle as a means of evaluating risk and the need for regulation: it provides no
guidance as to which risks deserve fear and those that do not. Sunstein (2005, 4)
argues that, in its strongest forms, the Precautionary Principle is "literally

incoherent" because risks are present on all sides of social situations: the Precautionary Principle is therefore paralysing, forbidding regulation, non-regulation, and everything in between. Sunstein (2005, 5) asserts this is so because people are averse to particular risks, not to risk in general; in other words, humans often act irrationally—or at least without statistical evidence—when choosing which risks to ignore and which to fear. Sunstein explores a variety of facets of human cognition (some of which are discussed below) that explain this tendency.

As previously noted, the weak versions of the Precautionary Principle border on common sense and are hardly objectionable; they can easily, and appropriately, be used to justify reasonable measures to address real risks such as human-induced climate change. People rarely demand absolute certainty regarding the danger posed by a risk before they are willing to take measures to guard against it (Sunstein 2005, 18). However, Sunstein (2005, 24, 120) argues that no reasonable person would advocate banning an activity merely because somebody argues, without evidence, that a risk is worth taking seriously due to some vague and undefined risk of harm; to avoid absurdity, Sunstein interprets the idea of a possible risk that justifies costly measures of risk avoidance as requiring at least a minimal threshold of plausibility, in the form of scientifically-supported suspicions or suggestive evidence of significant risk.

Sunstein (2005, 24) asserts that the threshold burden under the stronger versions of the Precautionary Principle is minimal; once it is met, there is a virtual presumption in favor of regulatory controls without meaningful risk analysis. It is in its strongest forms that the Precautionary Principle offers no guidance at all as to which risks actually should be regulated and those that should not; Sunstein (2005, 4) therefore limits his critique to these strongest forms. Sunstein (2005, 29) argues the Precautionary Principle fails because the stringent regulation of many risks, in the guise of precaution, itself creates other risks by ignoring the "opportunity benefits" of the regulated activity or by ignoring new risks created by the act of regulation itself.

One example Sunstein (2005, 29) offers to illustrate this problem is that the United States is more precautionary about the introduction of new human medicines than is most of Europe. While there are potential risks to any new medicine, by choosing to slow their introduction to the market the United States does not take precautions against certain illnesses that could be reduced by speedier approval procedures. By ignoring new risks that are substituted for the risk being regulated, the Precautionary Principle offers little actual guidance to policy makers (Sunstein 2005, 32–33).

3.2 Why the Precautionary Principle Fails

Sunstein identifies five facets of behavioral economics and cognitive psychology to explain both the appeal of the Precautionary Principle and the reasons that it fails to offer guidance. Three of them are particularly salient to our discussion here concerning Active SETI: the availability heuristic, probability neglect, and loss aversion. In addition to these three factors, another from the fields of risk communication and risk perception that Sunstein does not discuss seems particularly salient, which is the tendency of people to favor risks assumed voluntarily as opposed to those they view as imposed on them against their will.

The availability heuristic is easy to understand: it is a rule of thumb that makes some threats seem more likely to come to fruition than they really are because one can easily recall an example that represents the threat (Sunstein 2005, 36–39). This might apply to the Active SETI debate in the following way. If an example of one threat (for example, numerous works of science fiction which portray invading, predatory or hostile aliens) is more cognitively available than other risks associated with regulating or even banning Active SETI (such as fostering a political environment that is hostile to science), the risk that is less cognitively "available" may not be apparent.

The Precautionary Principle also encourages probability neglect. In evaluating risks, it is human nature to focus on the worst-case scenario and ignore the statistical probability of that scenario occurring. Sunstein's (2005, 39–41) claim here is that the Precautionary Principle appears to give guidance only because the probabilities of different outcomes, including worst-case scenarios, are ignored; rather than consider probabilities, people tend to zero in on one emotionally gripping outcome (e.g., in the Active SETI debate, alien invasion) that is only a discrete subset among a much larger set of probabilities.

Loss aversion makes the Precautionary Principle seem rational because people tend to fear a potential loss from their status quo more than they value a potential gain (Sunstein 2005, 42–43). People are more concerned with, and will be highly attentive to, any losses from an unfamiliar risk (again, alien invasion resulting from Active SETI) and will place a lower value on potential benefits that may be lost if they forego the same unfamiliar risk (for example, lost "opportunity benefits" that could result from contact with an altruistic and benevolent ETI). In other words, potential disadvantages of an unfamiliar risk matter more than potential advantages of the same risk, even if the potential advantages are at least as likely or even far more likely to occur than the disadvantages. Aggravating this problem is the tendency of people to be far more tolerant of *familiar* risks than *unfamiliar* ones, even if the unfamiliar risk is just as likely to occur than the familiar one, or even less likely to occur (Sunstein 2005, 42–43).

Sunstein (2005, 224) summarizes the impact of the availability heuristic, probability neglect and loss aversion on the Precautionary Principle as a plea for the "aggressive regulation of risks that are unlikely to come to fruition," at the cost of losing benefits that are at least as likely as some risks and often more far more likely; it is because of these various features of human cognition and risk perception, which "make certain potential hazards stand out from the background," that the Precautionary Principle seems to offer clear guidance. People tend to worry more, and take excessive precautions regarding, the most cognitively available risks—in other words, those risks that come quickly to mind; people tend to focus emotionally on worst-case scenarios regardless of how likely they are to occur (Sunstein 2005, 224–225).

This chapter adds another facet of human behavior to Sunstein's list of factors leading to the failure of the Precautionary Principle as a useful tool for evaluating risks, this factor also drawn from the fields of risk communication and risk perception. In these fields, it is understood that people accept a higher level of risk if that risk is taken on voluntarily (e.g., driving, skiing, smoking cigarettes) than they will accept from a risk that is imposed by others (e.g., asbestos, radon, pesticide residues) (Grabenstein and Wilson 1999). This fact is relevant to the Active SETI debate if people believe that those engaging in Active SETI are imposing an existential risk on others against their will. Of course, whether Active SETI actually imposes an existential risk is completely independent of what people believe.

3.3 Possible Alternatives: Maximin and the Anti-Catastrophe Principle

Sunstein suggests that in many cases, some form of the Precautionary Principle can be salvaged from the wreckage of its current formulations. Specifically, he suggests two possible substitutes for the Precautionary Principle in certain situations. The first is maximin, or choosing the policy with the best worst-case scenario; this is an option only if one can say with some degree of probability that one option indeed presents a worst-case scenario that is truly better (or worse) than the other policies that could be chosen (Sunstein 2005, 60–61).

This quandary points to the distinction between situations involving *uncertainty* (where probabilities cannot be assigned to different possible outcomes) and *risk* (where probabilities can be assigned to various possible outcomes); the problem becomes more complex when one considers that some situations originate in an environment of *uncertainty* and move gradually over time toward a situation of *risk* as more information because available that allows for the calculation of probabilities (Sunstein 2005, 60–61). The Active SETI debate may be one such situation.

To address some (but not all) situations involving uncertainty, Sunstein (2005, 109) also proposes an Anti-Catastrophe Principle that would guide policy makers to follow maximin by choosing the approach that presents the "best" worst-case scenario and that *also* completely avoids the risk of catastrophic and irreversible harm. However, the Anti-Catastrophe Principle has several parameters that greatly limit its utility in the debate over Active SETI. It is vulnerable to the same weakness as the Precautionary Principle, in that many situations present risks on all sides of the decision matrix. For the Anti-Catastrophe Principle to work, all relevant risks must be identified and taken into consideration: As Sunstein (2005, 114) points out, it makes no sense to take steps to avert one catastrophe if those very steps would create their own catastrophic risks. If a policy choice intended to avoid irreversible catastrophe would increase similar risks from an alternative source, then the Anti-Catastrophe Principle offers no guidance at all—particularly in cases of uncertainty, where policy makers are acting without the guidance of any calculable probabilities. In other words, the Anti-Catastrophe Principle is particularly unhelpful if all alternative courses of action carry with them similar potential worst-case scenario or varying outcomes that, while different in some respects, are all equally bad, and their relative probabilities of occurring cannot currently be calculated or even estimated. However, more options may become available as data are gathered and one moves from situations of uncertainty (with no available probabilities as to worst-case scenarios) to situations of risk (where probabilities can be calculated).

4 Active SETI and Risk Analysis: Egoism Versus Altruism?

4.1 Why the Precautionary Principle Is of No Help

John Billingham and James Benford (2011) cited Stephen Hawking's 2010 warning regarding the supposed dangers posed by Active SETI to justify both international consultations as well as a moratorium on Active SETI until an international consensus is reached. Their paper does not indicate whether this moratorium is intended to be self-imposed based on moral authority alone, or enforced via some mechanism of law. Others have previously sounded similar warnings (Shuch and Almár 2007).

While Billingham and Benford (2011) are by no means the first to say that either a voluntary or involuntary moratorium on Active SETI should be imposed, it is worth looking at the their justifications for this measure. Because any extant ETI civilization is likely to be vastly older and more advanced than the human race, they argue that such societies will be far beyond our current level of technological development. Billingham and Benford acknowledge we cannot know the societal, cultural, political and ethical make up of an unknown ETI. They could be altruistic (benign and benevolent), primarily motivated to benefit less-advanced humans through intellectual exchange. In other words, they may help us. They may behave altruistically toward us or at least regard us as moral equals. But we currently do not know the probability of that.

However, Billingham and Benford (2011) also suggest a more ominous outcome, citing Hawking's analogy (Khan 2010) to the detrimental effects suffered by indigenous populations in North and South America upon contact with Europeans, beginning with Christopher Columbus.[1] Acknowledging they cannot be specific about the possible risks involved, Billingham and Benford point to Hawking's concerns that ETI would probably have powers that would be beyond our own understanding, as well as ethics and morals that are presently unknown. In other

---

[1] It is worth nothing here that the strength and historical accuracy of this analogy/assumption has been questioned by some, including recently by Kathryn Denning (2010b) and, at least in the case of indirect contact between cultures—as may be the case if SETI succeeds—Paolo Musso (2012).

words, ETI may be egoistic or selfish and harm or even destroy us. But we currently do not know the probability of that outcome, either.

The arguments presented by Billingham and Benford (2011) illustrate precisely why appeals to the Precautionary Principle or reasoning similar to it are of little help in the debate over Active SETI. There are risks in both pursing Active SETI and in refraining from it. Indeed, there are risks (to be addressed at the end of this chapter) in terms of detrimental scientific, legal and public policy precedents that could flow from enjoining the practice through a moratorium that might be imposed despite the lack of even a minimal amount of evidence bearing on the *probability* (as opposed to the mere *possibility*) of risk. At its crux, the central question in this debate is whether ETI would behave egoistically or altruistically toward us. We can only speculate at this point. The risks posed by any outcome are unknown because we cannot currently calculate the relevant probabilities of extraterrestrial egoism versus altruism.

If we cannot currently calculate even tentative probabilities of the risk connected with either outcome (altruistic and benevolent ETI that want to help us versus egoistic and malicious ETI that want to hurt us), can maximin or Sunstein's proposed Anti-Catastrophe Principle be invoked to help us? Neither can because, in addition to the fact that we cannot calculate the *probabilities* of various outcomes, we also cannot determine which outcome truly presents the *worst-case scenario*.

At first blush, it would appear obvious that the threat of possible destruction, colonization or conquest by an invading ETI is a worse outcome than simply losing out on "a little help from our friends." The problem with that answer is we do not know the solution to either the Fermi Paradox or the value of L (representing the average duration of a technological civilization in our galaxy) in the Drake Equation (Denning 2005). Simply put, the human race's only chance of long-term survival might be "a little help from our friends." That's obviously highly speculative, but it is no less speculative (and no less *probable* given our current state of knowledge) than Hawking's suggestion that ETI may harm us.

There are possible dangers to both engaging in Active SETI (in terms of attracting the attention of an egoistic, predatory ETI) and to not engaging in it (at least in the form of lost "opportunity benefits" of contact with an altruistic ETI that may help us save us from ourselves, or from an asteroid, or from a distant supernova—pick your poison). But since we currently have no means of knowing what the probabilities are of either outcome, or even determining which outcome is truly the worst-case scenario, "precaution" toward only one side of the risk balance sheet while ignoring the other is simply of no help whatsoever.

Others have come to the same or similar conclusions. Regarding the risks of both traditional, passive SETI and Active SETI, Alan Penny (2012) has argued that both activities carry "existential risks" that may be real.[2] Penny argues for a

---

[2] Billingham and Benford (2011) disagree regarding passive SETI, claiming it carries no "innate risk"; others beside Penny take a contrary view or at least do not rule out the possibility that passive SETI may also present at least theoretical risks (Musso 2012; Baum et al. 2011).

variety of reasons that the detection by ETI of an Earth-originating Active SETI transmission could just as well lead to our salvation as to our destruction. In short, an ETI that detects an Earth-originating Active SETI signal might just as well egoistically or selfishly destroy us as altruistically save us, or even refrain from a previously-made decision to destroy us. Penny concludes that because we cannot predict what ETI might do is response to a message, we have no basis for deciding to transmit or not to transmit. Both listening and transmitting may be either dangerous or beneficial in terms of our continued existence. Musso (2012) has come to a similar conclusion, and holds that Active SETI is unlikely to be dangerous, but that we cannot completely exclude the possibility of danger.

Kathryn Denning (2010a, 1400–1401) has made the point succinctly: at some point, the debate over the risk of transmission leads us down a "blind alley" and we should focus, for now, on separating knowable from unknowable risks. Denning (2010a, 1401) further argues that for now the only calculable factor bearing on a risk–benefit analysis of Active SETI would be the quantification of exposure, i.e., the chances that any given transmission(s) have been or likely will be detected by ETI.

A vigorous debate exists over whether humans have already passed the point of no return in this regard, a debate which is beyond the focus of this chapter. Needless to say, opinions (and calculations) very widely, ranging from certainty that we have already irreversibly exposed ourselves to detection to certainty that we have not (Atri, DeMarines, and Haqq-Misra 2011). Penny (2012) argues that an ETI civilization only a few hundred years more advanced than us and within a few hundred parsecs of Earth would already have imaged our planet and detected oxygen in our atmosphere; all ETI would have to do is tune their radio telescopes to Earth and listen to our broadcasts (assuming they were close enough to detect them), which might tell them all they need to know. It is too late to keep quiet, Penny asserts. Billingham and Benford (2011) disagree and are certain we have not irreversibly exposed ourselves to detection through our intentional broadcasts and unintentional leakage. Ultimately, this question is not central to whether the Precautionary Principle can aid us, because the real uncertainty has to do with ETI intentions, capabilities and desires, in particular egoism versus altruism, not whether they have or even can currently detect our signals or will be able to do so in the near or distant future.

The likelihood of detection of our signals is only the first step in determining the probabilities needed to analyse whether to use either maximin or Sunstein's Anti-Catastrophe Principle as alternatives to the Precautionary Principle. Beyond the risk or likelihood of detection, one would need to calculate the probabilities of contact (response) of some kind, which Denning (2010a, 1401) asserts is simply unknown (and probably unknowable until it happens). According to Denning (2010a, 1402), contacts between civilizations on Earth can only illustrate the *range* of potential outcomes for human contact with ETI, rather than predict those outcomes that are most *likely*. Ultimately, Denning (2010a, 1402) argues we cannot "by any method" realistically calculate the risks or benefits of contact.

Seth Baum, Jacob Haqq-Misra, and Shawn Domagal-Goldman (2011) have explored the potential range of possible patterns of human-ETI contact. Their paper illustrates nicely the central problem with subjecting Active SETI to any kind of Precautionary Principle analysis: it is not just the *probabilities* of various outcomes of Active SETI that we cannot calculate. As pointed out earlier, we also currently have no way of knowing which course of action (transmitting or not transmitting) poses the most adverse worst-case scenario. This is because we simply cannot see into the future, and we have no precedents upon which we can reliably draw. Examples from human history do not inform us in any reliable way about what an encounter with ETI might mean for us—our destruction or our salvation or something in between. This quandary also renders analysis under alternatives to the Precautionary Principle, such as maximin or Sunstein's Anti-Catastrophe Principle, equally futile.

Critics of Active SETI who employ a Precautionary Principle-type analysis focus on the supposed adverse outcome of alerting an egoistic, selfish and dangerous ETI to our existence. However, as Baum et al. (2011) show, contact with ETI could play out in a wide variety of scenarios that are beneficial, neutral or harmful to us. While many of these contact scenarios fall into the (intentionally or unintentionally) "harmful" category, this does not allow us to conclude that contact is more likely to be harmful than not. This may stem simply from the fact that, as Baum et al. (2011) point out, we have compelling reasons to think that ETI, because of its likely great age and level of development compared to human technology, would be highly capable of destroying us, if it so desired.

But "highly capable" does not dictate "highly likely" or even "likely." Like Cass Sunstein, Baum et al. (2010) caution against a human tendency to jump to conclusions regarding matters that are highly uncertain and for which a broad range of outcomes is possible; while they acknowledge there are many scenarios in which ETI may be harmful to humanity, ranging from merely having a demoralizing cultural impact to destroying humanity (either intentionally or unintentionally), contact may also be essentially neutral to us under several possible scenarios.

But it is the scenarios involving contact's possible *beneficial* effects on humanity that seem most often to be forgotten or ignored in the debate over Active SETI. Baum et al. (2011) propose several beneficial scenarios resulting from contact with an altruistic ETI. For example, contact may be philosophically beneficial by uniting humanity in achieving positive goals. But the beneficial effects of contact may be more concrete. Cooperative extraterrestrials may endow humanity with useful knowledge regarding math, science and physics and biology. This information may enable the human race to overcome myriad obstacles that could ultimately cause our own destruction or set us back in our development: war, hunger, disease, political and religious strife, and even asteroids or distant supernova. Baum et al. also posit that an altruistic ETI may have achieved a sustainable mode of development that the human race could adopt to avoid exhaustion of our limited resources and the eventual collapse of our civilization through resource depletion, a fate some have suggested is possible (if not likely) if humanity continues its current course of development unaltered (Turner 2008).

These possible beneficial outcomes to contact with an altruistic ETI should not be discounted by those analyzing any risk posed by Active SETI. The many problems humanity currently faces are quite real, and far more demonstrable today than any imagined risk of destruction that an unknown and selfish ETI may pose to humanity. Baum et al. (2011) have not been alone in suggesting that Active SETI may lead to scenarios that benefit or even save humanity as opposed to endanger it. Douglas Vakoch (2011) has suggested that an Active SETI program may be a necessary prerequisite to establishing contact with ETI (whether that contact proves harmful, beneficial, or something in between). Vakoch (2011, 518) has also described what might be other potential "opportunity benefits" of Active SETI, pointing out that ETI may believe that it is the youngest civilizations, such as ours, that bear the burden of transmitting first since we have the most to gain from an interstellar conversation: "If other civilizations are waiting to reveal their presence until receiving an invitation from humankind, an Active SETI program may be a prerequisite to establishing communication with extraterrestrial intelligence."

This is all obviously speculation—but so is the fear that Active SETI will alert a hostile and selfish ETI to our presence, leading ultimately to our destruction, enslavement or other degradation. Worse, it is speculation in a vacuum of known facts regarding ETI. The bottom line for now is that, as Baum et al. (2011) assert, we cannot just assume that any one specific outcome would result from contact with an unknown ETI. Until contact actually occurs, there is no way to know what will happen. Baum et al. acknowledge the need for some kind of quantitative risk analysis of various contact scenarios, but also acknowledge the difficulty of assigning numerical qualities to an unknown ETI civilization of unknown qualities when it comes to egoism and altruism.

Like Billingham and Benford (2011), Paul Shuch and Iván Almár (2007, 142) have asserted that ardent proponents of Active SETI must admit that the "probability of negative outcomes" is not zero. True to a point: presumably, it would be zero only if we knew that ETI did not exist at all and we were indeed alone in the universe, but that is something we probably cannot ever know. In our current vacuum of facts regarding ETI, neither can we completely exclude the dangers of *not* engaging in Active SETI, something its ardent opponents really ought to admit. We also cannot completely exclude the dangers of engaging only in traditional Passive SETI. Again, while those dangers are speculative, they are no less speculative than the dangers assumed by some to be inherent in Active SETI.

In all of these questions, we are dealing with situations of *uncertainty* (where probabilities are unknown and possibly uncalculable) as opposed to *risk* (where probabilities can actually be calculated). Therefore, neither the Precautionary Principle, maximin, nor Sunstein's Anti-Catastrophe Principle can help resolve these questions given the current state of knowledge regarding ETI civilizations and their dispositions toward egoism and altruism.

4.2 The Factual Vacuum: A Reality Check

If and when we do detect a confirmed radio or other signal from an advanced extraterrestrial civilization, we will know at least one thing: ETI exists. We may know more. We may have a good idea of the location of the signal's origin (its direction and approximate distance from Earth), telling us at least where ETI was located when the signal was first broadcast. If the signal is more than a carrier wave and contains intelligible information we are able to decode and understand, we may learn a great deal more. We may have some idea of the civilization's age, level of technological development, physical constitution, habitat, culture, law and ethics, just to name a few potential properties. These would all be facts the human race could consider in debating and deciding the then-urgent question of whether and how to respond to such a message.

Contrast such a hypothetical scenario with the reality facing us now in the context of the debate over Active SETI and whether ETI would behave altruistically toward us if it learned of our existence. At the risk of overstating the obvious, anything we think we know about ETI is, at best, speculation, even if it is speculation informed by analogy from the only advanced technological civilization we know, the human one (Denning 2011).

To suggest that Active SETI poses a danger to humans is to make many unwarranted assumptions for which we currently have no evidence—in addition to the assumption that ETI even exists. The assertion of danger also assumes ETI, if it knew of our existence, would take some action that may harm us, intentionally or unintentionally. Therefore, the assertion of danger makes assumptions not only regarding ETI's intent and ethics but, implicitly, about the feasibility of ETI engaging in interstellar travel. While some scientists have proposed that an interstellar probe may be the most efficient means for an advanced extraterrestrial civilization to communicate with other civilizations (Rose and Wright 2004), the predominant SETI paradigm appears historically skeptical toward this suggestion (Shostak 2004). At the very least, there is a dearth of published scientific papers supporting the reality of interstellar travel by anyone.

The following brief set of "known unknowns" illustrates our lack of knowledge about the very existence of ETI and other issues bearing on the possible risks and benefits presented by Active SETI. As the term implies, these "known unknown" are merely facts or questions we know we do not yet have the answer to, but may acquire in the not-too-distant future. This list is sparse and incomplete (and rather obvious), but it is a starting point.

The first "known unknown" would be the discovery of the first extrasolar planet that is a true Earth analogue—a terrestrial planet of the approximate mass, volume and composition of the Earth located in the habitable zone of a star more or less like our Sun in its age and other relevant properties. As of this writing, such a planet has not yet been found, although speculation in the popular press has suggested that the announcement of such a discovery could come in the near future, perhaps as the fruit of the recently completed Kepler mission (O'Neill 2011).

The discovery of the first true Earth analogue may tell us that Earth-like planets are common in our galaxy, raising the chances that life, including intelligent life, may coexist with us.

A second "known unknown" would be the discovery in our solar system (or on an extrasolar planet) of signs of the existence of simple or microbial life forms that are unrelated in origin to life on Earth. This drives home an important fact: not only do we not know that complex (let alone intelligent) extraterrestrial life exists, we do not even know of a second origin of *simple* life anywhere in our own solar system, let alone our galaxy. Evidence of a second origin of simple life in our solar system might be found someday on Mars or on a moon such as Europa, Titan or Enceladus or it may be detected in the atmosphere of a distant extrasolar planet. It may even be found on Earth. The discovery of another origin of simple life in our solar system would tell us that life on Earth—at least simple life—is not a lone occurrence, if it can be determined that the new form of life is truly unrelated in origin to known life on Earth.

Finally, a third "known unknown" would be the discovery of evidence, such as that suggested by Paul Davies (2012), that ETI has existed in the distant past, even if we cannot determine from such evidence if it coexists in the galaxy with us today. (Even a radio signal broadcast from a thousand light-years away is evidence only of an ETI's existence a thousand years ago when the signal was sent; Davies suggests we may someday find physical evidence of the existence of ETI dating back millions, perhaps billions, of years ago.) Although we would probably have no idea of current status of such an advanced civilization, it would tell us that intelligent life has developed more than once in our galaxy.

Surely other "known unknowns" could (and should) be imagined, but the point is obvious. We are not only a long way from understanding the culture, motivation and ethics of any ETI that may exist, but we are in the dark as to even the most basic question: are we, or are we not, indeed alone in this universe? If and when the time comes that any one of these questions is answered, we should be alert to the possibility that our thoughts regarding Active SETI may need to be re-evaluated in light of the new information available.

5 What We Should Really Fear

The debate over Active SETI should continue. All reasonable voices deserve to be heard and listened to with respect. But to suggest restricting or stifling any human scientific, technological or intellectual activity (such as Active SETI) simply because we know *nothing* about ETI is an extraordinary proposition and a gross misapplication of Precautionary Principle-type thinking. If the human race had adhered to such a principle throughout its existence, we might not have modern science.

There are some issues bearing on Active SETI that scientists may be able to resolve in the near term, such as the probabilities of whether or not we have, already, irreversibly exposed ourselves to detection through our intentional broadcasts and unintentional radio leakage. Other issues are probably more intractable, at least in the short term, such as whether or not ETI is likely to behave in an altruistic manner toward us if contact ever occurs.

But the debate itself is fraught with a danger that may not be readily apparent to those engaged in it. That danger is to the SETI community specifically, and to science more generally. At least in the United States, science has been under political attack for decades. This is not an alarmist statement. One need look no further than the more than ten-year history of attacks and restrictions on embryonic stem cell research, or the now decades-long political war on climate scientists. These are attacks motivated, in many cases, by certain political and religious ideologies. Politically motivated attacks on other publicly funded science projects have become routine.

To date, at least in the United States, SETI's primary fear in the realm of public policy has been that of being regarded as silly. The late U.S. Senator from Wisconsin, William Proxmire, attacked NASA expenditures on SETI in the 1980s as being a waste of taxpayer money, until Carl Sagan persuaded him otherwise (Lemonick 2011). In 1993, Congress killed funding for the High Resolution Microwave Survey for much the same reason.

However, in our post-9/11 world, being regarded as a silly waste of taxpayer money may be the least that SETI has to fear from politicians. SETI cannot afford to be regarded as somehow threatening our security, physical or moral. Assertions of danger should not be tossed around casually and without hard, demonstrable evidence to back them up. Other authors have suggested various ways in which the discovery of extraterrestrial life may threaten the world views of religious conservatives around the globe (Tarter 2000). In the United States, such fears can have very concrete ramifications. In 2011, several Wisconsin state legislators introduced two bills intended to restrict the use of fetal tissue in scientific research (Finkelmeyer 2011). Neither bill passed. These efforts followed the successful passage of a bill in 2005 (ultimately vetoed by the governor) that would have banned both reproductive and therapeutic cloning in Wisconsin (Babe 2005). What makes these efforts particularly noteworthy is that the publicly funded University of Wisconsin-Madison has been the source of some of the most groundbreaking stem cell research. That has not stopped legislators in the state capitol building in Madison (just a few blocks from the university) from trying to kill that very research, in large part because it offends their view of the moral order.

So the danger to SETI isn't only from the Congress in Washington, DC. SETI efforts in state-funded institutions around the United States could be vulnerable to attack if conservative (or liberal) political and/or religious forces aligned in a manner similar to the way they have in attacking stem cell, cloning, and climate research. This warning should be kept in mind in view of a recent study indicating that trust in science among politically conservative Americans has declined markedly over the last 40 years (Gauchat 2012).

References

Atri, Dimitra., Julia DeMarines, and Jacob Haqq-Misra. 2011. "A Protocol for Messaging to Extraterrestrial Intelligence." *Space Policy* 27:165–169.
Babe, Ann. 2005. "Doyle Vetoes Human Cloning Ban." *The Badger Herald*, November 4. Accessed August 29, 2012. http://badgerherald.com/news/2005/11/04/doyle_vetoes_human_c.php.
Baum, Seth D., Jacob D. Haqq-Misra, and Shawn D. Domagal-Goldman. 2011. "Would Contact with Extraterrestrials Benefit or Harm Humanity?: A Scenario Analysis." *Acta Astronautica* 68 (11–12):2114–2129.
Billingham, John, and James Benford. 2011. "Costs and Difficulties of Large-scale 'Messaging,' and the Need of International Debate on Potential Risks." Accessed August 29, 2012. http://arxiv.org/abs/1102.1938v2.
Davies, Paul C. W. 2012. "Footprints of Alien Technology." *Acta Astronautica* 73:250–257.
Denning, Kathryn 2005. "'L' on Earth." Paper presented at the 56th International Astronautical Congress, Fukuoka, Japan, October 16–21.
Denning, Kathryn. 2010a. "Unpacking the Great Transmission Debate." *Acta Astronautica* 67:1399–1405.
Denning, Kathryn. 2010b. "The History of Contact on Earth: Analogies, Myths, Misconceptions." Paper presented at the 61st International Astronautical Congress, Prague, Czech Republic, September 27-October 1.
Denning, Kathryn. 2011. "Ten Thousand Revolutions: Conjectures about Civilizations." *Acta Astronautica* 68(3–4):381–388.
Finkelmeyer, Todd. 2011. "UW Officials Say Bill Would Have a 'Chilling Effect' on Biomedical Research." *The Capital Times*, August 6. Accessed August 29, 2012. http://host.madison.com/ct/news/local/education/article_4d7e69e6-bfa1-11e0-a631-001cc4c03286.html.
Gauchat, Gordon. 2012. "Politicization of Science in the Public Sphere: A Study of Public Trust in the United States, 1974 to 2010." *American Sociological Review* 77(2):167–187.
Grabenstein, John D., and James P. Wilson. 1999. "Are Vaccines Safe?: Risk Communication Applied to Vaccination." *Hospital Pharmacy* 34(6):713–729.
Khan, Amina. 2010. "Scientists Weigh in on Hawking's Alien Warning." Los *Angeles Times*, May 7. Accessed August 29, 2012. http://articles.latimes.com/2010/may/07/science/la-sci-hawking-aliens-20100508.
Lemonick, Michael D. 2011. "ET Call Us—Just Not Collect." *Time*, April 28. Accessed August 29, 2012. http://www.time.com/time/health/article/0,8599,2067855,00.html.
McIntyre, Owen, and Thomas Mosedale. 1997. "The Precautionary Principle as a Norm of Customary International Law." *Journal of Environmental Law* 9:221.
Michaud, Michael A. G. 2001. "If Contact Occurs, Who Speaks for Earth?" *Foreign Service Journal* 78(April):23–27.
Michaud, Michael A. G. 2003. "Ten Decisions That Could Shake the World." *Space Policy* 19:131–136.
Musso, Paolo. 2012. "The Problem of Active SETI: An Overview." *Acta Astronautica* 78:43–54.
O'Neill, Ian. 2011. "Big Question for 2012: Will We Find Earth 2.0?" *Discovery News*, December 21. Accessed August 29, 2012. http://news.discovery.com/space/big-question-for-2012-earth-20-111220.html.
Penny, Alan. 2012. "Transmitting (and Listening) May Be Good (or Bad)." *Acta Astronautica* 78:69–71.
Rose, Christopher, and Gregory Wright. 2004. "Inscribed Matter as an Energy-efficient Means of Communication with an Extraterrestrial Civilization." *Nature* 431:47–49.
Shostak, Seth. 2004. "Does ET Use Snail Mail?" *Space.com*, September 9. Accessed August 29, 2012. http://www.space.com/312-snail-mail.html.
Shuch, H. Paul, and Iván Almár. 2007. "Shouting in the Jungle: The SETI Transmission Debate." *Journal of the British Interplanetary Society* 60(4):142–146.

Sunstein, Cass N. 2005. *Laws of Fear: Beyond the Precautionary Principle*. New York: Cambridge University Press.

Tarter, Jill C. 2000. "SETI and the Religions of the Universe." In *Many Worlds: The New Universe, Extraterrestrial Life & the Theological Implications*, edited by Steven Dick, 143–149. Philadelphia and London: Templeton Foundation Press.

Turner, Graham. 2008. "A Comparison of *The Limits to Growth* with Thirty Years of Reality." *Global Environmental Change* 18(3):397–411.

Vakoch, Douglas A. 2011. "Responsibility, Capability, and Active SETI: Policy, Law, Ethics and Communication with Extraterrestrial Intelligence." *Acta Astronautica* 68(3–4):512–519.

Part III
Inferring Altruism

The Accidental Altruist: Inferring Altruism from an Extraterrestrial Signal

Mark C. Langston

Abstract This chapter discusses the likelihood of an alien race concocting and transmitting a computer virus to Earth, and explores the possibility of determining the intent of an extraterrestrial signal using the Prisoner's Dilemma as a framework. The fundamental question we must ask ourselves is this: How do we know whether the beings sending us the signal are friendly or hostile? Other questions follow from this premise: Is the signal meant as a greeting? A lure? A Trojan horse? Is the signal itself intended to disrupt our technology or society? Are any of these consequences possible despite the intent of the sender? Through an exploration of the costs and benefits of altruism, the author will attempt to answer these and other questions. The act of signaling will be explored separately from the content of the signal itself, and the impact of intent on each will be addressed.

Keywords Alien communication · Prisoner's Dilemma · Alien computer virus · Trojan horse virus · Extraterrestrial intelligence · SETI · Altruism · Selfishness · Intention

M. C. Langston (✉)
Infoblox, Santa Clara, USA
e-mail: seti@bitshift.org; mark@bitshift.org

M. C. Langston
1913 Baltimore Pike, Gettysburg, PA 17325-7013, USA

D. A. Vakoch (ed.), *Extraterrestrial Altruism*, The Frontiers Collection,
DOI: 10.1007/978-3-642-37750-1_9, © Springer-Verlag Berlin Heidelberg 2014

1 Alien Computer Viruses

Early one morning in the not-too-distant future, SETI locates and archives a bona-fide artificial signal of extraterrestrial origin.[1] When the alien signal is loaded into the researcher's computer for analysis, the computer appears to crash. Shortly thereafter, computers across the Internet begin to crash in a suspiciously similar manner. Within a matter of hours, the Internet has been crippled. Stock markets are unable to function. Global communication is destroyed. Power grids begin to falter. Travel and commerce have been brought to a halt, and panic ensues. The cause is ultimately traced back to executable code embedded in the alien signal that managed to penetrate the SETI researcher's computer and spread worldwide.

This type of scenario is similar to that envisioned by Richard Carrigan (2006). Though the possibility of this happening is remote it is, as Carrigan argues, not zero. Let us consider, therefore, the conditions that must hold for such a thing to occur. First, the extraterrestrial intelligences (ETIs) must have some knowledge of the current state of our computer technology. That includes knowing that our computers operate on binary data, knowing that those computers use instructions defined by strings of binary data of a certain length, that those instructions take specific operators, that the computer processor uses a memory pointer and an instruction pointer, that the computer uses a form of temporary and a form of persistent storage, and how those two types of storage are accessed and used.

Assume that an alien civilization inhabits a planet orbiting Proxima Centauri, a star just over 4 light years distant. This civilization is near enough to Earth to have received information regarding relatively current terrestrial computer technology via our noisy broadcast television or radio. Ignore for the moment our rapid adoption of satellite and fiber-optic transmission technologies that obsolete much of our old broadcast system, thus reducing the likelihood that any such information would ever reach Proxima Centauri. This civilization would quite possibly possess all of the information I list above as required for such an attack to occur. At least, they would possess the signals containing that information.

Also assume that this alien civilization is not only quite a bit more advanced than ours, but also that these aliens are more intelligent on average than the typical human. They are able to decode the signals, decipher whatever human language the information has been presented in, and comprehend everything conveyed therein in a relatively short period of time: one Earth year. They have the material, financial, and political resources to react to and subsequently act on this alien signal in a global, coordinated manner without significant societal impact or

---

[1] In this chapter, we will be discussing the reception and interpretation of a signal from an intelligent alien civilization. In so doing, we have already assumed that such a civilization exists and is at least as technologically advanced as our own. It is not therefore unreasonable to also assume, for purposes of this chapter, that other such civilizations exist, and that those civilizations may be far in advance of our level of technology. We will, however, impose a few limits to our assumptions: the speed of light is still inviolate, time travel does not exist, and wormholes may not be used for travel.

upheaval. They see us as a threat to their continued existence in and dominance of this sector of the galaxy. Thus, they devise a plan to leave us defenseless in the face of an alien invasion. They construct the facilities and equipment necessary to transmit a signal containing their computer virus back to Earth, hidden in what seems to be a message of peace and goodwill. The signal is sent, coincident with the launch of an armada of alien vessels that travel quite near the speed of light, designed to arrive in the solar system very soon after the signal reaches Earth.

Four more years pass, and the aliens have arrived in our solar system. They park their battleships behind Jupiter and begin surveillance of Earth, waiting for the telltale signs of societal collapse. Eventually, a SETI researcher or research team picks up the extraterrestrial signal containing the cleverly-hidden virus, and…nothing happens. The virus relied on information almost a decade out of date. It also relied on information based on a computer chip architecture completely different from the one the researcher is using: The aliens received information about the top-of-the-line computer processor of 2002: the Intel Pentium 4, while the researcher is using a modern SPARC T3 processor. The age isn't as of much significance as the two CPUs in question: One is a CISC-based processor, the other, a RISC-based processor. The acronyms aren't particularly relevant here: the salient point is that the two are as different as night and day when it comes to the instructions they are designed to execute. Trying to get CISC code to run at the binary level on a RISC processor would be akin to expecting someone who only speaks Swahili to discuss the finer points of Shakespeare with someone who only speaks Tagalog.

The time difference does become an issue if we instead discuss alien civilizations on more distant stars. Suddenly, the number of years that pass between the transmission of the information the extraterrestrial intelligences (ETIs) rely on to build their virus and the subsequent reception of said virus back here on Earth can span entire evolutions and revolutions in computer technology and processor architecture. Go from just 4 light years out to 10, and our malevolent neighbors are suddenly at least two decades behind in computer knowledge. That's long enough to mean the difference between a modern PC and a 486-based machine without any recognizable modern form of Internet access.

But let's get back to our Centauran virus: it's been loaded on a SETI researcher's Sun workstation, and has done precisely nothing. Does this invalidate Carrigan's scenario? Not necessarily. That data could be sent to other machines, ones using a processor architecture more similar to that used by the ETIs when designing their virus. Will it therefore wreak untold havoc on our modern, technology-reliant society? Well, assuming the ETIs were working at the binary level—that is, assuming they sent a long string of ones and zeroes that represented a sequence of instructions for the processor they learned about—and assuming the virus eventually landed on a computer using an Intel processor sufficiently old enough, or sufficiently compatible enough, that those instructions were still valid they may or may not do anything (the x86 processor architecture goes out of its way to preserve as much backwards compatibility as possible in its instruction set, though this can't always be done; this is one reason why higher-level programming languages exist, to abstract away from having to deal with the actual processor instructions). They would still need to be executed: the virus would have to be "run" on the computer. This is usually done via direct user intervention: clicking on an icon or typing a command. However, this allows the computer's operating system (OS) to decide what to do with the file. This is usually dictated by the file's type, which is in turn usually defined by the file's extension. SETI researchers aren't in the habit of trying to execute files containing signals received from other stars, and even if they were, the OS wouldn't treat it as an executable file.

There is another way that computer viruses can be executed on a computer without human intervention: the virus tricks the computer into running it. This, however, requires a detailed knowledge of the computer hardware, the operating system, and the particular piece of software being run within the operating system (a web browser, a signal detection and analysis program, a statistical package, etc.). The virus would exploit weaknesses in that piece of software, typically discovered by benevolent or malevolent researchers via painstaking months of investigation and reverse-engineering, using knowledge that relies on a detailed understanding of the various types of weaknesses typically found in computer software. Those weaknesses would be leveraged to execute the virus payload—in this case, the malicious Centauran code.

Let's say this happens, as vanishingly unlikely as this scenario has become: the Centauran virus finds a suitable host and executes. It still has to do something more malignant than crash the one computer it's on. It needs to spread, and spread quickly, if it's going to cause the kind of worldwide catastrophe the Centaurans are counting on. Relying on random manual infection of this sort is far too inefficient to be any real threat. It needs to get onto the Internet without human intervention. It needs either to be autonomous, or hide itself in something so wildly popular that it will entice people in all major sectors of our infrastructure to install and run it themselves. The latter would require even more knowledge about us in terms of economics, sociology, politics, and so forth, and that knowledge would need to be much more current than what the Centaurans possessed when they created the virus. It would be wonderful if the Centaurans could figure out a way to get a virus to execute simply by viewing captioned images of cats, but if that were a viable attack vector, I can promise you it would already be in use by much more domestic sources. So let's examine the possibility of autonomy.

For the Centauran virus to propagate, it needs to be able to detect and use modern networking hardware. And here the Centauran virus is likely to be stopped in its tracks: the odds of any two computers having an identical hardware configuration, including the networking hardware, are small. The odds of enough computers on the planet sharing that same configuration are essentially zero. The reason why this is relevant is because we assumed at the beginning that this virus was working at the level of "bare metal": that is, it was using raw, binary data to represent a string of instructions hard-wired into the computer's processor. At this level, the virus authors would need to know about things like networking drivers, and how to interface with them. It's not out of the question that such an advanced tidbit of knowledge could have been broadcast in 2002. What is beginning to stretch credulity is that said knowledge would still be relevant a decade later. Not only has the hardware in use changed in that period of time, so too have the drivers necessary to interface with them. In addition, there are many different makes and models of networking hardware in use at any given time throughout the world. So the Centaurans couldn't possibly hope to have their virus communicate directly with the networking hardware in the computer at that level. That means they would instead have to rely on the computer's operating system.

The operating system of a computer provides a useful abstraction layer for such tasks as communicating with a network card. This allows software writers to ignore all of the details I outlined above, and focus on the more important aspects of their task. Rather than walking into a deli and instructing the employees in every minute detail of building the sandwich you'd like, from slicing the meat, cheeses, and bread to placing them along with your chosen condiments and additions in the proper order on a plate, you simply give the counter person the name of the sandwich you'd like, and get on with the more important task of lunch. The Centauran virus could conceivably tell the OS it wants to connect to the Internet, assuming several more critical events have occurred: The Centaurans have received enough low-level programming information about an OS to accomplish this, the OS in question is still the dominant OS back on Earth, and the Centaurans have received an expert-level education in how the Internet operates.

In 2002, Windows XP was just becoming the dominant PC OS. Windows 2000 was still quite prevalent in business environments, and Windows 98 or Windows ME was still to be found in great number in homes. Not to mention the resurgence of Apple with OS X, and the vast array of Unix-based OSes in universities and research organizations around the world. A decade later, it's Windows 2003 and Windows 7, and newer versions of OS X and all those Unix OSes. The diversity of operating systems alone precludes the type of worldwide takeover this virus was meant to achieve. Even if we were a monoculture of a single Windows operating system, and that OS remained dominant for the decade necessary for the Centaurans to achieve their aim of accessing the networking hardware in computers worldwide, they would still need to understand that the Internet existed, let alone its inner workings. We take all these things for granted, and with the entire Internet at our fingertips, most of us would struggle with learning enough of this information to achieve what the Centaurans intend. Now imagine trying to do it on a sparse collection 10-year-old broadcasts you cannot control, or search, or request follow-up information on.

As unlikely as this scenario is, it raises an interesting question: If we did receive a signal from an ETI, would we be able to tell or determine the sender's intent? Would we somehow be able to tell that the Centaurans had hidden a virus in the signal, meant to bring society to its knees, or are we at the mercy of our assumptions about extraterrestrial altruism?

2 Inferring Altruistic or Selfish Intent

> We are all here on earth to help others. What I can't figure out is what the others are here for.
>
> W. H. Auden (2002, 347)

What is altruism and how does it apply to SETI? Altruism is commonly defined in two ways: the belief in or practice of disinterested and selfless concern for the well-being of others, and the behavior of an animal that benefits another at its own expense. Either definition could be applied to a signal sent into the cosmos by an alien race. But would we be correct in assuming any such act was an altruistic one? The concept of altruism may be universal (Minsky 1985), so it's not unreasonable to assume that an alien civilization would be acquainted with the idea.

A computer is completely trusting of its input, and disambiguates mercilessly. Its interpretation of input is based on a finite set of well-understood rules. Thus, the sender's message is interpreted plainly within that ruleset: if an executable instruction is found in the proper context, the computer assumes it was meant to be executed, and that instruction is totally unambiguous to the computer: there can be only one way in which that particular instruction may be interpreted in a given context. This property allows the creation of astoundingly complex software. However, it is exactly this clarity and brittleness that gives rise to "bugs" in computer programs: unintended effects of well-intentioned but poorly-structured code. The benefit of computers is that they do exactly what you tell them to do. The drawback of computers is that they do exactly what you tell them to do.

In his paper "Science and Linguistics" Benjamin Lee Whorf (1940, 229) wrote, "We cut nature up, organize it into concepts, and ascribe significances as we do, largely because we are parties to an agreement to organize it in this way—an agreement that holds throughout our speech community and is codified in the patterns of our language. The agreement is, of course, an implicit and unstated one, but its terms are absolutely obligatory; we cannot talk at all except by subscribing to the organization and classification of data which the agreement decrees." Computers may interpret a signal in one way. Humans, on the other hand, may have quite a different interpretation of that same signal. In addition, we humans have the added benefit of considering the intent behind the signal itself; we can take the signaler into consideration.

With respect to the content of a signal, communication as consensus presents interesting problems when applied to interstellar distances. In the beginning of this chapter, we presented a scenario in which an alien race was able to communicate with our own terrestrial computer systems in their own language. This scenario makes many assumptions, including that the signal was intended only for receivers on Earth. A more likely scenario may posit a signal that is not language-specific, species-specific, or even location-specific. How does one decode a signal written in a language so totally foreign that no assumptions whatsoever can be made about its content? Various solutions have been proposed (Freudenthal 1960; Sagan 1973) in which the foundations for this agreement are present within the message, defining the terms through which the sender is attempting to communicate with the recipient. In this situation, the sender assumes both parties will come to consensus and agree to use the terms in the primer properly. This technique has been demonstrated in various works of fiction, perhaps most notably the book *Contact* (Sagan 1997). This allows us some hope that we may be able to establish the linguistic common ground necessary to interpret the contents of a signal. However, such a primer would by necessity provide only the bare minimum necessary to interpret the contents of the message. What we would lack is the cultural contexts in which the concepts being communicated arose, and knowledge of the socio-political forces that led to the transmission of the signal. A simple, "Hi, we're over here!" could mean anything from "WARNING: Avoid at all costs. Imminent collapse into ultra massive black hole in progress," to "We just wanted to let you know where we were before we annihilate you. In our society, it's considered polite." We cannot rely on the content of the signal by itself when determining whether or not the signal is intended to be altruistic.

To explore the question of altruism in an extraterrestrial signal, the signal may be considered with respect to both the act of signaling itself, and the signal's content. Either component may be altruistic or selfish, resulting in four possibilities:

Beacon altruism: The act of signaling is intended as an act of altruism, to inform other civilizations of the signaler's existence.
Content altruism: The signal contains a message designed to impart some benefit to the recipients.
Beacon selfishness: The signaling act is intended to harm the recipient. Perhaps it was meant as a lure to entice other civilizations to expend significant resources attempting to contact and/or visit the signaler, disrupting the recipient's civilization and leaving them vulnerable to attack. Or perhaps it was meant to throw the civilization's religious beliefs into question or incite panic in the populace.
Content selfishness: The signal contains a message designed to harm the recipient. Perhaps the scenario described at the beginning of this chapter was the intended effect of the data contained within the signal. Perhaps the signal contains instructions for building a doomsday device that, when activated, destroys the recipient's world thereby eliminating the signaler's competition in the long term.

The time and distance involved in interstellar communication makes any such act a one-way affair. Therefore, it may be assumed that the signaler will not reap any immediate benefit nor incur any immediate harm due to the actions of the signal recipients.

Under these assumptions, acts of beacon altruism would seem to provide the least benefit to the signaler. There can be no short-term gains to the signaling civilization, except perhaps social and political benefits to certain individuals derived from the decision and act. The costs of beacon altruism are quite high in comparison, due to the strength and duration of signal necessary to produce a significant likelihood of reception. Content altruism would seem to suffer from the same problem: high costs with little to no short-term gain. Furthermore, altruistic signaling may produce long-term harm, in that a recipient civilization may view the signal source as a threat, and act to minimize or eliminate that threat.

Selfish acts, on the other hand, may provide a greater benefit to the signaling civilization. The ability to remotely hobble or destroy a potential competitor at a comparatively negligible cost may seem attractive to a xenophobic and/or highly-competitive civilization. As described above, these effects are possible either through the content of a signal, or through the act of signaling itself. With this potentially greater reward, however, comes increased risk. If the signal recipient is similarly competitive but more technologically advanced than the signaler, there is the possibility that the signaling civilization could be located and destroyed, using the signal as a means to locate the signaler.

Given the relative costs and benefits of altruistic versus selfish signaling, which is more likely? The question presents a classic Prisoner's Dilemma scenario (Axelrod 1984). The basic Prisoner's Dilemma structure is one based on cooperation versus betrayal: two parties are faced with possible harm. If one party cooperates and the other party betrays them, the betrayer is not harmed, while the cooperator receives the greatest amount of harm. If both parties cooperate, both parties are minimally harmed. If both parties betray the other, both parties are harmed slightly more than they would be if both had cooperated, but neither is harmed to the full extent possible. If we consider alien signaling in the Prisoner's Dilemma framework, it would be of greatest benefit to the signaling civilization to be selfish, while assuming that the recipient would infer altruism on the part of the signaler. This isn't an unreasonable assumption to make. Fehr and Fischbacher (2003) have demonstrated that we humans tend towards cooperation in such situations, even though logic dictates the best outcome for both parties would be to act in their own interests.

Based on this information, should we, the recipient of an extraterrestrial signal, view that signal and its contents as an altruistic act, or an act of selfishness? Should we trust that the signaling civilization has our best interest at heart, or should we think the worst and act accordingly? We have no means of evaluating the cultural context which gave rise to the alien signal. There is no knowledge of the social and political pressures that drove the signaling civilization to send a signal in the first place, nor those that shaped the signal's content. Indeed, there is no knowledge of whether social interaction as we know it exists on the signalers' world; we merely assume that concepts like "social pressure" and "politics" exist as we know them.

We must therefore infer the signaler's intent due to a lack of context in which to place the act. Is the signal meant as an interstellar "Hello, neighbor"? Is it a tantalizing lure meant to exhaust our resources or to provide the signalers' descendants with flash-frozen culinary delights? Are the contents of the signal an *Encyclopedia Galactica*? Or are they an insidious computer virus designed to destroy any sufficiently advanced technology with which it comes into contact? We must ask these and similar questions once the initial excitement regarding signal discovery has died down.

The problem with doing this is that we may never know the correct answers. The theory of cultural relativity states that concepts like morality, hostility, and so forth vary among cultures. As such, what the recipients deem an immoral act (e.g., sending a signal that contains instructions for incubating a fast-acting virus that destroys all biological life on the recipients' world, disguised as a cure for all disease) may be viewed as not only moral, but highly desired by the senders (e.g., a culture whose religion holds in the highest regard the act of introducing another to the afterlife quickly and efficiently). Western culture is full of examples of these erroneous inferences. In J. Michael Straczynski's *Babylon 5* television series, an interplanetary war is predicated on a cultural misunderstanding: An alien race makes first contact with a human warship with its gun ports open—a sign of strength and respect in the alien culture. The humans mistake this gesture as an act of aggression and open fire. In American culture, the "thumbs-up" gesture (fist outstretched and vertical with a thumb extended upward) is typically interpreted as "I'm okay" or "everything is great." In many Arab cultures, however, the gesture is considered offensive, equivalent to the "middle finger" gesture in American culture (Axtell 1998).

Without knowledge of the signalers' culture, it is impossible to intuit the original intent of the signal. Even an instance of beacon altruism—where the signalers' only intent was a friendly, content-free "hello"—could be construed as an untoward act by the recipients. If the signal were such that it caused significant problems with the receiving equipment or sensitive but otherwise completely unrelated equipment, the recipients may interpret the act as hostile (e.g., perhaps the aliens deliver messages using small amounts of hyper-dense material accelerated to nearly the speed of light, meant to be deciphered at impact after being decelerated and captured by a special orbiting apparatus that we sadly lack. It would be the interstellar equivalent of tying a message to a rock and hurling it through a priceless stained-glass window!). If public knowledge of the signal were to cause a social and/or political upheaval in the recipient's civilization, there is no way to determine whether those effects were intended or accidental.

On the other hand, a signal originally intended as selfish may be interpreted as altruistic by the recipients. An act of beacon selfishness may be viewed as nothing more than a friendly "hello," and the recipient civilization fails to demonstrate the anticipated reaction the signalers had hoped for. In this instance, it is the signaler that made faulty assumptions about the recipients' culture. Likewise, an act of signal selfishness, though it may contain instructions for constructing a dangerous and ultimately deadly device, may provide the recipient civilization with precious knowledge of advanced technology or manufacturing techniques, and they may determine the true nature of the device before it becomes active. Hostile computer code meant to wreak havoc with the recipients' computers may in fact alter them in such a way as to make them self-aware, or may provide the recipients with enlightening insights into new programming techniques. In each case, the signalers' intended ill effects not only fail to manifest, but the recipient civilization benefits in some way from the content of the message.

These acts of accidental altruism highlight the difficulties a recipient civilization may face once an extraterrestrial signal has been detected. The odds of correctly interpreting the sender's intent and full meaning are small. According to Wiio's First Law, communication usually fails, except by accident (Wiio 1978). Ultimately, the determination of intent must be made with respect to our own culture's concepts, values, and mores, while remembering that we know nothing of the social, political, or cultural contexts in which the alien signal originated.

In the face of all this uncertainty, how should we proceed? The Prisoner's Dilemma suggests that we should be trusting only if we believe the signaling civilization's intent is altruistic. However, the Prisoner's Dilemma holds only if the actions of both entities have known interpretations by and effects on the other party. As demonstrated here, this is not the case: it is entirely possible for the signaling civilization to act altruistically yet have that act interpreted as selfish (in this case, dangerous or untrustworthy) by the recipients. It is also possible for the signalers to act selfishly only to be viewed as interstellar altruists by grateful recipients. Thus, the Prisoner's Dilemma scenario cannot be relied upon to inform our reactions to an alien signal. Perhaps the best course of action is a prudent one that embodies guarded optimism: prepare for the worst and hope for the best.

References

Auden, W. H. 2002. *The Complete Works of W.H. Auden: Prose. Volume II. 1939–1948.* Princeton, NJ: Princeton University Press.

Axelrod, Robert M. 1984. *The Evolution of Cooperation.* New York: Basic Books.

Axtell, Roger. 1998. *Gestures: The Do's and Taboos of Body Language Around the World.* New York: Wiley.

Carrigan, Richard. 2006. "Do Potential SETI Signals Need to Be Decontaminated?" *Acta Astronautica* 58(2):112–117.

Fehr, Ernst, and Urs Fischbacher. 2003. "The Nature of Human Altruism." *Nature* 425:785–791.

Freudenthal, Hans. 1960. *LINCOS: Design of a Language for Cosmic Intercourse.* Amsterdam: North-Holland Pub. Co.

Minsky, Marvin. 1985. "Communication with Alien Intelligence." In *Extraterrestrials: Science and Alien Intelligence*, edited by Edward Regis, 117–132. New York: Cambridge University Press.

Sagan, Carl, ed. 1973. *Communication with Extraterrestrial Intelligence (CETI).* Cambridge, MA: MIT Press.

Sagan, Carl. 1997. *Contact.* New York: Simon & Schuster.

Whorf, Benjamin Lee. 1940. "Science and Linguistics." *Technology Review* 42(6):229–231, 247–248.

Wiio, Osmo A. 1978. *Wiionlait—javähänmuidenkin.* Espoo: Weilin+Göös.

Interstellar Intersubjectivity: The Significance of Shared Cognition for Communication, Empathy, and Altruism in Space

David Dunér

Abstract What kind of indispensable cognitive ability is needed for intelligence, sociability, communication, and technology to emerge on a habitable planet? My answer is simple: *intersubjectivity*. I stress the significance of intersubjectivity, of shared cognition, for extraterrestrial intelligence and interstellar communication, and argue that it is in fact crucial and indispensable for any successful interstellar communication, and in the end also for the concepts that are the focus of this volume, empathy and altruism in space. Based on current studies in cognitive science, I introduce the concept of intersubjectivity as a key to future search for extraterrestrial intelligence, and then explain—leaning on phylogenetic, ontogenetic, and cultural-historical studies of cognition—why intersubjectivity is a basic requisite for the emergence of intelligence, sociability, communication, and technology. In its most general definition, intersubjectivity is the sharing of experiences about objects and events. I then discuss what "intelligence" is. I define it as cognitive flexibility, an ability to adjust to changes in the physical and socio-cultural environment. Next, I discuss sociability and complex social systems, and conclude that we probably can expect that an extraterrestrial civilization which we can communicate with has a high degree of social complexity, which entails a high degree of communicative complexity and high degree of cognitive flexibility. Concerning communication, I discuss intention, attention and communicative complexity. I also stress three socio-cognitive capacities that characterize advanced complex technology: a sustainable,

Special thanks to Göran Sonesson, Per Lind and Erik Persson for valuable comments on my chapter, and to Dainis Dravins, Peter Gärdenfors, Mathias Osvath, and Jordan Zlatev who generously shared their knowledge and articles in preparation. Finally, thanks to all participants of the seminar at the Centre for Cognitive Semiotics, Lund University, for exciting discussions about cognition seen from a true universal perspective.

D. Dunér (✉)
Inst. för kulturvetenskaper, Lund University, Biskopsgatan 7, 223 62 Lund, Sweden
e-mail: David.Duner@kultur.lu.se

D. Dunér
Centre for Cognitive Semiotics, Lund University, Lund, Sweden

complex social system, with a regulated system for collaboration, such as ethics; complex communication for collaboration and abstract conceptualization; and a high degree of distributed cognition. Finally, if we conclude that intersubjectivity is a fundamental requisite, we then have some options for future interstellar communication. We should target Earth analogues, monitor them, and finally initiate an interstellar intersubjective interaction.

Keywords Altruism · Astrobiology · Cognition · Cognitive science · Cultural evolution · Empathy · Evolution · History of technology · Intelligence · Interstellar communication · Intersubjectivity · Joint attention · Philosophy of mind · SETI · Sociability

1 Introduction

When we read texts like this one, when we write, or talk to each other, we interact with someone with intentions, we socialize, have joint attention, try to reach the other's inner world. What makes this endeavor possible is that we to some extent share experiences. What we need is intersubjectivity.

When we monitor the skies, listen to the stars, or analyze electromagnetic waves from outer space in the search for extraterrestrial intelligence, we are searching for something that we can understand, can communicate with, that has intentions, is self-conscious and social, has advanced civilization and technology. In return, in interstellar communications, we have to show that we are alive, intelligent and self-conscious. We must recognize that they have attentions and intentions, and they have to recognize that we have it. What is needed is intersubjectivity.

Interstellar message construction and the analysis and decoding of extraterrestrial signals are about searching for "something" or "someone" we can exchange information with. That is, this transmitter or receiver should be "something" that we can recognize as "intelligent." However, it is not enough to say that it has technology to transmit electromagnetic waves. It should also be able to understand and decode our messages, believing that there are intentions behind them, that there are meanings hidden in them; thus, it should have an ability to communicate. So the "thing" we are searching for is in some respects something similar to us, something that we recognize as intelligent and with which we can communicate and exchange knowledge and experiences: i.e., a social and communicating intelligent being with advanced technology. At least four characteristics of such an extraterrestrial life form have thus been presumed in our search: intelligence, sociability, communication, and technology. These characteristics, I would argue, are interrelated and depend on more fundamental cognitive skills.

So the question here is what kind of indispensable cognitive ability is needed for intelligence, sociability, communication, and technology to emerge on a habitable planet? My answer is simple: *intersubjectivity*. In this chapter, I will stress the significance of intersubjectivity, of shared cognition, for extraterrestrial intelligence and interstellar communication, and argue that it is in fact indispensable for any successful interstellar communication, and in the end also for the concepts that are focus of this volume, empathy and altruism in space. Empathy, the knowing of other feelings, and altruism, the knowing of other needs leading to a subsequent unselfish action, as well as intelligence, sociability, communication, and technology, rest on intersubjective abilities of the terrestrial minds, and presumably also the extraterrestrial minds in universe.

Based on current studies in cognitive science, I will introduce the concept of intersubjectivity as a key to future search for extraterrestrial intelligence, and then explain, leaning on phylogenetic, ontogenetic, and cultural-historical studies of cognition, why intersubjectivity is a basic prerequisite for the emergence of intelligence, sociability, communication, and technology. Finally, I will propose an interstellar intersubjective interaction for future detections of Earth analogues.

The following argument rests on the belief that we need to focus on the cognitive foundations of interstellar communication, and that cognition and communication are results of the biocultural coevolution. The mathematical, logical and technological constructions of interstellar messages are of limited use if we do not take into account the cognitive basis of intelligence in space, the emergence and evolution of cognitive capacities, how interstellar messages can be cognitively understood, and the cognitive and societal requisites for a sustainable advanced technology. From a general cognitive perspective, I aim to propose new strategies for future interstellar communication. Elsewhere I have discussed the significance of cognitive science as a tool to understand the cognitive challenges the human mind faces in space, and the cognitive foundations of interstellar communication (Dunér 2011a; b; c; 2012; 2013). In this chapter I will explore how and in what way intersubjectivity is fundamental for intelligence, sociability, communication, and technology to evolve in space.

2 Intersubjectivity

In contemporary cognitive science, intersubjectivity has become a key concept for understanding not only empathy and altruism, but also intelligence, sociability, and communication (Gillespie 2009; Hrdy 2009; Gillespie and Cornish 2010; Tylén et al. 2010; Fusaroli, Demuru, and Borghi 2012; Gentilucci et al. 2012). In its most general definition, intersubjectivity can be explained as the "*sharing of experiences* about objects and events" (Brinck 2008, 116). To be more precise, Zlatev et al. (2008b, 1) describe intersubjectivity as "the sharing of experiential content (e.g., feelings, perceptions, thoughts, and linguistic meanings) among a plurality of subjects." The intersubjectivity originates then from a sharing of experiences through actions, and these shared experiences can basically be of three kinds: emotions, attention, and intention (Stern 1985). Through a process of observations and imitations, the individuals share object-directed actions. According to Peter Gärdenfors, intersubjectivity includes five capacities: representing other beings' emotions (empathy), attention, desires, intentions, and beliefs and knowledge. These capacities have emerged gradually in that order through an evolutionary process (Gärdenfors 2013). Empathy, the ability to share others' emotions, according to Stephanie Preston and Frans de Waal (2002), is available to most mammalian species and some birds, which indicates that empathy has old evolutionary roots. Higher order intersubjective skills that we find among humans, such as representing beliefs and knowledge of other beings, are particularly relevant for this present study concerning communication with extraterrestrial intelligence.

The ability to share and represent others' mentality, i.e., intersubjective ability, is an important part of our inner worlds (Thompson 2001; Zlatev et al. 2008a). In the phenomenological tradition, empathy and intersubjectivity play a significant role in the experiencing of another person as a subject and the world as a shared world (Husserl 1973; Stein 1917; Gallagher 2004; Moran 2004; Zahavi 2001). Empathy, the representing of other human beings' emotions, motives, intentions and desires, bodily expressions of emotions, beliefs and knowledge, are impossible without a rich inner world. Other species on Earth have varying degrees of cognitive skills, including awareness of their own subjectivity, their own existence and mental processes, but they seem not to have to the same extent as the species *Homo sapiens sapiens* an awareness of other minds, of other beings' subjectivity. It seems that to be human, or in other words, to be intelligent, is not only to be aware of our own thoughts, but also be aware of others' thoughts, feelings, intentions, etc. According to Zlatev et al. (2008b) the quintessence of the human mind is to be a shared mind. Intersubjectivity is what makes us human. I would add that beings, which we would recognize as intelligent in space, would have intersubjective skills, be aware of themselves and of other minds, and be able to share experiences, actions, information and mental content. If they are conscious of other minds they would also be self-conscious, be able to ponder on their own thoughts and existence. We can probably conclude that an extraterrestrial being technologically capable of transmitting and receiving interstellar messages has intersubjective skills. That said, this does not mean that they actually will think like us. There are species-specific capacities for intersubjectivity, based on biological factors, but also as a consequence of ecological, societal and cultural factors. Most likely these extraterrestrial intelligent species would have other bodies, social organization and cultural history.

Even when intersubjective ability is present, an interchange of information will not be easy. Due to our totally different biological and cultural attributes, future encounters with aliens will face severe problems concerning intersubjectivity, in coordinating our inner worlds, feeling empathy, etc. A human and an extraterrestrial will probably also have trouble perceiving the same target, aligning their attention, adjusting their actions, and imitating each other. Because of our divergent evolutions, empathy and intersubjectivity toward extraterrestrials would probably be even more problematic than in the case of inter-species communication on Earth.

What we can hope for is that we might be capable of sharing attention. Among the shared experiences (emotions, attentions, intentions), it is probably primarily sharing attention, joint attention or attentional intersubjectivity that we can hope for. It would be more difficult to understand each other's emotions and intentions, which require more knowledge about the context and history of the other.

3 Cognition

The field of *astrocognition* deals with cognitive processes in space, the evolution of cognitive abilities, and cognition in extraterrestrial environments (Dunér 2011a; Osvath 2013). If we are discussing the existence of extraterrestrial intelligence in space, I maintain, we seriously have to take into account the research within cognitive science and affiliated research areas in order to get satisfactory answers to the question: What is needed for higher cognitive skills to evolve? What physical, biological, societal, cultural and other environmental factors shape cognition? What cognitive abilities are needed for a living organism to be able to manipulate its environment or, in another words, to develop technology?

3.1 Cognitive Flexibility

What is intelligence? In the search for extraterrestrial intelligence, we should at least have some ideas of what kind of phenomena we are looking for (Regis 1985). I do not think we need to have, or even could have, an Aristotelean definition, a finite set of necessary and sufficient qualities, because we do not know what we will encounter and an openness in our definition would be a necessary strategy when encountering the unknown. Instead, in line with a prototype theory (Rosch 1975; 1978), we should discuss what kind of qualities we are looking for, qualities that we can recognize. I think we should not avoid this question, as it is sometimes in the interstellar communication literature. The famous Drake Equation for estimating the number of civilizations in our galaxy with which communication is possible, does not define what we mean by that "intelligence" we are searching for. Rather, it has an operative definition and just looks for a civilization able to transmit electromagnetic waves. "Intelligence" is, in that sense, "the ability to transmit electromagnetic waves." This is of course not what we mean when we recognize something as "intelligent" in ordinary life. Nor is it sufficient. A satellite can contact us and receive messages from us, but it is not intelligent. Rather, it is a cultural product or extended tool of an intelligent being. If we want to exchange information with an extraterrestrial civilization, it is not enough that the "intelligent" being is able to construct advanced devices; it should also be able to communicate, to share experiences through a medium. This kind of intelligent being needs to have intersubjective skills.

The definition of intelligence has been the subject of a lively debate (Sternberg 2002) and has often been connected to problem solving. Intelligence as the ability to solve problems, to make rational choices, to reason logically, to handle the constraints and limitations of time, space and materials, is not enough to explain the development of technology for interstellar communication. The social constraints are missing. Intelligence is an adaption to the physical *and* social environment. As a broader concept, cognition includes not just the abilities that we call rational, logical or intelligent. An important part of what it is to be intelligent is to have emotional skills, a capacity to emotionally appraise the relevant environment with attraction, disgust, etc., and to respond to socializing, bounding, coupling, etc., in a social group. Emotions are shortcuts through competing options, a faster way to decide, instead of calculating all of them (Frank 1988; Damasio 1994). As Darwinian creatures, extraterrestrials would probably have emotions and feelings, longing for some things, dislike other things, just like us.

I would pinpoint just two features (among many other possible ones) that we find in that we call "intelligent." First, it can imagine things not existing—things, events, etc., not present in time or space right in front of the thinking subject. An intelligent being can test various options or "simulate" events in its mind, instead of doing it in the world outside the brain. Second, an intelligent being is also able to engage in intersubjective interactions, understand other minds, imagine and envision what they will do, what they feel and reason. To be intelligent is to have intersubjective skills, to be able to understand and make interferences about other minds. If the extraterrestrial being that we encounter is lacking these two abilities, it would probably not have complex communication and advanced technology, and we would not be able to communicate with it.

To conclude, in the most general sense I would say that "intelligence" is cognitive flexibility, an ability to adjust to changes in the physical and sociocultural environment. Intelligence can be seen as evolved mental gymnastics required to survive and reproduce within a specific environment. This includes the capability of representing activities and being able to make inner models of reality and other minds. "Intelligence," and in our case "extraterrestrial intelligence," is a rather misleading and narrow concept, too connected to problem solving and rational reasoning, and does not include other mental abilities that are indispensable for a life form to have civilization, culture, and technology. Thus, instead of searching for extraterrestrial intelligence, we should search for extraterrestrial cognitive flexibility, especially extraterrestrial intersubjectivity.

3.2 The Evolution of Cognition

It is evident that cognitive flexibility has evolutionary benefits and can be regarded as a good strategy for the adaptation to a changing environment. "Intelligence is an adaptation," Jean Piaget wrote, "To say that intelligence is a particular instance of biological adaptation is thus to suppose that it is essentially an organization and that its function is to structure the universe just as the organism structures its immediate environment" (Piaget 2001 [1963]). The mental life of an organism is an accommodation to the environment.

Intelligence or cognitive flexibility has emerged through an evolutionary process due to its benefits for survival, orientation and adaptation to a variable environment in a Darwinian struggle for existence. Intelligence is thus the ability to respond to changes in the environment with flexibility and success, and a plasticity of learning, i.e., to be able to learn from experience. As we know it, we can presuppose that the phenomenon we call "life" and are searching for in outer space, has experienced a Darwinian evolution, including variation, heredity, and selection. With their senses, Darwinian creatures explore their environment, orientate in it and search for sources, under a selection pressure leading to different capacities for its specific ecological niche (Gibson 1979; Gärdenfors 2003). This cognitive flexibility, it has to be emphasized, is also a fruit of the pressure from the socio-cultural environment. In other words, intelligence, or rather cognitive flexibility, is a result of the biocultural coevolution of the embodied minds that are interacting with their physical and socio-cultural environment. Extraterrestrial minds, like terrestrial minds, have adapted to their specific environment and the specific social interactions between the minds of their species.

Living creatures have to tackle a variable and changing environment, cope with the abundance of information in the environment, sift through the mass of data, make decisions of what is relevant or not, and find out how these pieces of data relate to each other (Thornton, Clayton, and Grodzinski 2012). In complex environments, it is advantageous not just to take statistical co-occurrences into account but also to try to find general rules and then be flexible and able to solve various problems in different contexts, and further to be able to make mental representations or models of how the world works. By simulating events in their brains, living creatures prepare for actions not yet occurred.

There are reasons to believe that under the right circumstances this environmental pressure will lead to more complex and flexible cognitive abilities. We cannot, though, presuppose that intelligence is a necessary outcome of evolution. Ernst Mayr (1985) expected the probability of intelligent life to evolve to be very low. The benefits of acquiring improved cognitive abilities have to be balanced against the costs of having an energy-consuming brain. But on Earth we find that intelligence seems, like vision and other abilities, to have emerged several times in the course of evolution and in separate evolutionary lines, i.e., convergent or parallel evolution (Seed, Emery, and Clayton 2009; Osvath 2013). The more intelligent or cognitive flexible species on Earth, such as primates, dolphins, and corvids, seem to have some qualities in common: First, they are social and have a high degree of social complexity; and second, they are "all-round," multi-adaptable to very different environments and diets (Osvath 2013).

If we can come closer to an understanding of the processes behind the rapid brain evolution that began a few million years ago on Earth—the encephalization in the Phanerozoic (Carter 2012)—we can use this knowledge to formulate astrocognitive theories. The evolution of human intelligence is part of a general process of greater encephalization (Bogonovich 2011), or the increase brain size relative body size, over time. The social-brain hypothesis (Dunbar 1996; 1998) says that there is a correlation between the size of an animal's social group and the size of its brain, leading to the conclusion that social behavior drives encephalization.

The complex social structure of the group is probably a very important drive for the emergence of intelligence. The brain has increased in capacity in order to tackle different kinds of social relations. The difference between humans and other primates is the state of the socio-cognitive capabilities. Humans engage with others in joint activities that share goals and attention, which facilitate the use of linguistic symbols and the creation of cultural norms (Tomasello et al. 2005; Herrmann et al. 2007; Thornton, Clayton, and Grodzinski 2012). But must we suppose that the extraterrestrials also are social creatures? Could it be possible that the intelligent "it" we are communicating with is a solitary phenomenon, a "Solaris," an end of an evolutionary process? Not likely. For the selective evolutionary process to give rise to higher cognitive flexibility, we have to suppose a variation of distinct genetic units that are flexible enough to quickly adapt to environmental changes.

The biocultural coevolution, in respect to long-term processes and different environmental pressures, will probably make extraterrestrial cognitive abilities and intelligence very different from ours. But if there are intelligent (or rather cognitive flexible) beings in space, we can then probably suppose that they are social and multi-adaptable to different environments, that through the course of their biocultural coevolution have acquired capacities to handle and orientate themselves in their environment, both in their physical environment, but perhaps even more importantly in their social environment. They need to be able to handle complex social relations, to understand other individuals' feelings, thoughts, attentions, intentions, etc. In short, they need intersubjectivity. At least among themselves they would most likely show various degrees of empathy and altruistic behavior in order to orient themselves as social beings in the group. The question is then, would they show empathy towards other forms of life?

4 Sociability

Intelligent species are social species. Human society is a complex environment that requires cognitive flexibility in order to survive. Individuals need to understand and keep a check on what the others are doing, thinking and feeling. In return, sociability and the social context enhance the adaption to the physical environment, and make the individuals less vulnerable to a hostile environment. The sociability is advantageous to the individuals themselves as well as the group in its entirety. These social skills cannot subsist without a cognitive capacity to understand, feel and share experiences of other minds.

4.1 Cultural Evolution

Characteristic of this human social interaction is the ability to learn from others, i.e., *culture*, the transmission of learned behavior and knowledge that is not biologically encoded, or in other words, the ability to transfer information from generation to generation that does not use the genetic code for the transfer but is learned, taught, and transferred by a multitude of communicative and cultural devices and artifacts, like language, signs, pictures, sounds, objects, etc. Culture presupposes enduring joint beliefs or common knowledge. Significant for human cognition is the infant's prolonged period of dependency on its parents, in order to learn skills, behaviors, attitudes, knowledge about the environment, etc., that are not contained in the genetic code, but are indispensable for a flexible and less-vulnerable existence as an adult. Culture is also dependent on an alloparental care, a system where individuals others than the biological parents take the parental role, which can be regarded as an altruistic behavior for the benefit of new generations. What makes us human is to a large extent our propensity for imitation and seeing motivations of others (Calcagno and Fuentes 2012). That is our ability to see things from the perspective of our fellows.

We probably have to assume that social and cultural skills, and in the end a complex social system, play a significant role if we want to explain the evolution of intelligence, communication, and advanced technology in space. Theories discussing the cultural evolution in space are very much needed (Dick 2003; Vakoch 2009). Biological theories of the evolution of cognition are of course important, as I have maintained earlier in this chapter, but we should also discuss the "post-biological" cultural evolution leading to technological civilizations. The social, cultural, educational skills are fundamental prerequisites for the survival and technological development of an extraterrestrial civilization. We can expect that an extraterrestrial intelligent civilization has developed intricate social interactions, a complex social system, that enhances its chances of survival, not just for the individuals, but more importantly for the entire biosphere. And furthermore, if we succeed in interstellar communication, we will in fact socialize with extraterrestrial social beings, and this will demand that we both have advanced social skills and a flexibility to handle and understand very different ways of organizing the social interactions.

4.2 Complex Social Systems

Humans and other intelligent species on Earth have more or less complex social networks. There are reasons to believe that social complexity has been the driving force behind the emergence of intelligence, brain size and communicative complexity. Complex social worlds are like selective environments, driving species towards increased cognitive processing ability, that in its turn leads to higher social complexity, and when social complexity increases, it gives rise to a greater selection pressure on individuals for cognitive skills, that feedback producing even more social complexity and so on (Bogonovich 2011).

In complex social systems "individuals frequently interact in many different contexts with many different individuals, and often repeatedly interact with many of the same individuals over time." (Freeberg, Dunbar, and Ord 2012a, 1785; Freeberg, Ord, and Dunbar 2012b). A complex society contains a large number of interacting individuals. These individuals are of different types and take different social roles. Furthermore, the interaction between individuals has a high degree of diversity. A large number of individuals interact in many different contexts and often with the same individuals.

An "advanced" society, or a civilization with advanced technology, is a complex social system. We can probably expect that an extraterrestrial civilization, which we can communicate with, has a high degree of social complexity, which entails a high degree of communicative complexity and a high degree of cognitive flexibility. Such a socially complex extraterrestrial civilization would have (1) many individuals rather than few; (2) a high rather than low density; (3) many different member roles rather than few roles; and (4) an egalitarian structure rather than a hierarchical structure.

These four characteristics of social complexity will enhance the emergence of advanced technology. Many individuals entail greater collective brain power. A high density entails more frequent and faster interactions between individuals. Many different member roles entail a distributed and specialized cognitive processing. And finally, egalitarian societies have greater diversity of directional relations and more reversals and agonistic interactions. In hierarchies, there are fewer relationships between individuals, and the directional relations are severely limited. Because of this greater diversity in relations, the communicative complexity of egalitarian societies is greater than in despotic systems.

It is more likely that a long-lasting advanced extraterrestrial society is egalitarian than despotic. First, an egalitarian society with distributed decision-making is better adapted to the physical constrains of the universe. To be dictator in the huge time and space-frames of the universe is rather difficult. Hierarchical systems are impractical in space, where the huge distances will make orders from despot to subordinates take a very long time to arrive, in contrast to egalitarian, distributed decision-making. In the history of colonial empires on Earth we find many examples of the difficulties of holding power and the empire together over longer distances. A society with distributed decision-making is more flexible, better able to change and less vulnerable to disturbances in the communication network. Second, if an extraterrestrial civilization has survived for a longer period of time, it must have found ways of dealing with agonistic behavior and conflicts (cf. Pinker 2011), and must have developed reliable collaborative systems for overcoming physical and societal threats. In order to deal with destructive behavior, these civilizations must have advanced intersubjective skills to understand other subjects, must have a high degree of communicative complexity to sustain and strengthen the intersubjective interactions between its members, including long experience of communicating with a diversity of groups and species, and must have arrived at some sort of reliable "ethics" or regulation system for behavior.

5 Communication

Communication is in its quintessence social, something communal rather than private. Some authors have supposed that social complexity leads to communicative complexity. According to the "social complexity hypothesis" for communication, "groups with complex social systems require more complex communicative systems to regulate interactions and relations among group members" (Freeberg, Dunbar, and Ord 2012a, 1785). If we receive a complex message, or a complex artificial signal, from outer space, we then have reasons to believe that the transmitting civilization has a complex social structure (and accordingly also intelligence).

Interstellar communication is, like spoken human languages, intersubjective, as a system for sharing information and for socializing. Communication can be regarded as a sharing of mental states, and the expression as information about a mental state (Østergaard 2012). The semantics is based on a "meeting of minds," as Gärdenfors puts it: "the meanings of expressions do not reside either in the world or (solely) in the mental schemes of individual users, but emerge from *communicative interactions* between the language users" (Gärdenfors 2013). The evolution of semantics could be seen as a coevolution of intersubjectivity, cooperation, and communication (Gärdenfors 2008a; b). In linguistics and cognitive science, probably no one would deny that intersubjectivity plays a critical role in the acquisition of language, but it has still not been discussed in the context of interstellar communication.

Elsewhere I have discussed the cognitive foundations of interstellar communication (Dunér 2011b), and maintained that communication is based on cognitive abilities embodied in the organism that has developed through an evolutionary and socio-cultural process by interacting with its specific environment (See also Arbib 2013; Holmer 2013). In the following I will dig into this further, and discuss one of the most crucial cognitive abilities for language acquisition, i.e., intersubjectivity. Communication presupposes shared knowledge, or perhaps better, shared experiences.

5.1 Intention

I want you to react in a certain way. That is why I send this message. And I have, like we all have in ordinary everyday communication between humans, some ideas of what kind of response our messages might get. There are intentions behind our messages. A communicative intention means that the sender's utterance is meant to produce a particular response, and that the receiver recognizes that the sender intends (Grice 1989 [1957]).

By sending a message, the sender has some idea about how other minds will receive it. The problem is, when we transmit a message to an extraterrestrial civilization, we do not know if they will respond to it. Neither do we have any idea of how they will respond. Intersubjectivity is lacking. We want to achieve something more than just sending a message and not knowing if it is received. My intention in writing this chapter is not just to send a message to an unknown addressee, but to elicit a response, even though I am well aware of that few of us are lucky enough to get response. Thus, at least a third order intention is what is needed in true interstellar communication—in other words, we should know that they know that we know them, by transmitting a message saying: "We are doing this communicative act to you, just you," that they believe is an intentionally sent message aimed to generate a response.

5.2 Attention

All successful communication requires intersubjectivity, an idea of the interlocutor's mental state. The communicating subjects need to know what the other is referring to. The interpreter has to make assumptions about the state of the interlocutor's knowledge, attention, feelings, etc., and then adjust to what he or she thinks the other thinks. These intersubjective interactions are based on common experiences or shared actions. When communicating, communicators need to have shared devices for sharing and manipulating attention. In conversations between humans, we constantly monitor each other's attentional status. There are strong arguments, according to Michael Tomasello (2005), that an infant can understand a symbolic convention only if it understands its communicating partner as an intentional agent with whom one may share attention toward something. A linguistic symbol can in that case be said to be a marker for an intersubjective and shared understanding of a situation. Sign use is in other words social, intersubjective, a sharing of meanings. Imitation and shared attention proceed, phylogenetically and ontogenetically, other more complex communication systems involving iconic and symbolic signs (cf. Kita 2003; Oller and Griebel 2008; Andrén 2010; Lenninger 2012).

To achieve mutual understanding in interstellar communication, we need to establish an intersubjectivity that could lead to the possibility of entering the others' inner thoughts and views of reality. It is crucial to find out whether the others are, like ourselves, intentional agents, so that we in that case could relate to their world, and have perspectives on our worlds that can be followed, directed, and shared.

5.3 Complex Communicative Systems

In order to reach more complex communication, we first need not just attention, imitation and iconic signs, but also have to use symbolic signs—conventional or arbitrary signs that are detached representations and, as such, dependent on culture and human interaction. If the extraterrestrials are intelligent, they probably have some kind of symbolization abilities and abstract thinking detached from the environment, with which they can reason about things not existent, things that are not right in front of them, facing their senses, in a specific moment in time. In other words, to reach a higher degree of communicative complexity, they need signs where the expression is separated from the content. The fact that symbolic signs are characterized by this constraint, as Göran Sonesson (2013) has shown, makes it nearly impossible to use symbols for interstellar communication. Even if we agree with the content (for example about mathematics, physics, chemistry, etc.), we could have very different ways of expressing this content. So, even though we cannot decode and understand their symbolic messages, they must show that they have it. Second, complex communicative systems contain "a large number of structurally and functionally distinct elements [...] or possess a high amount of bits of information." (Freeberg, Dunbar, and Ord 2012a, 1787). A complex communicative system shows a large repertoire of distinct signals.

To conclude, in the presumably artificial signals from outer space we have to look for: (1) symbolic signs, and (2) signaling complexity. These two characteristics of complex communicative abilities can indicate that we are receiving a message from an extraterrestrial intelligent civilization.

6 Technology

Advanced technology can be defined heuristically as technology for interstellar communication. Technology in general could be described as ways of manipulating the environment, using objects in the environment outside the body in order to strengthen the genetically given capacities, such as body strength, perception, and cognition.

Technology is not just applied science. It is a misconception that technology is an application of scientific theories, a product of the rational, inventive mind. An innovation needs a larger innovation system, including many people with different roles, technicians, designers, investors, lawyers, marketers, etc. It is not enough to have a new bright idea. To become an innovation, the technological invention has to be used, to be an answer to the needs of a society. Technology is to a large extent a social phenomenon, a product of the cultural evolution. One of the most challenging factors in the Drake equation is f_c, the fraction of intelligent civilizations that develop technology for interstellar communication (Sagan 1973). This factor actually concerns the cultural evolution of technical civilizations; or, to put it in another way, what is needed for an intelligent life form to evolve advanced technology? To answer this question we have to turn to studies in the evolution of cognition, how hominids began using and manipulating their environment, and to studies in the history of technology, how the cultural evolution of *Homo sapiens sapiens* made it possible to achieve higher technology (Donald 1991; Tomasello 1999; Steels 2004; Richerson and Boyd 2005). The rise of civilization involved closeness, interaction of many individuals, exchanges of ideas, products, and experiences that paved the way for a technological society.

In order to achieve advanced technology extraterrestrial life forms need to have complex social structure, complex communicative skills, high degree of distributed cognition, sustainable society, and well-developed intersubjective abilities. Technology is not a matter of science, rational reasoning alone; intersubjective skills are completely indispensable.

6.1 Advanced Technology

If we are searching for extraterrestrial civilizations (Vakoch and Harrison 2011) with advanced technology, i.e., technology for interstellar communication or other devices for reaching beyond their home planet, we also have to discuss what makes advanced technology possible. I would like to stress three socio-cognitive capacities that characterize advanced complex technology and that are crucial for the development of it: (1) a sustainable complex social system, with a regulated system for collaboration, such as ethics; (2) complex communication for collaboration and abstract conceptualization; and (3) a high degree of distributed cognition. All these three capacities require intersubjective skills.

Cooperation is what makes higher technology possible. Cooperation in its turn requires some fundamental cognitive and communicative functions (Leimar and Hammerstein 2001; Bowles and Gintis 2003; Richerson, Boyd, and Henrich 2003; Fehr and Fischbacher 2004; Stevens and Hauser 2004; Kappeler and van Schaik 2006; Lehmann and Keller 2006; Gärdenfors 2008b; Tomasello 2009; Cheney and Seyfarth 2012). All the three factors mentioned above (complex social system, complex communicative system, and distributed cognition) enhance such cooperative behavior towards higher technology. In many cases basic cooperation—for example, flocking behavior and the way of increasing cooperation by separating members of the in-group from the out-group among non-human animals on Earth—requires just limited cognitive capacities and no communicative skills. More advanced forms of cooperation, as we find among humans, as Gärdenfors (2012, 165) argues, "presuppose a high awareness of future needs, a rich understanding of the minds of others, and symbolic communication."

First, concerning a sustainable complex social system: Cooperation about detached non-present goals requires advanced coordination of the inner worlds of the individuals. In order to achieve advanced technological skills, the individuals have to cooperate in joint activities where they are sharing goals and attentions. By coordinating their roles when working towards a specific goal, they achieve a joint intention. To this we can add that they must be able to engage in prospective planning, to anticipate the future, i.e., have the capacity to represent future needs, a prospective thinking or "mental time travel" (Suddendorf and Corballis 1997; Roberts 2007). An extraterrestrial civilization capable of transmitting messages for a long period of time, and which has perhaps more advanced technology than we do, must have survived and be capable of avoiding disasters and crises in their history, such as those that we face—nuclear war, global warming, pollution, decreasing biodiversity, etc. This indicates that they have a functioning social structure that can handle and avoid crises, a complex social system that regulates risks and destructive behavior. It is easier to develop advanced technology for destruction than an ethics for survival. It is easier to understand the laws of physics and chemistry than to understand and predict the human mind and the complex social and cultural interactions of humans. The question of the L-factor, the life span of advanced technological civilizations, deals with this bottleneck (Sagan 1973; Shostak 2009; Denning 2011). Longevity, sustainability and technological growth as well as regulatory systems for behavior (ethics) are linked together. Civilizations develop technology for destruction before they develop a sustainable ethics. If a civilization has advanced technology, which needs cooperation between large numbers of individuals, they have to be able to trust each other; some sort of ethical consent has been established.

Second, a complex communicative system is needed in order to handle the social complexity, to facilitate collaboration, to transfer information between the individuals, which in its turn are indispensable for the development of technology. The communicative system must enable the users to construct abstract concepts and symbols, to generalize, to discuss things and events not existent, that have ceased to exist, and have not yet have come into being.

A third factor important for the emergence of higher technology is distributed cognition, the ability to use external objects and minds to enhance thinking. The ability to construct external cognitive artifacts is significant in human cognition (Norman 1991; 1993; Malafouris and Renfrew 2010; Malafouris 2012). These organism-independent artifacts compensate for the limitations of the biological memory. As such, they are part of our distributed cognition. Gradually through human cultural history, the externalization and materialization of memory have increased, from the art of writing, to the printing press and Internet (cf. Donald 1991; 2008). We have invented more devices strengthening our inborn sensory equipment and more devices for thinking. We are using our environment to think, and distributing our thinking to physical objects, but also to other minds. A diversified, specialized distributed cognition has some radical advantages for developing complicated and thought challenging technology. Constructing advanced technology involves a large number of people. The famous Manhattan Project that led to the first atomic bomb included more than 130,000 people. A technologically-advanced civilization cannot rest on the brain power of a few individuals, but needs as many as possible sharing their knowledge, cooperating, and completing different specialized tasks.

To conclude, if there are intelligent beings able to communicate with advanced technology, they would probably have a complex social system, complex communication, and a high degree of distributed cognition, and needless to say, intersubjectivity.

7 Conclusion

If we succeed in transcending the insurmountable moats of the immense distances of light years between the stars, we will confront the next problem, the cognitive and semiotic problem of interstellar communication. Interstellar communication is a dyadic interaction. The question is if this interaction will be an asymmetrical and hierarchical relationship. This is what Stephen Hawking and others have feared. In other words, are there different ethical standards, different scopes of inclusion in the ethical community (Persson 2012), and an asymmetry in respect to knowledge, science, and technology? Recent phylogenetic, ontogenetic, and cultural-historical studies clearly show that intersubjectivity is an indispensable requisite for the evolution of human intelligence, sociability, and communication, as well as advanced technology. If so, then intersubjectivity is important for any future interstellar communication. In the following, I will explain how this knowledge can be used in an interaction with a future extraterrestrial interlocutor.

7.1 Future Candidates

Earlier I suggested that an interstellar message should not be a general, abstract message, but a concrete message tied to the spatiotemporal setting (Dunér 2011b). In this chapter I also suggest that we should search, for heuristic reasons, for an extraterrestrial being similar to us. To begin with, to enhance the feasibility of communication with an extraterrestrial intelligence, we should choose a habitable planet with similar physical appearance. In this section, I will give some suggestions of how we can focus our search, and what targets we could select as future candidates for communication.

Our chances for functioning communication increase if we search for and construct messages aimed for an intelligent extraterrestrial species similar to us, a social creature adapted to an equivalent physical environment. This would probably also enhance our ability to recognize the signal as an artificial message. What I propose is not an analogical argument, suggesting that they necessarily are like us. Intelligent beings in outer space might be very different. Searching for extraterrestrial cognitive flexibility could instead be seen as a heuristic approach. The probability for a fruitful exchange would be greater if we formulate our message for a species that has evolved under similar circumstances, and is similar to us. In addition (Dunér 2011b), we should not only tie the message to the spatiotemporal situation, and formulate a non-symbolic message, but also try to communicate with something similar to us in order to increase the possibility to mutual comprehension.

I would suggest three steps in a road map towards an interstellar intersubjective interaction. First, we should search for Earth analogues, i.e., Earth-like exoplanets within the habitable zone (Kane and Gelino 2012) around solar-type stars (a G2 main sequence star of 4.5 Gyr of age)—in other words, exoplanets with similar magnitude, physical and orbital characteristics (period, eccentricity, inclination, etc.), gravitation, atmosphere, chemistry, temperature, etc., as Earth. Hopefully, such candidates would be detected in the near future (Fridlund et al. 2010). If we believe that cognition is somehow and to some extent related to and adapted to the physical environment, we could expect that similar environmental pressures would lead to similar cognitive adaptions. That is why Earth analogues would be particularly interesting. The extraterrestrial intelligent beings' perceptual system and cognition have adapted to regularities in their physical environment such as planetary motion, gravitation, light, radiation, atmospheric conditions, chemistry (water, minerals), etc. The organisms are adapted to their specific environment; their sensory equipment has evolved in order to facilitate the orientation in and interaction with it. This environment-specific sensory ability leads to environment-specific cognitive abilities and conceptualization of the world. In other words, different environments mean different sensory modalities, which in its turn mean different conceptualizations of the world (Jonas and Jonas 1976). The possibility of sharing experiences with an extraterrestrial intelligence will be enhanced if we share similar physical environments, and thus similar selection pressures. If we are very different from each other, with brains adapted to completely different environments, we would to some extent have different views of the "reality." The "reality" is not objective, in the sense that our view of it is independent of our bodily constitution, our senses, cognition, etc. The sensory impressions from the "reality" must in some way be processed before it becomes something we can think about. The view that there is only one "reality," might be too simplistic, defending that there is a "reality" that is objective and intelligible, where the laws of physics and mathematics are objective, transcendental and universal, everywhere the same, no matter who is actually perceiving the "reality" that transcends our existence as embodied, perceiving, thinking beings living in the world. As Douglas Vakoch (2011) points out in a captivating mountain metaphor, different civilization might have taken different paths to the summit. There is not one single way of conceptualizing the world. "Science" or conceptual representations of the "reality" might have taken very different paths in different contexts. In line with this challenge to the universality principle of science (Rescher 1985), I think that the reason for this divergence is the human embodiment, the enactive mind's interaction with the environment, which means that our views and conceptualization of the world cannot be separated from our idiosyncratic biology, ecology, and culture—whom we are, where we are, and where we come from. By communicating with an extraterrestrial intelligent civilization and getting their view on reality, we will find out if our "objective" view of reality is shared with thinking beings other than our own species.

Second, we should monitor these Earth analogues for a period of time, analyzing their (1) electromagnetic leakage, and (2) the absorption spectrum of its atmosphere. First of course we turn our radio telescopes towards these candidates and listen. If electromagnetic leakage is detected, we have to determine if it might be artificial, i.e., a voluntarily transmitted message of a cognitive flexible life form; and second we have to develop methods for detecting communicative or signaling complexity, i.e., search for a great diversity of elements of the signaling system. The second monitoring strategy would be absorption spectroscopy (Kaltenegger et al. 2010). First of all, of course, would be to look for biomarkers, and by analyzing the chemistry and composition of these candidates' atmospheres, we might detect signs that reveal the environmental conditions of that exoplanet. Life on Earth exists in equilibrium with the rest of the planet in, for example, the case of the biomarkers oxygen and methane. A non-periodic rapidly-changing chemical equilibrium might indicate a decline in habitability. Granted, of course, that we know what the normal conditions are like at the monitored planet, we could accordingly develop methods for determining the sustainability and stability of its atmosphere, if there are signs of environmental disasters, rapid climate change, greenhouse effects, low biodiversity, radioactive leakage and other indications of unfriendly circumstances for life. In other words, we need methods of detecting artificial, unstable pollution in the spectrum of their planetary atmosphere or other "technomarkers" that are indicators of artificial non-periodic chemical processes. If we can find out if they have a sustainable and stable atmosphere (at least not polluted in a way similar to our planet)—this could indicate that they either have not evolved advanced technology yet or they have already gone through the bottlenecks of their civilization. In both cases they should be rather benign. If so, we should not fear to continue to step three, to construct an interstellar intersubjective interaction directed to just them, tightly connected to the situation in time and space. On the other hand, if it turns out that they have an artificially affected unstable environment, this can indicate that they have rather primitive altruism. Silence might then be recommended.

7.2 Interstellar Intersubjective Interaction

Instead of formulating an abstract, universal, symbolic message, I suggest that we do something together. To be more precise, I propose interaction tightly connected to the spatiotemporal setting, i.e., trying to establish *joint attention* (Tomasello 1999; Moore and Dunham 1995) by context-specific signals. We would then develop a mutual referential behavior, directed gaze or mutual gaze, by pointing to an object or spatial location, for example with a laser pointer, and checking if the recipient attends to the same object or location. In its turn, the recipient checks if the pointer notices the recipient's attention. The option is thus to try to tune in our spatial organizations, and together observe things observable to both terrestrials and extraterrestrials. For example, we can use certain astronomical landmarks or periodic events in their neighborhood, or in our own near environment, to which we can direct our joint attention—for example, as has been earlier suggested with reference to known pulsars in the neighborhood, as the Pioneer plaque represented the Sun's relationship to 14 known pulsars. Or the nearby Andromeda galaxy, which, as Carl Sagan noted, would be the only object that both the recipients and we could see first-hand. Even better, I would say, is to choose closer targets for joint attention that cannot be confused. Everett M. Hafner proposed transmissions simulating astronomical objects—for example the fluctuation of the Sun's cycle back and forth between the stars (Hafner 1969; Vakoch 1998). The sounds of geological activity, such as volcanoes, earthquakes, thunder, and ocean waves, included in the Voyager recording (Vakoch 2009), are something we both might experience if we both hear in the same frequency range under same atmospheric conditions. By using such indexical references toward some concrete phenomena in the physical environment, we do not need to presuppose a universal science that we should have in common, and do not have to point to our models of the phenomena. Instead, we firmly connect our interaction in physical reality. If we succeed in this, we will have taken a crucial step toward an interstellar intersubjectivity.

Joint attention or "attentional intersubjectivity" is the case when "two or more subjects simultaneously focus their attention on the same target" (Brinck, Zlatev, and Andrén 2006, 1; Brinck 2001). The target can be an object, an entity that has a position in space and time, or in a wider sense, events in space and time. A target can also be a spatial location (for example, a place to go to), or a direction (for example, the way to go). What characterizes a target is that it is an object for an undivided attention of one or both subjects. This attention can be symmetric (the target is noticed by both subjects) or asymmetric (the target is initially noticed by just one of the subjects and then the other subject aligns its attention). According to Brinck, Zlatev and Andrén (2006, 4), there are three levels of attentional inter-subjectivity of increasing complexity: synchronous, coordinated, and reciprocal. Synchronous attentional intersubjectivity occurs when participants are "performing similar individual actions relative to a single target in the same spatiotemporal context." This is not a social behavior, because their individual actions are performed independently of each other. Coordinated attentional intersubjectivity occurs when the subjects are "adjusting their actions relative to a single target." This behavior is social and can be interactive, i.e., their actions directly affect the other. Finally, reciprocal attentional intersubjectivity is when subjects are "mutually matching their actions relative to a single target." This behavior is a more complex interactive behavior.

This is what we can hope for in an interstellar intersubjective interaction, an attentional intersubjectivity of various sorts following the typology of Brinck, Zlatev and Andrén (2006). A simple case could be an asymmetric synchronous attentional intersubjectivity: We, the terrestrial intelligence, T, focus our attention on the target, t. The extraterrestrial intelligence, E, is attracted by our attention

Fig. 1 Interstellar
intersubjective interaction.
The terrestrial intelligence, T,
focus its attention on the
target, t. The extraterrestrial
intelligence, E, is attracted by
T's attention focusing. E then
follows T's orientation to t,
with the result that both E and
T focus their attention on
t. The goal would be to
engage in a reciprocal
attentional intersubjectivity,
where E attends not just to
T's attention to t and vice
versa, but to a third-order
attention in which E attends
to T's attending to E's
attention and vice versa.
Artist's rendition by Sten
Dunér

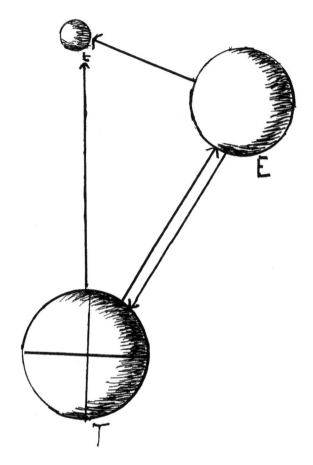

focusing. E then follows our orientation to t, with the result that both of us focus
our attention on t (see Fig. 1). The case could also be the other way around where
E initiates the attentional intersubjective episode: that E focuses its attention on t;
we notice that E focus their attention on t; and we follow their orientation to
t. Even better would be a social coordinated attentional intersubjectivity: We focus
our attention on t, and ostensibly focus on t and engage in referential behavior
towards t. Then E notices our referential behavior, which leads to E following our
attention and referential behavior and notices t. The goal would be to engage in a
reciprocal attentional intersubjectivity, where E attends not just to our attention to
t and vice versa, but to a third-order attention (Zlatev 2008) in which E attends to
our attending to E's attention and vice versa. By a referential behavior towards a
target that only they can see from their position, and which is hidden to us—for
example an eclipse or a transit of a planet between their planet and their star—they
will know that we know what they observe.

The benefit of this interstellar intersubjective interaction is that it will create (1)
a common experience about an object or event. We both will have an unambiguous

conception of the very same object, a common reference, that we both have turned our attention to, and we both will know that the other know what that object is. From this shared experience, we can begin a conversation about the object we direct our attention to; we can compare our views, learn from the other's view, and find what we have in common or not. Interstellar intersubjective interaction would hopefully be an unambiguous starting point for future interstellar communication. An interstellar intersubjective interaction will have the further advantage of focusing on (2) the prelinguistic semiotic stages in the phylogenetic and ontogenetic development of communication, an intersemiotic communication of attentional intersubjectivity, of iconic and indexical signs, rather than the more complex sign use, such as using abstract, arbitrary and detached symbolic messages. (3) The interaction will not be an information transfer as such, and thus be a rather "ecological" encounter that just triggers a response. Finally, (4) the content of the message will not be some presupposed universal abstract conception or "fact," but solely the intention to communicate.

7.3 Altruism

In this chapter, I have argued that there is a certain cognitive ability, namely intersubjectivity, that is fundamental for intelligence, sociability, communication, and technology. Intersubjectivity is a *sine qua non* for interstellar communication. We can expect that if we find a civilization with advanced technology, its members would have some sort of intersubjective skills. If they have intersubjectivity, they would also show empathetic and altruistic behaviors towards at least some of their conspecifics. Some basic *reciprocal altruism* (Trivers 1971; 2006)—"you scratch my back, and I'll scratch yours"—would be expected, a strategy for cooperation that needs the ability to recognize individuals and a memory of previous outcomes. But among humans we find much more advanced forms of altruism, such as *indirect reciprocity* (Leimar and Hammerstein 2001; Nowak and Sigmund 2005), in which a person can help another person, without expecting help from that person, but from someone else. Humans also use commitments, contracts, and conventions in cooperation. The question is, will we be included in their altruistic domain? If they have survived, have been able to develop a sustainable environment, a stable society, and an ethics, they might have achieved an empathetic and altruistic behavior towards other life forms, a behavior that can handle agonistic and destructive behavior, inequalities and asymmetry. An altruistic or "ecological" behavior might be to leave the other less-developed civilization alone, not engage in interstellar communication, and let the other develop without interference. They might say to themselves that our species is not ready for intergalactic integration, as has been suggested as a solution to the Fermi paradox, the so-called "zoo hypothesis" (Ball 1973; Baxter 2001; cf. Haqq-Misra and Baum 2009). We can be rather sure that if a message comes through, it will change the other culture. From our own history, we learn that cultural encounters and information transfers always include an asymmetric relation due to different experiences. And that is what communication is about, to share experiences.

Another question is this: are we in our present state able to send an altruistic message? Mathematical and scientific messages do not entail signs of altruism, and the level of our civilization shows obvious problems in altruistic behavior. So it can be argued that is not we (and Hawking), but the extraterrestrials, who should be afraid. We cannot just claim that we are altruistic—we have to show it, we have to prove it—and that we cannot, I am afraid, in the present time, with agonistic behavior, environmental problems, decreasing biodiversity, etc. Even though our situation might be improving, we are not there yet. This terrestrial intersubjective altruism is something we can work on, whether or not any extraterrestrial intelligent civilization exists.

References

Andrén, Mats. 2010. *Children's Gestures from 18 to 30 Months*. PhD diss. Lund: Travaux de l'institut de linguistique de Lund.

Arbib, Michael A. 2013. "Evolving an Extraterrestrial Intelligence and Its Language-Readiness." In *The History and Philosophy of Astrobiology: Perspectives on Extraterrestrial Life and the Human Mind*, edited by David Dunér. Newcastle upon Tyne: Cambridge Scholars Publishing.

Ball, John A. 1973. "The Zoo Hypothesis." *Icarus* 19:347–349.

Baxter, Stephen. 2001. "The Planetarium Hypothesis: A Resolution of the Fermi Paradox." *Journal of the British Interplanetary Society* 54(5–6):210–216.

Bogonovich, Marc. 2011. "Intelligence's Likelihood and Evolutionary Time Frame." *International Journal of Astrobiology* 10(2):113–122.

Bowles, Samuel, and Herbert Gintis. 2003. "The Origins of Human Cooperation." In *The Genetic and Cultural Origins of Cooperation*, edited by Peter Hammerstein, 429–443. Cambridge MA: MIT Press.

Brinck, Ingar. 2001. "Attention and the Evolution of Intentional Communication." *Pragmatics and Cognition* 9(2):255–272.

Brinck, Ingar. 2008. "The Role of Intersubjectivity in the Development of Intentional Communication." In *The Shared Mind: Perspectives on Intersubjectivity*, edited by Jordan Zlatev et al., 115–140. Amsterdam: Benjamins.

Brinck, Ingar, Jordan Zlatev, and Mats Andrén. 2006. "An Applied Analysis of Attentional Intersubjectivity." In *Deliverables of Stages in the Evolution and Development of Sign Use* 7, 1–21. Lund: Lund University. Accessed November 1, 2012. http://lup.lub.lu.se/luur/download?func=downloadFile&recordOId=931481&fileOId=2303466.

Calcagno, James M., and Agustín Fuentes. 2012. "What Makes Us Human?: Answers from Evolutionary Anthropology." *Evolutionary Anthropology* 21:182–194.

Carter, Brandon. 2012. "Hominid Evolution: Genetics Versus Memetics." *International Journal of Astrobiology* 11:3–13.

Cheney, Dorothy L., and Robert M. Seyfarth. 2012. "The Evolution of a Cooperative Social Mind." In *Oxford Handbook of Comparative Evolutionary Psychology*, edited by Jennifer Vonk and Todd K. Shackelford, 507–528. Oxford: Oxford University Press.

Damasio, Antonio R. 1994. *Descartes' Error: Emotion, Reason, and the Human Brain*. New York: G.P. Putnam.

Denning, Kathryn. 2011. "'L' on Earth." In *Civilizations Beyond Earth: Extraterrestrial Life and Society*, edited by Douglas A. Vakoch, and Albert A. Harrison, 74–86. New York: Berghahn Books.

Dick, Steven J. 2003. "Cultural Evolution, the Postbiological Universe and SETI." *International Journal of Astrobiology* 2(1):65–74.

Donald, Merlin W. 1991. *Origins of the Modern Mind: Three Stages in the Evolution of Culture and Cognition*. Cambridge, MA: Harvard University Press.

Donald, Merlin W. 2008. "A View from Cognitive Science." In *Was ist der Mensch?*, edited by Detlev Ganten, Volker Gerhardt, Jan-Christoph Heilinger, and Julian Nida-Rümelin, 45–49. Berlin: de Gruyter.

Dunbar, Robin. 1996. *Grooming, Gossip and the Evolution of Language*. Cambridge, MA: Harvard University Press.

Dunbar, Robin. 1998. "The Social Brain Hypothesis." *Evolutionary Anthropology* 6(5):178–190.

Dunér, David. 2011a. "Astrocognition: Prolegomena to a Future Cognitive History of Exploration." In *Humans in Outer Space—Interdisciplinary Perspectives*, edited by Ulrike Landfester, Nina-Louisa Remuss, Kai-Uwe Schrogl, and Jean-Claude Worms, 117–140. Wien: Springer Verlag.

Dunér, David. 2011b. "Cognitive Foundations of Interstellar Communication." In *Communication with Extraterrestrial Intelligence (CETI)*, edited by Douglas A. Vakoch, 449–467. Albany, NY: State University of New York Press.

Dunér, David. 2011c. "Die Evolution kognitiver Prozesse im Universum." *Sagenhafte Zeiten* 13(6):14.

Dunér, David, ed. 2012. "Introduction: The History and Philosophy of Astrobiology." In The History and Philosophy of Astrobiology, a special issue of *Astrobiology* 12(10):901–905.

Dunér, David, ed. 2013. *The History and Philosophy of Astrobiology: Perspectives on Extraterrestrial Life and the Human Mind*. Newcastle upon Tyne: Cambridge Scholars Publishing.

Fehr, Ernst, and Urs Fischbacher. 2004. "Social Norms and Human Cooperation." *Trends in Cognitive Science* 8:185–190.

Frank, Robert H. 1988. *Passions Within Reason: The Strategic Role of the Emotions*. New York: Norton.

Freeberg, Todd M., Robin I. M. Dunbar, and Terry J. Ord. 2012a. "Social Complexity as a Proximate and Ultimate Factor in Communicative Complexity." *Philosophical Transactions of Royal Society. Series B: Biological Sciences* 367:1785–1801.

Freeberg, Todd M., Terry J. Ord, and Robin I. M. Dunbar. 2012b. "The Social Network and Communicative Complexity: Preface to Theme Issue." *Philosophical Transactions of Royal Society. Series B: Biological Sciences* 367:1782–1784.

Fridlund, Malcolm, et al. 2010. "The Search for Worlds Like Our Own." *Astrobiology* 10(1):5–17.

Fusaroli, Riccardo, Paolo Demuru, and Anna M. Borghi. 2012. "The Intersubjectivity of Embodiment." *Journal of Cognitive Semiotics* 4(1):1–5.

Gallagher, Shaun. 2004. "Situational Understanding: A Gurwitschian Critique of Theory of Mind." In *Gurwitsch's Relevancy for Cognitive Science*, edited by Lester Embree, 25–44. Dordrecht: Springer.

Gärdenfors, Peter. 2003. *How Homo Became Sapiens: On the Evolution of Thinking*. Oxford: Oxford University Press.

Gärdenfors, Peter. 2008a. "Evolutionary and Developmental Aspects of Intersubjectivity." In *Consciousness Transitions—Phylogenetic, Ontogenetic and Physiological Aspects*, edited by Hans Liljenström, and Peter Århem, 281–385. Amsterdam: Elsevier.

Gärdenfors, Peter. 2008b. "The Role of Intersubjectivity in Animal and Human Cooperation." *Biological Theory* 3(1):1–12.

Gärdenfors, Peter. 2012. "The Cognitive and Communicative Demands of Cooperation." In *Games, Actions and Social Software: Multidisciplinary Aspects*, edited by Jan van Eijck and Rineke Verbrugge, 164–183. Berlin and Heidelberg: Springer.

Gärdenfors, Peter. 2013. "The Evolution of Semantics: Sharing Conceptual Domains." In *Evolutionary Emergence of Language*, edited by Rudolf Botha. Oxford: Oxford University Press.

Gentilucci, Maurizio, Claudia Gianelli, Giovanna Cristina Campione, and Francesca Ferri. 2012. "Intersubjectivity and Embodied Communication Systems." *Journal of Cognitive Semiotics* 4(1):124–137.

Gibson, James J. 1979. *The Ecological Approach to Visual Perception*. Boston, MA: Houghton Mifflin.

Gillespie, Alex. 2009. "The Intersubjective Nature of Symbols." In *Symbolic Transformations: The Mind in Movement Through Culture and Society: Cultural Dynamics of Social Representation*, edited by Brady Wagoner, 23–37. London: Routledge.

Gillespie, Alex, and Flora Cornish. 2010. "Intersubjectivity: Towards a Dialogical Analysis." *Journal for the Theory of Social Behaviour* 40:19–46.

Grice, Paul. 1989 [1957]. "Meaning." In *Studies in the Way of Words*. Cambridge, MA: Harvard University Press.

Hafner, Everett M. 1969. "Techniques of Interstellar Communication." In *Exobiology: The Search for Extraterrestrial Life*, edited by Martin M. Freundlich, and Bernard M. Wagner, 37–62. Washington, DC: American Astronautical Society.

Haqq-Misra, Jacob D., and Seth D. Baum. 2009. "The Sustainability Solution to the Fermi Paradox." *Journal of the British Interplanetary Society* 62(2):47–51.

Herrmann, Esther, Josep Call, María Victoria Hernàndez-Lloreda, Brian Hare, and Michael Tomasello. 2007. "Humans Have Evolved Specialized Skills of Social Cognition: The Cultural Intelligence Hypothesis." *Science* 317:1360–1366.

Holmer, Arthur. 2013. "Greetings Earthlings!: On Possible Features of Exo-language." In *The History and Philosophy of Astrobiology: Perspectives on Extraterrestrial Life and the Human Mind*, edited by David Dunér. Newcastle upon Tyne: Cambridge Scholars Publishing.

Hrdy, Sarah Blaffer. 2009. *Mothers and Others: The Evolutionary Origins of Mutual Understanding*. Cambridge, MA: Belknap Press of Harvard University Press.

Husserl, Edmund. 1973. *Zur Phänomenologie der Intersubjektivität*, 3 vols., edited by Iso Kern. Haag: Nijhoff.

Jonas, Doris, and David Jonas. 1976. *Other Senses, Other Worlds*. New York: Stein and Day.

Kaltenegger, Lisa, et al. 2010. "Deciphering Spectral Fingerprints of Habitable Exoplanets." *Astrobiology* 10(1):89–102.

Kane, Stephen R., and Dawn M. Gelino. 2012. "The Habitable Zone and Extreme Planetary Orbits." *Astrobiology* 12(10):940–945.

Kappeler, Peter M., and Carel P. van Schaik, eds. 2006. *Cooperation in Primates and Humans: Mechanisms and Evolution*. Berlin: Springer.

Kita, Sotaro, ed. 2003. *Pointing: Where Language, Culture and Cognition Meet*. Mahwah, NJ: Laurence Erlbaum.

Lehmann, Laurent, and Laurent Keller. 2006. "The Evolution of Cooperation and Altruism: A General Framework and a Classification of Models." *Journal of Evolutionary Biology* 19(5):1365–1376.

Leimar, Olof, and Peter Hammerstein. 2001. "Evolution of Cooperation Through Indirect Reciprocity." *Proceedings of the Royal Society of London. Series B: Biological Sciences* 268(1468):745–753.

Lenninger, Sara. 2012. *When Similarity Qualifies as a Sign: A Study in Picture Understanding and Semiotic Development in Young Children*. PhD diss. Lund: Lund University.

Malafouris, Lambros. 2012. "Linear B as Distributed Cognition: Excavating a Mind Not Limited by the Skin." In *Excavating the Mind: Cross-sections Through Culture, Cognition and Materiality*, edited by Niels Johannsen, Mads Jessen, and Helle Juel Jensen. Aarhus: Aarhus University Press.

Malafouris, Lambros, and Colin Renfrew, eds. 2010. *The Cognitive Life of Things: Recasting the Bounderies of the Mind*. Cambridge: The McDonald Institute Monographs.

Mayr, Ernst. 1985. "The Probability of Extraterrestrial Intelligent Life." In *Extraterrestrials: Science and Alien Intelligence*, edited by Edward Regis, Jr., 23–30. Cambridge, UK: Cambridge University Press.

Moore, Chris, and Philip J. Dunham. 1995. *Joint Attention: Its Origins and Role in Development.* Hillsdale, NJ: Lawrence Erlbaum.

Moran, Dermot. 2004. "The Problem of Empathy: Lipps, Scheler, Husserl and Stein." In *Amor Amicitiae: On the Love That Is Friendship*, edited by Thomas A. Kelly, and Phillip W. Rosemann, 269–312. Leuven: Peeters.

Norman, Donald A. 1991. "Cognitive Artifacts." In *Designing Interaction: Psychology at the Human-Computer Interface*, edited by John M. Carroll, 17–38. Cambridge, UK: Cambridge University Press.

Norman, Donald A. 1993. "Cognition in the Head and in the World." *Cognitive Science* 17:1–6.

Nowak, Martin A., and Karl Sigmund. 2005. "Evolution of Indirect Reciprocity." *Nature* 437(7063):1291–1298.

Oller, D. Kimbrough, and Ulrike Griebel, eds. 2008. *Evolution of Communicative Flexibility: Complexity, Creativity, and Adaptability in Human and Animal Communication.* Cambridge, MA: MIT Press.

Østergaard, Svend. 2012. "Imitation, Mirror Neurons and Material Culture." In *Excavating the Mind: Cross-sections Through Culture, Cognition and Materiality*, edited by Niels Johannsen, Mads Jessen, and Helle Juel Jensen. Aarhus: Aarhus University Press.

Osvath, Mathias. 2013. "Astrocognition: A Cognitive Zoology Approach to Potential Universal Principles of Intelligence." In *The History and Philosophy of Astrobiology: Perspectives on Extraterrestrial Life and the Human Mind*, edited by David Dunér. Newcastle upon Tyne: Cambridge Scholars Publishing.

Persson, Erik. 2012. "The Moral Status of Extraterrestrial Life." *Astrobiology* 12(10):976–984.

Piaget, Jean. 2001 [1963]. *The Psychology of Intelligence.* New York: Routledge.

Pinker, Steven. 2011. *The Better Angels of Our Nature: The Decline of Violence in History and Its Causes.* London: Allen Lane.

Preston, Stephanie D., and Frans B. M. de Waal. 2002. "Empathy: Its Ultimate and Proximal Bases." *Behavioral and Brain Sciences* 25(1):1–72.

Regis, Edward, Jr., ed. 1985. *Extraterrestrials: Science and Alien Intelligence.* Cambridge, UK: Cambridge University Press.

Rescher, Nicholas. 1985. "Extraterrestrial science." In *Extraterrestrials: Science and Alien Intelligence*, edited by Edward Regis, Jr., 83–116. Cambridge, UK: Cambridge University Press.

Richerson, Peter J., and Robert T. Boyd. 2005. *Not by Genes Alone: How Culture Transformed Human Evolution.* Chicago, IL: University of Chicago Press.

Richerson, Peter J., Robert T. Boyd, and Joseph Henrich. 2003. "Cultural Evolution of Human Cooperation." In *Genetic and Cultural Evolution of Cooperation*, edited by Peter Hammerstein. Dahlem workshop reports 90, 357–388. Cambridge, MA: MIT Press.

Roberts, William A. 2007. "Mental Time Travel: Animals Anticipate the Future." *Current Biology* 17(11):R418–R420.

Rosch, Eleanor. 1975. "Cognitive Representations of Semantic Categories." *Journal of Experimental Psychology: General* 104:192–233.

Rosch, Eleanor. 1978. "Principles of Categorization." In *Cognition and Categorization*, edited by Eleanor Rosch and Barbara B. Lloyd, 27–48. Hillsdale, NJ: Erlbaum.

Sagan, Carl, ed. 1973. *Communication with Extraterrestrial Intelligence (CETI).* Cambridge, MA: MIT Press.

Seed, Amanda M., Nathan J. Emery, and Nicola S. Clayton. 2009. "Intelligence in Corvids and Apes: A Case of Convergent Evolution?" *Ethology* 115:401–420.

Shostak, Seth. 2009. "The Value of 'L' and the Cosmic Bottleneck." In *Cosmos and Culture: Cultural Evolution in a Cosmic Context*, edited by Steven J. Dick, and Mark L. Lupisella, 399–414. Washington DC: National Aeronautics and Space Administration.

Sonesson, Göran. 2013. "Preparations for Discussing Constructivism with a Martian." In *The History and Philosophy of Astrobiology: Perspectives on Extraterrestrial Life and the Human Mind*, edited by David Dunér. Newcastle upon Tyne: Cambridge Scholars Publishing.

Steels, Luc. 2004. "Social and Cultural Learning in the Evolution of Human Communication." In *Evolution of Communication Systems*, edited by D. Kimbrough Oller and Ulrike Griebel, 69–90. Cambridge, MA: MIT Press.

Stein, Edith. 1917. *Zum problem der Einfühlung*. Halle: Buchdruckerei des Waisenhauses.

Stern, Daniel N. 1985. *The Interpersonal World of the Infant: A View from Psychoanalysis and Developmental Psychology*. New York: Basic Books.

Sternberg, Robert J. 2002. "The Search for Criteria: Why Study the Evolution of Intelligence." In *The Evolution of Intelligence*, edited by Robert J. Sternberg and James C. Kaufman, 1–7. Mahwah, NJ: Lawrence Erlbaum Associates Publishers.

Stevens, Jeffrey R., and Marc D. Hauser. 2004. "Why Be Nice?: Psychological Constraints on the Evolution of Cooperation." *Trends in Cognitive Sciences* 8(2):60–65.

Suddendorf, Thomas, and Michael C. Corballis. 1997. "Mental Time Travel and the Evolution of Human Mind." *Genetic, Social and General Psychology Monographs* 123(2):133–167.

Thompson, Evan, ed. 2001. *Between Ourselves: Second-person Issues in the Study of Consciousness*. Thorverton: Imprint Academic.

Thornton, Alex, Nicola S. Clayton, and Uri Grodzinski. 2012. "Animal Minds: From Computation to Evolution." *Philosophical Transactions of Royal Society. Series B: Biological Sciences* 367:2670–2676.

Tomasello, Michael. 1999. *The Cultural Origins of Human Cognition*. Cambridge, MA: Harvard University Press.

Tomasello, Michael. 2005. "Uniquely Human Cognition is a Product of Human Culture." In *Evolution and Culture: A Fryssen Foundation Symposium*, edited by Stephen C. Levinson, and Pierre Jaisson, 203–217. Cambridge, MA: MIT Press.

Tomasello, Michael. 2009. *Why We Cooperate?* Cambridge, MA: MIT Press.

Tomasello, Michael, Malinda Carpenter, Josep Call, Tanya Behne, and Henrike Moll. 2005. "Understanding and Sharing Intentions: The Origins of Cultural Cognition." *Behavioral and Brain Sciences* 28:675–691.

Trivers, Robert L. 1971. "The Evolution of Reciprocal Altruism." *The Quarterly Review of Biology* 46(1):35–57.

Trivers, Robert L. 2006. "Reciprocal Altruism: 30 Years Later." In *Cooperation in Primates and Humans: Mechanisms and Evolution*, edited by Peter M. Kappeler and Carel P. van Schaik, 67–83. Berlin: Springer.

Tylén, Kristian, Ethan Weed, Mikkel Wallentin, Andreas Roepstoorf, and Chris D. Frith. 2010. "Language as a Tool for Interacting Minds." *Mind and Language* 25:3–29.

Vakoch, Douglas A. 1998. "Constructing Messages to Extraterrestrials: An Exosemiotic Perspective." *Acta Astronautica* 42(10–12):697–704.

Vakoch, Douglas A. 2009. "Encoding Our Origins: Communicating the Evolutionary Epic in Interstellar Messages." In *Cosmos and Culture: Cultural Evolution in a Cosmic Context*, edited by Steven J. Dick, and Mark L. Lupisella, 415–439. Washington, DC: National Aeronautics and Space Administration.

Vakoch, Douglas A. 2011. "Asymmetry in Active SETI: A Case for Transmissions from Earth." *Acta Astronautica* 68:476–488.

Vakoch, Douglas A., and Albert A. Harrison, eds. 2011. *Civilizations Beyond Earth: Extraterrestrial Life and Society*. New York: Berghahn Books.

Zahavi, Dan. 2001. "Phenomenology and the Problem(s) of Intersubjectivity." In *The Reach of Reflection: Issues for Phenomenology's Second Century*, edited by Steven Crowell, Lester Embree, and Samuel J. Julian, 265–278. West Hartford, CT: Electron Press.

Zlatev, Jordan. 2008. "The Co-evolution of Intersubjectivity and Bodily Mimesis." In *The Shared Mind: Perspectives on Intersubjectivity*, edited by Jordan Zlatev, Timothy P. Racine, Chris Sinha, and Esa Itkonen, 215–244. Amsterdam: Benjamins.

Zlatev, Jordan, Timothy P. Racine, Chris Sinha, and Esa Itkonen, eds. 2008a. *The Shared Mind: Perspectives on Intersubjectivity*. Amsterdam: Benjamins.

Zlatev, Jordan, Timothy P. Racine, Chris Sinha, and Esa Itkonen. 2008b. "Intersubjectivity: What Makes Us Human?" In *The Shared Mind: Perspectives on Intersubjectivity*, edited by Jordan Zlatev, Timothy P. Racine, Chris Sinha, and Esa Itkonen, 1–14. Amsterdam: Benjamins.

Other Minds, Empathy, and Interstellar Communication

Tomislav Janović

Abstract If an extraterrestrial intelligence should have the technological capacity to decode an interstellar message, or at least to receive our signal, then it is highly probable that its society would be based on a reasonably high degree of cooperation among its members. Cooperation, in turn, is hardly conceivable without an ability to understand and express emotions and intentions—ability indispensible for setting off a communication process, even in the absence of a common code. This is the role of empathy—affective understanding of other minds. As a psychological mechanism underlying complex types of cooperative behavior, empathy might thus be a psychological universal—a fairly widespread characteristic of intelligent life. In standard communicative situations on Earth, empathy is essential to both of the participants in the communication process. To optimize this process with respect to the resources employed, the sender is typically required to foresee what the receiver already knows. That is, one usually wants to structure a message so that only the necessary information get explicitly encoded, leaving everything else—the potentially redundant part of the information content—implicit. However, in case of interstellar communication, even an impoverished message, leaning heavily on the common context, might fail to get across. Overestimating decoders' decoding potentials—being too optimistic about aliens' cognitive abilities or the commensurability of their representational system with ours—may prove fatal for our project. In order to forestall this risk, I propose, and try to justify, the following guideline: if our communicants are incapable of understanding the *informative intention* behind our message they might still be able to understand our *communicative intention*—the intention to simply reveal our presence as intentional beings. For it is much more likely that they will be able to empathically recognize such an intention than to interpret a signal embodying an explicit representational content.

T. Janović (✉)
Department of Philosophy, University of Zagreb, Croatian Studies Center, Borongajska cesta 83d 10000 Zagreb, Croatia
e-mail: tjanovic@inet.hr

D. A. Vakoch (ed.), *Extraterrestrial Altruism*, The Frontiers Collection, 169
DOI: 10.1007/978-3-642-37750-1_11, © Springer-Verlag Berlin Heidelberg 2014

Keywords: Empathy · Alien mind · Intention · Communicative intention ·
Context · Representation · Code · SETI

1 Introduction

A natural way to begin a discussion on interstellar communication is to speculate
on the characteristics of a civilization capable of intercepting an interstellar signal
and understanding its message. Some writers on the subject, however, find such a
speculation superfluous, given the constraints on the conditions of interstellar
communication. From a certain point of view, that is a useful strategy: treating the
constraints as obvious enough, and more-or-less universally agreed upon, enables
one to avoid the "speculative" and focus on the "technical" issues—the choice of
the information channel, code type, or the encoding/decoding procedure.

Contrary to such an approach, I will begin my chapter by trying to put forward
and look into what I take to be our basic assumptions/expectations regarding the
mental and the social characteristics of the putative recipient(s) of our signal. I will
then turn to some more specific—albeit (as I will try to show) not less justifiable—
assumptions bearing on the main topic of my chapter: the concept of empathy. The
logic behind this strategy should be obvious enough: since our presuppositions
about the nature of extraterrestrials and their societies constrain our choice of both
the content of our signal and the type of the code used, they deserve to be
explicated and scrutinized. For overestimating decoders' decoding potentials—
being too optimistic about aliens' cognitive abilities or the commensurability of
their representational system with ours—may prove detrimental to the fate of our
project. In order to forestall this risk, I propose, and try to justify, the following
principle as a guideline for our attempt at interstellar contact: if it turns out that our
communicants are incapable of understanding the *informative intention* behind our
message, they might at least be able to understand our *communicative intention*—
our wish to reveal our presence as intentional, communicative agents.

2 Attributes of Extraterrestrial Minds

To begin with, I propose a list of attributes expressing what I take to be our most
general assumptions regarding the nature of beings capable of interstellar
communication:

(1) Mindlike Behavior: reaction potential that is *functionally equivalent* to possession of a human-like mind[1];

(2) Representational (Conceptual) Powers: the capacity to represent things internally (through "percepts," "concepts," "beliefs," "goals," "reasons," etc.) and apply them in execution of cognitive tasks (typically, tracking and re-identifying units of the environment)[2];

(3) Practical Intelligence (Rationality): high degree of consistency between represented input states ("perceptions," "beliefs"), represented needs ("desires," "goals"), and output (behavioral) states ("actions");

(4) Social Skills: capacity for representing, and responding to, challenges posed by the creature's *social* environment (capacity to "understand" their in-groups and to respond to their needs by communicating with them in direct, non-symbolic ways);

(5) Coordinated Action: capacity to act collectively, in accordance with a common goal;

(6) Extelligence[3]: capacity for non-genetic transmission of knowledge (e.g., by imitation or some form of social learning);

(7) Language Competence: capacity for symbolic communication;

---

[1] Of course, if we were to encounter an alien creature exhibiting behavioral plasticity of such a degree, it would be irresistible to assume that it *does* possess a human-like mind. A more cautious attitude, however, would demand that we suspend this assumption and remain neutral to the question of the nature of our creature's "inner life"—if it had one, that is. (Why not also leave open the possibility that our communicant is a zombie?) This is why, in my initial determination of alien mind, I want to remain as economical as possible, i.e., avoid any reference to potentially anthropomorphic notions, including "consciousness," "experience," "feelings" etc. One way to do this is to refrain from mentalistic vocabulary altogether and speak of "mindlike behavior." Another way is to make use of Daniel Dennett's notion of "intentional system." By intentional system Dennett (1996, 34) means any entity "whose behavior is predictable/explicable from the intentional stance"—by treating it as an agent, i.e., by "attributing to it beliefs and desires on the basis of its perception of the situation and its goals and needs." The point is that an entity, in order to be an intentional system, need not *literally* possess any beliefs and desires, i.e., be *aware* of its goals and needs as humans typically are; it needn't, in fact, possess any distinctively *mental* features in the sense of our folk-psychological attributions. It needn't even be a sentient being. For no one would *explicitly* attribute an agency-level mind to a jellyfish or a vacuum cleaner, notwithstanding the fact that we routinely understand and predict behavioral reactions of such pseudo-agents by implicitly treating them as real (i.e., minded) agents. (Think of the usefulness of predicting the next move of a chess playing computer by "reading" its "mind.") This is not contradicted by the fact that, when prompted to weigh in on a creature's mental status, we are often unable to make up our minds. (Do mice have minds? What about parrots? Industrial robots or chess playing computers certainly not. Or?) The advantage of Dennett's instrumentalist notion is that it spares us from taking a stance on such issues—it spares us from formulating conjectures about the (unknowable) mental underpinnings of someone's/some thing's behavioral characteristics. However, in the case of our interstellar communicants we do have a good reason to take such a stance.

[2] Types of mental representation are here given in quotation marks in order to avoid their literal, i.e., anthropomorphic reading.

[3] The term was invented by Stewart and Cohen (1999).

(8) Theoretical Intelligence and Creativity: ability to develop and advance "abstract" representations (concepts and beliefs that are independent of one's immediate needs and goals);

(9) Trans-specific Curiosity: interest in other types of minds and societies ("non-ethnocentric" and "non-xenophobic" attitude).

Assumptions (1), (2), (4), (7) and (9) are relevant primarily, though not exclusively, for the creature's ability to decode/interpret an interstellar message, while (3), (5), (6) and (8) are more "technological" assumptions—assumptions bearing on the civilization's reaching a technological level necessary for receiving/emitting an interstellar signal.

Obviously, the proposed list is not to be taken as complete, still less as consistent, since most of the cited attributes appear to be mutually *inclusive*, not exclusive. That is, if a creature instantiates a property F from the list, it seems necessary—by the very manner we understand and use our key concepts—that it also instantiates some other property, G. Does this tell us something about the nature of intelligent life *as such*? Hardly. For, as is well known, the key concepts appearing on the list—"mind," "evolution," "internal representation," "symbolic communication," "intelligent behavior"—are either too vague or too idiosyncratic to be used in formulating testable generalizations about intelligent behavior universe-wide. This is not to say that there are no biological, psychological or sociological theories and insights that might be relevant for our speculations about extraterrestrial minds and the ways they form societies. The problem is rather that it is fully unclear—unlike in the case of physics or chemistry—whether these theories and insights apply to extraterrestrial conditions and how their generality is to be assessed. Far from being peculiar to interstellar communication, this problem also permeates current interdisciplinary debates on mind, cognition, and consciousness. Since this is not the right place to review the complex ramifications of these debates, I will illustrate my point by considering just a few of the listed properties.

Take the "basic" assumptions (1) and (2). Surely, what else could it mean for a creature to manifest mindlike behavior if not that it possesses internal states/processes systematically controlling its overt reactions? But what does "internal" exactly mean? This is open to very different interpretations. Are internal states equivalent to representational states, and what does it take for a representational system to produce outputs functionally similar to actions produced by a human agent? What does it mean for such a system to "represent" the states of its environment and its own states? Under what conditions do *phenomenal properties* emerge and how does a creature become *aware of itself*? Is self-awareness no more than a manifestation of a subject's meta-representational powers, the latter, in turn, being representational powers of a higher order? Or are all phenomenal properties (including self-awareness) *qualitatively* different from the representational ones, making (1) logically independent from (2)?

Or take some other assumptions from my list. It is hard to imagine a population reaching a stage of a high-tech society—a society capable, among other things, of interstellar signal emission/reception—without at least some of its members making use of a mechanism for cognitive schematization—a systematic way of organizing sensory inputs into "percepts," "concepts," "beliefs." But is the creature's capacity to use such cognitive devices for various purposes (including communication) the *only possible* evolutionary path leading to rational management of resources, technical innovations and expertise knowledge—the type of knowledge necessary for interstellar communication? Or is this conjecture merely a product of our biased ways of thinking about our own mind and its achievements? Likewise, high technology seems to require not only theoretical intelligence and creative thinking but also the capacities necessary for establishing and maintaining a well-organized society—social skills, the ability to act collectively, and the ability to acquire, store, and transmit knowledge in non-genetic, cultural ways (i.e., independently of individual minds). But can the possession of any of these attributes be treated as a necessary condition for technological progress? I doubt it. For the assumption, however unlikely, that high technology can also evolve in an "individualistic" society—among creatures with no or very limited degree of social interaction—cannot be excluded on a priori grounds.[4]

What are we to conclude from this? On the one hand, it is obvious that such quandaries blur the prospects of interstellar communication because it is not clear what kind of mind we should attribute to our interstellar communicants and, consequently, what kind of message will best serve this mind's communicative potentials and needs. On the other hand, notwithstanding the limits of our present knowledge of extraterrestrial minds and societies, I take it that most of us would be inclined to treat the listed attributes as almost *logically required* by the conditions for receiving an interstellar signal and understanding its message. In other words, whether there be anyone out there or not, and whatever the nature of this being, it ought to possess characteristics more or less equivalent to the listed ones if it is to receive our signal let alone understand the embedded message.

Is there a way out of this impasse? Can the plausibility of our assumptions be tested independently from intuitive appeal and deductive reasoning? Can we rely on any kind of real-world data that would constrain our speculations? An obvious place to look for an answer to these questions is the mainstream research program of studying extraterrestrial life. Despite the lack of direct evidence of non-Earthly life forms, indirect evidence—evidence of *life-supporting environments* (at least for the Earthlike kind of organisms)—is piling up. In addition, the number of discovered

---

[4] Think of a Platonic, totalitarian society ruled by an intellectual elite, or by a single individual, responsible for strategic decision-making, innovations, and technological progress. Imagine that this ruling class or ruling individual is served by all other members of the society, each obediently fulfilling its limited task (like drones in an ant or a bee colony) to the benefit of the population as a whole, but without there being any noteworthy interaction with other society members. In such a case, we would have a high level of behavioral coordination and social division of labor without real cooperation between individual members. See Barkow (2013, in this volume) for this kind of scenario. Cohen and Stewart (2002, 284) warn that "if an alien had sufficiently great intelligence, then the relevant store of know-how would *not* be beyond the capacity of any individual, and extelligence would be unnecessary."

exoplanets is growing at an unpredictable rate[5] and some of the data provided by these discoveries offer fresh grain for the speculative grinding mill. Can these new circumstances somehow help the student of extraterrestrial mind—and not only the student of extraterrestrial *life*—to leave the philosopher's armchair, at least for a moment, and make a move beyond what may be known a priori?

Now even if such a move were possible and justifiable, i.e., even if new insights provided by astronomers and astrobiologists could work as objective constraints on our speculations about alien minds and the ways they form societies, this might confront us with another, opposite danger: the constraints might be too tight. That is, actual scientific scenarios of alien life, leaning on very limited sources of evidence that are currently available, might prove too insular to be of any use to interstellar communication. It could even be counterproductive.

Since the very beginning of the search for extraterrestrial intelligence (SETI) program, there have been those who have warned precisely against this kind of error—the error of *unrepresentativeness*. Some of them go a step further and pro- pose a thoroughly different approach embodied in a new science—"xenoscience." The rationale for this radical proposal is a methodological flaw allegedly inherent to astrobiology—the "official" research program of exploring alien life. For,

> [a]strobiologists tend to… assume that Earth's story is the only story, and that alien life *must* be very similar to ours. And so they search for aqueous planets, and assume that oxygen is a must and that the only way to get oxygen is via chlorophyll in plant-like organisms, and so on. SF authors realised long ago that far more fanciful scenarios can be conceived, even if we do not yet know whether they could really exist (Cohen and Stewart 2002, 98).

Therefore, according to the critics, a new, transdisciplinary approach is needed: one that will fruitfully combine up-to-date science with creative science fiction, that will take seriously what scientists have to say about the most general con- straints on the possible, but that will also—*within these constraints*—enable the imagination to go wild. Xenoscientists thus follow in the footsteps of those science fiction writers who envision worlds *truly* beyond human experience but, at the same time, manage to "get their science right"—up to the smallest detail.

But even if delimited by scientific reason, isn't the possibility space—the space open for creative imagination—still unrealistically huge for this kind of cooper- ation between science and science fiction to be of any use to interstellar com- munication? This is where xenoscientists make their most promising contribution. They try to apply a more subtle filter than the one typically applied in scientific thinking about extraterrestrial life. Its purpose is to separate the *physically possible* from the *biologically probable* or, as xenoscientists would express it, "parochials" from the "universals":

---

[5] Thanks to the Kepler mission, a space-based telescope, the number of discovered exoplanets in 2011 exceeded the number of such planets discovered ever before. Data gathered during Kepler's mission from March 2009 to May 2013 yielded more than 3500 planet candidates, of which more than 130 have already been confirmed as planets (NASA Exoplanet Archive 2013).

> The difference between parochials and universals is that we cannot confidently expect
> parochials to occur again at some other place or time. However, we *can* expect universals
> to recur: the fact that they have evolved independently on several occasions implies that
> they are *likely*. If we re-ran Earth, the default would be that they would happen again
> (Cohen and Stewart 2002, 102–103).

Xenoscientists propose and discuss a huge variety of potential biological universals, from ecological to anatomic—water-and-oxygen ecosystems, carbon-based life, silicon-based life, genetic inheritance, genetic variability, natural selection, homeostasis, homeorhesis, land life, water life, eukaryotic cells, photosynthesis, "synthesizers" (plant-like organisms), "exploiters" (animal-like organisms), recombination, sexual reproduction, the K-reproductive strategy (greater investment in smaller amount of offspring), the r-reproductive strategy (smaller investment in greater amount of offspring), skeletons, jointed limbs, fur, central nervous systems, digestion, copulation, eyesight, flight, etc. They also consider several socio(bio)logical traits. Some of these traits are more general, such as intelligence, tool use, cooperation, communication, extelligence, and high technology. Other traits are more specific, such as interstellar communication or the production of non-replicating and self-replicating machines. But they are conspicuously silent about other potential characteristics of alien minds—the ones that (as I will try to show) people engaged in interstellar communication should be particularly interested in: How likely is it that aliens are able to feel anything? How likely is it that they will recognize our intent to communicate with them? Should we expect them to react to our revealed presence—not only behaviorally, but "inwardly" too? And if so, in what ways?

The silence on these—admittedly difficult—issues is hardly surprising. For almost any kind of mind one can think of seems compatible with whatever present-day life sciences—let alone physics and chemistry—have to say. Or to put it the other way around: it is hard to imagine alien minds and alien societies that would *not* fit into the possibility frame envisaged by xenoscientists[6]—however duly they set up their constraints. For, how *probable* a specific kind of mind is, measured by the criterion of universality (i.e., possibility of recurrence)—they are not able to tell. Nor is anybody else. For all we know, the most peculiar aspects of our minds (including the minds of our closest mammalian ancestors) are products of only a *single* evolutionary line of Earth's history. This does not mean that properties such as self-consciousness or rational thought are necessarily parochial—because there are traits of which it can plausibly be argued that they might have evolved elsewhere despite the fact that they evolved only once on Earth[7]—but they are certainly not *obvious universals*.

Our dilemma thus remains: are the basic attributes of minds and the ways they build societies (when they do)—exemplified in the attributes listed above—

---

[6] They speak of a "phase space of imaginative possibilities," borrowing the idea from Henri Poincaré (Cohen and Stewart 2002, 18).

[7] Extelligence is the case in point. See Cohen and Stewart (2002), Chap. 12.

psychological universals? Are they, considering the conditions of their occurrence on Earth, likely to recur elsewhere in the universe? And can a positive answer to these questions be corroborated *independently* of the limited sources we are presently stuck with.

Analogous to xenoscientists' treatment of biological traits, a satisfactory answer to these questions would require two things: an evolutionary comparative analysis of the types of minds that we know of *and* an assessment of the degree to which the results of these analyses can be generalized, i.e., extended to outer-space conditions, discounting the possible idiosyncrasies of terrestrial evolution as a one-time event (in order to forestall xenoscientsts' worries about "parochial thinking" and non-representativeness).

As to the first part of this strategy, it very much resembles the "paradigm shift" advocated by the pioneers of evolutionary psychology in the 1990s. According to the proponents of the radically Darwinian science of the human mind (Tooby and Cosmides 1992, 108), "we cannot understand what it is to be human until we learn to appreciate how truly different nonhuman minds can be, and our best points of comparison are the minds of other species and electronic minds." The purpose of such comparisons was to make us sensitive to an "entire class of problems and issues that would escape us if we were to remain 'ethnocentrically' focused on humans, imprisoned by mistaking our mentally imposed frames for an exhaustive demarcation of reality."

Indeed, in the past 20 years or so—since the founders of the Darwinian science of the mind issued their methodological warning—many revealing insights have been gained by comparative interdisciplinary analyses of known (terrestrial) types of minds. But what about the second part of the proposed strategy—the one that is supposed to make these findings bear on interstellar communication?

Although nobody would deny that many original ideas and insights have helped us realize "how truly different nonhuman minds can be," and will certainly continue to shape our understanding of other minds, I cannot think of any methodologically justified means of extending this knowledge to extraterrestrial conditions. After all, we still lack almost *any* data relevant for the probability assessment of biogenesis—let alone the psycho- and sociogenesis—in extraterrestrial conditions, including even a rough estimate of the frequency of life-supporting planets. This is why most factors in the Drake Equation, even in its amended versions, still wait to be substituted by minimally reliable figures. The same applies to the famous Fermi Paradox, not to mention the diverse evolutionary paths that psychogenesis might take, as a consequence of the—possibly very subtle—variations in the initial environmental conditions.[8] And lacking these data, we cannot really know *how* different the "truly different nonhuman minds" can be and how *probable* is their re-appearance on the time-and-space scale of the universe—conditions necessary for our conjectures to be more than intuition-based guesswork.

---

[8] See Lem (1974), Chap. III ("Cosmic Civilizations").

The most we can do in this epistemically unfavorable situation is to submit particular assumptions—like the ones made explicit by my list—to philosophical thought experiments, *probing their plausibility in light of our specific goal*—constructing an interstellar message that could be recognized as such. To apply this kind of conceptual analysis to all properties listed above, considering all interesting—let alone possible—ways of their instantiation, would be unrealistically ambitious and far beyond the purpose of this chapter. In the speculations to follow I will therefore take up only a few characteristics—the ones I find particularly relevant for interstellar communication. My aim is to give these characteristics a *more specific reading* and to show how this move might forward our project.

3 Social Intentions, Empathy, and Altruism

So, if in envisaging alien minds and societies for the purposes of interstellar communication we cannot count on almost any other empirical evidence apart from the knowledge gained by studying terrestrial minds, and are thus forced to rely on conjectures and deductive reasoning, then we might have a reason to be bolder—*more specific and more imaginative*—in our hypothesizing. This reason has to do with the minimal communicative potentials of our alien communicants, bearing in mind the minimal goal of interstellar communication (as I see it)—to signal our presence to a creature capable, at least, of *understanding such a signal as an intentional (purposeful) act.*

Motivated by this line of reasoning let me propose three further hypothetical characteristics of alien minds:

(10) Phenomenal Consciousness ("inner" perspective);
(11) Empathic Powers (cognitive-affective understanding of others' mental states in respect to a situational context[9]);
(12) Capacity for Altruism (ability to act cooperatively, against one's own short-term interest).

First of all, note that these three items are actually more specific characterizations of the three properties from my original catalog—properties (1), (4), and (5), respectively. The capacity to react phenomenally (experientially), and not just behaviorally, to environmental stimuli is one of the ways—for us humans the default way—to be a minded creature[10]; empathy is the key social skill, and altruistic behavior a plausible evolutionary solution leading from individual

---

[9] This definition is adopted from Janović et al. (2003, 811).

[10] There are authors that take phenomenal properties to be ontological primitives. According to David Chalmers (1996), even the smallest ingredients of the physical world (elementary particles) can have non-physical properties that are proto-constituents of consciousness ("protophenomenal properties"). When elementary particles combine into larger structures, protophenomenal properties also combine, producing fully structured conscious experience.

intelligence to extelligence and, possibly, to complex societies characterized by high technology and high degree of social coordination. Note further how the three characteristics are conceptually (logically) interconnected[11]: being able to react to the world "phenomenally" (in the sense that it "feels like something" to be affected or acted upon) is a necessary condition for experiencing empathy (no experiential properties → no empathic experience[12]), while taking an empathic stance can plausibly be interpreted as a trigger of, and thus a sufficient condition for, altruistic behavior—the kind of behavior that is sufficient (albeit not neces- sary) for group selection to take place, systematically favoring those groups whose members are able to coordinate their actions in line with a common goal. It is plausible to expect that such coordination/cooperation will be much more pre- valent in extelligent societies—in societies that have developed "memetic" means of storing, retrieving, and transferring knowledge. In addition to the "formal" rules of memetic transfer (language rules), this knowledge will typically consist of rules of conduct, practical know-how, theoretical knowledge, and other forms of soci- ety's "cultural capital."

I am aware that the proposed scenario is a rough-and-ready one, with a flavor of an anthropomorphic just-so story. However, it has one important virtue (consid- ering the aim of interstellar communication in its most modest formulation): it does not require our hypothetical communicants to be endowed with intellectual powers that would qualitatively match our own in the sense that their conceptual schemes, interests, and values would be minimally congruent with ours. It is therefore not vulnerable to the well-known Rescherian objections (Rescher 1985; 1987; Vakoch 1999; Basalla 2006), unlike many other hypothetical scenarios of alien life. However, I am also aware that this *negative* virtue is not enough to make it work. Anybody advocating such a fanciful theory would at least have to offer a more detailed account of the connection between its three key elements: the ability to act intentionally (to have intention-like mental states), empathy (ability to recognize and react to intention-like states of others), and cooperative (altruistic) action. This is what I will try to do in the rest of this section. I will thereby lean on some well received ideas and theories (primarily of Michael Tomasello) of human development—both phylogenetic and ontogenetic.[13]

Among various intention-like states let us consider a specific subclass of these states and call them (following Tomasello 2008, 97) *social intentions*. Unlike other intention-like states, directed towards simple means for fulfilling individual goals (like an intention to pick up a stone in order to crack nuts or to step into sunlight for the purpose of warming up after rain), social intentions are directed towards other individuals, typically in-groups or conspecifics, as more complex and more

---

[11] They are also *causally* connected, as exemplified by at least one evolutionary line of terrestrial evolution.

[12] For an elucidation of the relation between conciousness, empathy, and its rudimentary social aspects (intersubjectivity) see Thompson (2001). David Dunér (2013) makes an interesting attempt in this volume at extending Thompson's arguments to extraterrestrial conditions.

[13] For an overview of some of these theories and their historical roots see Janović et al. (2003).

efficient means for satisfying individual goals and needs. Now the critical point is that in order for A's social intention (for instance, that B helps him crack nuts) to be fulfilled, B (in order to comply with A's intention) has to possess: (1) intention-like inner states herself, (2) a way of identifying such states in others (e.g., of A's intention that B helps him crack nuts), and (3) a way of recognizing others as apt candidates for a joint action (e.g., cracking nuts).[14] This is where *empathy* comes into play. As a process of recognizing others' mental states in respect to a particular situation,[15] empathy enables potential cooperators to fulfill conditions (2) and (3) and thus set off a recursive process of mutual recognition of intentions. Taken in the broadest possible sense (regardless of the way they are materially realized), empathic skills of some sort seem to underlie all types of complex altruistic behavior, so they might be a psychological universal. Namely, unlike simple cases of nepotistic or reciprocal altruism that do not require more than a simple mechanism for identifying kin or recent cooperators on the grounds of overt (perceptive) stimuli, more complex forms of altruism—characterized by a postponed, uncertain or indeterminate reward for acting cooperatively—require a more sophisticated mechanism for recognizing the other party's hidden motives ("intentions").

A specific type of complex cooperative behavior, characterized by expression and empathic recognition of social intentions, has been widely recognized as a proxy and a precursor of advanced symbolic behavior. John Searle thus invites us to

> ... imagine a class of beings who were capable of having intentional states like belief, desire and intention but who did not have a language. What more would they require in order to be able to perform linguistic acts?... The first thing that our beings would need... is some means for externalising, for making publicly recognisable to others, the expressions of their intentional states. A being that can do that on purpose, that is a being that

---

[14] Searle (1990, 414–415; cited in Tomasello 2008, 73) speaks of a "background sense of the other as a candidate for cooperative agency." He takes this special kind of sense—functionally equivalent to empathy—to be a "necessary condition of all collective behavior," including communication.

[15] Even when used in this-worldly contexts—i.e., when applied to known types of minds— "empathy" is a vague term referring to an unspecified variety of mental phenomena. This is why there are so many definitions of empathy. Batson (2009) tries to bring some order into this chaotic field by identifying eight basic uses of the term. He sees each of these uses as motivated by researchers' need to answer at least one of two crucial questions: "How can one know what another person is thinking and feeling? What leads one person to respond with sensitivity and care to the suffering of another?" (Batson 2009, 3). My use of the term is closest to Batson's first and most general definition of empathy: "knowing another person's internal state, including his or her thoughts and feelings." Some authors call this type of empathy "cognitive empathy," while others speak of "mindreading." But the term itself is irrelevant for the point I am trying to make. What is important is that empathic recognition of intentions (and other mental states) need not function as an inference from a hidden knowledge store (implicit "theory of mind"), requiring special cognitive abilities. It needn't be a conceptually mediated process at all. The alternative way to achieve the same goal is to simulate ("mirror") another mind when prompted by relevant behavioral/situational cues. In fact, it is this simulation model that best fits recent empirical findings (Gallese 2001; Gallese and Goldman 1998).

does not just express its intentional states but performs acts for the purpose of letting others know its intentional states, already has a primitive form of speech act (Searle 1979, 193–194).

Of course, for a rudimentary communication process to take place, in addition to such a capacity—the capacity to have intentions and to make them "publicly recognizable to others"—a complementary capacity is needed: a capacity to recognize "externalized" expressions of other's intentions. When realized simultaneously and recursively, these two capacities/roles transform ordinary social intentions (in the sense explained above) into communicative intentions, making communication possible even in the absence of conventional linguistic devices (meaningful signs and grammatical rules).

At this point, it is important to sweep aside a common objection to the scenario I am arguing for. This objection, often attributed to the proponents of an alternative model of communication, boils down to the following question: isn't it natural to assume a *common code* ("language") for any kind of communicative behavior to take place?[16] The purpose of the common code is to "translate" overt stimuli to internal representations, and thus enable all kinds of messages—not only the simplest "social intentions"—to be expressed ("encoded") and understood ("decoded"). But how do the participants in communication get hold of the code? There are only two possible answers: the code is either hardwired, an inherited component of the cognitive make-up of organisms sharing the same genome, or prearranged and passed down through generations by way of cultural learning— essential part of the community's extelligence. While the first (genetic) solution sounds highly implausible, the second one is susceptible to a well-known objection: how is this prearrangement to come about if not by means of another code, which would itself require another one, and so ad infinitum? That is to say, the common code solution to the puzzle of the emergence of symbolic conventions suffers from the "bootstrapping problem," originally recognized by the philosopher Ludwig Wittgenstein.[17] The only cogent way to circumvent this problem is to presume a capacity for recognition of others' intentions independent of, but necessary for, any type of coded communication.

An obvious virtue of this solution is its applicability to all instances of communicative behavior lacking a pre-arranged code, including the case currently under discussion—communication with an unknown extraterrestrial intelligence

---

[16] This is the main idea of the "code model" or the "information-processing" approach to communication. According to Shannon and Weaver's (1949) seminal account, all communication is a kind of encoding–decoding activity governed by a system of rules ("code") shared by the participants in the process. The rules enable "messages" (internal representations of objects or states of affairs) to be paired with "signals" (modifications of the external environment) in a systematic way. As a consequence, the same message (mental representation) occurs at both ends of the communication process. This mediating procedure is necessary for obvious reasons: the messages themselves, as they are defined, cannot travel through space–time, and therefore cannot be directly conveyed. No telepathy is possible.

[17] See Wittgenstein (1953) and Tomasello (2008, 58–59).

that might be so different from us that any communicative attempt by use of a code will be doomed to fail. Of course, we could take a more optimistic stance and presuppose some minimal congruence between our two types of mind that would enable our communicants to divine the conceptual basis of our encoding procedure. But, as I have already stressed, it would be methodologically wiser *not* to take this avenue in the "phase space of imaginative possibilities"; that is, *not* to assume that our communicant will be able to reconstruct the code from the signal itself (or even understand the very idea of a code).

Bearing this in mind and endorsing the idea of codeless communication—communication by means of empathical recognition of social intentions—it is unavoidable to ask about the possible interests and motives that a creature might have for engaging in such an interaction. Tomasello (2008, 82–88) cites three basic motives: requesting help or information, offering help or information, and sharing emotions or beliefs. He takes these motives to be phylogenetic and ontogenetic precursors of advanced social interactions between intentional agents. But they can also be taken as psychological universals (in the sense explained above), at least for those kinds of minds that satisfy the basic requirements from my list (most importantly, having the ability to express/recognize intention-like mental states). To these motives, it is important (concerning my main topic) to add another, more specific one: *deception* or sharing false beliefs or emotions. Of course, this kind of motive would presuppose a respective ability, on the part of the communicator, to disguise his real intent—to misinform (deceive) the communicant. In appropriate circumstances, this could result in communicants believing and/or behaving in accordance with the communicator's hidden intent (Janović et al. 2003, 815–817).

In cases of more specific and/or more complex communicative motives, a more complex means for their recognition (including the recognition of deceitfulness) would be required. However, this would not necessarily require a common code. For, the more *implicit knowledge* is shared between the communicants, the less explicit (symbolically coded) the message has to be—the less has to be put "in" the signal itself (Tomasello 2008, 79). Think of any typical situation in which two people—e.g., two soccer players belonging to the same team—are able to understand each other's intentions by use of very simple gestures (eye or head movements, for instance), without relying on linguistic symbols. Note that such non-verbal ways of expressing/understanding intentions needn't even be agreed upon in advance, (like in cases of conventional communication). Moreover, by use of spontaneous, non-conventional iconic gestures, or analogous ways of "pointing" to the same referent, one might be able to successfully refer to complex, abstract, or absent objects, like aspects of things, past or future states of affairs, or other types of referent that one would normally consider as communicable only through conventionalized means (Tomasello, 2008, 82, 107). Apart from the empathic ability and the shared implicit knowledge, the only requirement for this kind of communication to get off the ground is a *tacit* agreement between the communicating parties to the effect that they will be mutually cooperative in facilitating their social undertaking.

According to the proposed model, such cooperation can be analyzed in two logically-distinct steps[18]: (1) after indicating a *communicative* intention to the communicant, by sending her a "for you" signal (i.e., attracting her attention), the communicator (a "he" henceforth) will (2) help the communicant (a "she") infer his *informative* (*referential*) intention. This can be done by pointing to a common object of attention—the referent of the communicative act. Should the communicant decide that it is in her interest to be cooperative, she will comply with the communicator's request. The recognition of the communicative intention—that the communicator would like to share some piece of information—is straightforward and requires nothing more than basic empathic skills. The recognition of the informative intention—of *what* piece of information is to be shared—is a more intricate matter: the only way for the communicant to understand this kind of intention (say a pointing gesture) is to infer its referent from the *contextual knowledge* shared with the communicator: "when the context—the shared conceptual ground—is set up in enough detail, however that is done, a pointing gesture can refer to situations as complex as one wants" (Tomasello 2008, 99).

But what if there isn't any common conceptual ground? Although this issue is critical for my main point, before I address it I want to elucidate the respective roles of the two essential elements of intentional communication—mutual recognition of intentions and contextual knowledge—in cases of *conventionalized* communicative acts.

4 Inferring Intentions

Granted that some system of rules (syntactic and semantic), shared by all parties in the communication process, is necessary for this process to come about, it certainly isn't sufficient. In order to understand the coded signal, the addressee has to have some idea about what the sender could have "meant." She has to have some way of retrieving his "meaning-intention" ("message") coded in the signal. One way of doing this—in fact the only way, according to the received view of linguistic communication—is through *inference*. One takes the explicit part of the message encoded in the signal as a premise[19] and, with help of some additional evidence, arrives at the meaning-intention as a conclusion. However, this procedure is hardly ever straightforward. Typically, messages cannot be recovered without the support of manifold auxiliary hypotheses—relating to the verbal context, to behavioral and/or situational cues, to relevant memories, and to all sorts of other things including body skills and the most general type of knowledge about the world

---

[18] These two steps need not be *psychologically* distinct: neither the communicator nor the communicant need experience them as distinct.

[19] It is thereby assumed that this very premise (the explicit part of the message) is simply decoded. In other words, encoding/decoding processes cannot be entirely replaced by inference.

("common sense")—all this constituting an implicit, temporary, and situation-bound cognitive structure called "context."[20]

Note that the same mechanism that is at work in cases of non-conventionalized communication works in cases of exchanging information through any kind of conventional (pre-existing) code. In trying to figure out the meaning of the message—what information the sender has intended to convey—the addressee is led by a general tacit assumption that it was not only the sender's social intention to convey the particular piece of information, but that he also intended *to make very intention evident* to her (i.e., to the addressee). Generally, making one's intention to convey a particular message evident contributes essentially to understanding of the message itself.

This somewhat speculative idea presents the core of one of the most influential (and the most disputed) model of communication—the alternative to the code model (Grice 1957). In the "intention model"—as I have called it[21]—for a piece of behavior to be described as "communication" it is not only necessary that some informative effect in the addressee be produced (by the addressee's understanding of the content embedded in the signal), but also that the *communicator's intention to produce this effect be recognized*. As I have shown in the previous section, the fulfillment of the communicative intention—a successful recognition of the communicator's social intent to communicate a particular piece of information—helps fulfill the informative intention—retrieving the information itself.

The importance of the distinction between the communicative and the informative intentions should be obvious enough. Consider an interstellar signal—the type of communicative act we are particularly interested in. It is one thing for an extraterrestrial being to understand the information embodied in the signal, and another thing to understand the same information *as being intended* to be conveyed (and understood). In the former case the information could be interpreted as

---

[20] Contexts thus relate not only to wider bits of discourse, or to physical contingencies of communication process (time, place, properties of the natural or social environment, etc.), but also to everything else involved in this process. Sperber and Wilson (1986, 15) highlight this broader meaning of context:

> [A] context is a psychological construct, a subset of the hearer's assumptions about the world. It is these assumptions, of course, rather than the actual state of the world, that affect the interpretation of an utterance. The context in this sense is not limited to information about the immediate physical environment or the immediately preceding utterances: expectations about the future, scientific hypotheses or religious beliefs, anecdotal memories, general cultural assumptions, beliefs about the mental state of the speaker, may all play a role in interpretation.

[21] One might also call it—following Sperber and Wilson (1986)—the "inferential model." The difference is just one of emphasis. In their idiom, it is the means (inference) that is pointed out. In the expression I favor, it is the goal (recognition of intention) that is given priority (bearing in mind that inference is not the only means for achieving this goal, as some recent findings strongly suggest).

unintended, say mistaken for a byproduct of Earth's informational pollution. So recognizing our intent to communicate a particular piece of information is a prerequisite for our alien communicant's successfully retrieving this information from the signal. Without the awareness of a particular physical structure ("signal") being a product of a social/intentional/communicative act, the retrieval of information from this structure seems impossible.

Now the intentional model of communication has several virtues that have implications for my main topic. The first one is that putting an intention constraint on information exchange has appeal to common sense, because it rules out informative effects that we would consider as byproducts of purely physical processes.[22] We often receive information about things and events from our surroundings—both natural and social—without any kind of agency being involved (except on our part). To be sure, it is not the one-way character of such information-flow (surrounding → agent) that disqualifies it from the category of communication. There are fascinating examples of pseudo-intentional processes of information-exchange ("messenger-RNA" sending a "chemical signal" to the ribosome; the infamous bee dance, etc.), which clearly involve two parties. What is lacking, however, in these examples—and what excludes them from the class of genuine communicative acts—is the intentional (phenomenal) element. This is why it was crucial to add this element to our hypothetical list of alien's attributes.

The second virtue of the intention model has to do with its universal applicability. Although originally designed to apply primarily to the cases of human verbal communication, it is easily extendable to all forms of communicative behavior, independent of the nature of the signal or the code being used (if any). This is advantageous because it treats verbal communication as a special case of its evolutionary predecessor, pre-verbal communicative behavior, and not the other way around. As I have shown, understanding intentions and letting them be understood by others is a much older mechanism, much more general in scope and function, and much more widespread—presumably not only in terrestrial terms!—than is the mechanism that achieves the same goal through a developed system of mental representations, including linguistic signs (Gallese 1999; Rizzolatti and Arbib 1998). Needless to say, the advantage of the latter lies in its unsurpassed efficiency in communicating all sorts of contents (messages). However, such sophisticated communicative means require minimally developed and mutually congruent representational systems on both sides of the communication channel— a requirement not to be taken for granted, especially in the specific case of communication with an unspecified alien addressee.

---

[22] Grice (1957) calls the "meaning" conveyed in this way "natural meaning." Black clouds "mean" rain in this, and only in this sense. A being capable of intentions can also exhibit the non-intentional, "natural" kind of meaning: a bruise on my forehead "means" injury (as long as I am not showing it to somebody deliberately).

5 Communicative Intent Alone: A Less Ambitious Goal

As I have speculated, if the addressee of our interstellar signal will have the technological potential to decode, or at least to receive, our message—a condition that we are forced to take for granted—it is then highly probable that its society will be based on, and its achievements the result of, a reasonably high degree of cooperation among its members. Such cooperation, in turn, is hardly conceivable without the potential cooperators having the ability to make their inner states known to others and to empathically interpret overt signs of others' inner states.

However, even if my speculation about aliens' alleged empathic skills is sound, why suppose that our signaling will be successful? After all, in all known or imaginable cases of intentional communication without a common code, the bodily signs—a direct (visual or tactual) contact with an empathizing agent—seem to be the only means of making one's intentions known to others. Can the signal itself somehow "indicate" its intentional origin? And what kind of signal/message could serve this purpose? Being unable to provide a clear and convincing answer, I will give a few hints based on several of my previous points.[23]

In standard communicative situations, empathic ability is essential to *both* participants in the communicative process. Apart from playing a role in the communicant's recognition of the communicator's informative intentions, empathy is indispensable for the communicator too: it enables the sender to anticipate the "informational needs" of the receiver (Baron-Cohen 1995, 29). To optimize the communication process in respect to the resources employed, the sender is thus required to foresee what the receiver already knows. That is, one usually wants to structure the message in such a way that only the necessary information gets explicitly encoded, leaving everything else—the potentially redundant part of the information content—implicit (contextually available).

The elliptical art of expression, taking for granted a great deal of shared contextual knowledge, is typical of everyday human interactions. Even highly ambiguous, isolated messages—such as "Don't do it!"—get across safely despite the lack of explicitly encoded additional information. Obviously, what compensates for the absent verbal (explicit) context are implicit ("assumed," "taken for granted") features of the shared context. Yet, these can be treated as reliable auxiliary sources of information only if they really meet the communicant's informational needs. If they don't, the communicant's interpretive task will be doomed to failure. Clearly, the solution to this problem can only lie in the communicator's ability to anticipate what the communicant will anticipate about their

---

[23] David Dunér (2013, in this volume) has proposed a very interesting solution for at least some of these hard problems. His solution is based on Tomasello's concept of "joint attention." In many respects, the conceptual foundations of Dunér's proposal are the same as mine, as is his choice of the issues relevant (but often neglected as such) for thinking about interstellar communication and its principal constraints.

mutual knowledge (context) in the moment she receives the signal in his meta-representational ability.[24]

But how are we to know what is known to our interstellar communicant? What should we anticipate about the intersubjective features of our supposed shared context—features that are cognitively available to both parties *and* represented by their respective representational systems in mutually congruent ways? Bodily cues (facial, vocal, gestural, postural), so typical of our everyday encounters (both verbal and non-verbal), can function as indicators of internal goings-on under very restrictive conditions. What about the "situational" cues based on our supposedly common "background assumptions" (Searle 1978)—the common aspects of our world knowledge ("common sense")?

There has been a lot of speculation, both in the SETI community and among independent scholars, about possible conceptual commonalities and epistemic overlaps between us and our alien communicants. Marvin Minsky (1985, 117), for instance, argues that due to the universality of natural selection, "all problem-solvers... are subject to the same ultimate constraints—the limitations of space, time, and materials." To deal with these constraints aliens will, according to Minsky, use the same two principles of representing the world that we use: "the principle of economy" and "the principle of sparseness." The former principle will generate "symbol-systems for representing objects, causes, and goals"; the latter will guarantee that "every evolving intelligence will eventually encounter certain very special ideas—e.g., about arithmetic, causal reasoning, and economics—because these particular ideas are very much simpler than other ideas with similar uses." Many authors agree with Minsky. They are persuaded that the universality of natural laws and regularities (natural selection!) will ensure that all evolving minds will sooner or later arrive at more or less the same set of universal ideas and represent them in more or less the same ways. Ideas like three-dimensionality of space, direction of time flow, physical laws, chemical elements and compounds, items of the macro-world (e.g., planets, stars, and galaxies), gravity and its consequences, perhaps even some biological, behavioral, or social features seem universal enough and therefore good candidates for the shared communicative context.

However, despite these and other interesting proposals no decisive argument has ever been offered that would justify an optimistic attitude in respect to the shared knowledge problem. There is no way to dispel the worries—voiced by Nicholas Rescher (1985; 1987) and other critics of the "one world–one science" argument—concerning the commensurability between the interstellar communicants' representational systems.[25]

---

[24] As Sperber (1995, 198) put it, "[y]ou have to be doubly intelligent to see the intelligence in others. You need the ability to represent in your own mind the mental representations of other creatures. You need, that is, the ability to entertain representations of representations, what, in our jargon, we call 'meta-representations'."

[25] As Cohen and Stewart (2002, 285–291) show, even mathematical concepts might not be universal—they might be *invented*, not *discovered*.

So it might still be wrong to assume "that evolving alien brains will carve up the universe in the same way that ours do, and order what they find into the same patterns and logical schemes that we use" (Cohen and Stewart 2002, 287). Regarding interstellar communication, this means that the context simply cannot do the job it is designed to do in order to facilitate the decoding procedure—compensate for the lack of a prearranged code. So the only context one could confidently rely on is the one consisting of information deducible from the communicative situation itself, taken in the narrowest possible way. That is, the only thing that the addressee will anticipate upon receiving an interstellar signal is that the signal might embody a message and that this message has been emitted over a great distance by a civilization technologically equipped for such an undertaking. But even this minimal context might not be available if our signal fails to reveal our communicative intention. For it might be mistaken for a non-intended piece of information!

In light of this difficulty, we should think of a less ambitious goal for interstellar communication. In some specific cases of terrestrial behavior classified as communication, there is no other content of the message besides the very intent to communicate it. Or, to put it more technically, the content of the message is the communicative intention itself. The effect the sender thus intends to produce is solely the recognition, by the receiver, of her informative intention. My point is that to understand such a reduced message—barren of any specific informational content—will require minimal cognitive capacities of the receiver as compared to understanding messages with a genuine informational (representational) content. So being unsure about what is known both to ourselves and to our alien communicants, and lacking both a common code and ways to make this code deducible from the signal itself, we should satisfy ourselves with a message that would simply reveal our communicative intent as such. This would in effect be identical to a natural sign—in the sense of Grice's natural meaning—indicating an intentional mind and a potentially cooperative agent. Of course, the crucial requirement for understanding such a message would be the possession of a rudimentary mindreading/empathizing ability on the part of our unknown addressee.

What is more, a pure communicative intent, barren of any further informative content, might arouse less suspicion about possible *hidden motives* of our communicative attempt than a richly structured signal carrying a more informative message. In other words, our alien communicant might take our disclosure as intentional agents for our only communicative motive, thus reducing the chances that our contact be interpreted as a threat or an intent to deceive, increasing the chances for a rejoinder.

To conclude, it might be more advantageous for us to seek for a communicative soft-spot in the "social-affective" makeup of our putative communicants than in their intellectual abilities. For it is more likely that they will be able to empathically recognize a sheer communicative intent than to understand—let alone correctly interpret—a signal embodying an explicit representational content (i.e., an intent to convey a specific piece of information). If so, the receivers of our signal might be able to experience us as conspecifics—kin creatures capable for empathic understanding and intentional communication—irrespective of the very probable, and possibly insurmountable, differences between two types of intelligence and, consequently, two types of extelligence—the differences in general "worldviews," interests, representational systems, and symbolic practices between the two civilizations.

In its structural features, such a simple interstellar gesture would resemble our ordinary expressions of communicative intent—either verbal (like "hello" or "excuse me") or non-verbal (like eye contact with an unknown passenger sitting in a parallel train, or knocking on the wall that divides a prisoner from his unknown inmate in the adjacent cell). But the specific circumstances of its expression—its originating from a possibly altruistic and empathic being from other part of the universe—might, for the recipient of such a gesture, make the whole difference in the world.

References

Barkow, Jerome H. 2013. "Eliciting Altruism While Avoiding Xenophobia: A Thought Experiment." In *Extraterrestrial Altruism: Evolution and Ethics in the Cosmos*, edited by Douglas A. Vakoch, 37–48. Heidelberg: Springer.

Baron-Cohen, Simon. 1995. *Mindblindness: An Essay on Autism and Theory of Mind*. Cambridge, MA: MIT Press.

Basalla, George. 2006. *Civilized Life in the Universe: Scientists on Intelligent Extraterrestrials*. New York: Oxford University Press.

Batson, Daniel C. 2009. "These Things Called Empathy: Eight Related But Distinct Phenomena." In *The Social Neuroscience of Empathy*, edited by Jean Decety and William Ickes, 3–16. Cambridge, MA: MIT Press.

Chalmers, David. 1996. *The Conscious Mind*. New York: Oxford University Press.

Cohen, Jack, and Ian Stewart. 2002. *What Does a Martian Look Like?: The Science of Extraterrestrial Life*. Hoboken, NJ: John Wiley and Sons.

Dennett, Daniel. 1996. *Kinds of Minds: Towards an Understanding of Consciousness*. London: Weidenfeld and Nicolson.

Dunér, David. 2013. "Interstellar Intersubjectivity: The Significance of Shared Cognition for Communication, Empathy, and Altruism in Space." In *Extraterrestrial Altruism: Evolution and Ethics in the Cosmos*, edited by Douglas A. Vakoch, 141–167. Heidelberg: Springer.

Gallese, Vittorio, and Alvin Goldman. 1998. "Mirror Neurons and the Simulation Theory of Mindreading." *Trends in Cognitive Sciences* 12:493–501.

Gallese, Vittorio. 1999. "From Grasping to Language: Mirror Neurons and the Origin of Social Communication." In *Towards a Science of Consciousness*, edited by Stuart R. Hameroff, Alfred W. Kaszniak, and David J. Chalmers, 165–178. Cambridge, MA: MIT Press.

Gallese, Vittorio. 2001. "The 'Shared Manifold' Hypothesis: From Mirror Neurons to Empathy." *Journal of Consciousness Studies* 8:33–50.

Grice, Paul. 1957. "Meaning." *Philosophical Review* 66:377–388.

Janović, Tomislav, Vladimir Ivković, Damir Nazor, Karl Grammer, and Veljko Jovanović. 2003. "Empathy, Communication, Deception." *Collegium Antropologicum* 27:809–822.

Lem, Stanisław. 1974. *Summa Technologiae*. Krakow: Wydawnictwo Literackie.

Minsky, Marvin. 1985. "Why Intelligent Aliens Will be Intelligible." In *Extraterrestrials: Science and Alien Intelligence*, edited by Edward Regis, Jr., 117–128. Cambridge, UK: Cambridge University Press.

NASA Exoplanet Archive. 2013. Planets Count Page. Last modified July 31. http://
    exoplanetarchive.ipac.caltech.edu/docs/counts_detail.html.
Rescher, Nicholas. 1985. "Extraterrestrial Science." In *Extraterrestrials: Science and Alien
    Intelligence*, edited by Edward Regis, Jr., 83–116. Cambridge, UK: Cambridge University Press.
Rescher, Nicholas. 1987. *Scientific Realism: A Critical Appraisal*. Dordrecht: Reidel.
Rizzolatti, Giacomo, and Michael A. Arbib. 1998. "Language Within Our Grasp." *Trends in
    Neuroscience* 21:188–194.
Searle, John. 1978. "Literal Meaning." *Erkenntnis* 13:207–224.
Searle, John. 1979. "Intentionality and the Use of Language." In *Meaning and Use*, edited by
    Avishai Margalit, 181–197. Dordrecht: Kluwer.
Searle, John. 1990. "Collective Intentions and Actions." In *Intentions in Communication,* edited by
    Philip R. Cohen, Jerry Morgan, and Martha E. Pollack, 401–416. Cambridge, MA: MIT Press.
Shannon, Claude, and Warren Weaver. 1949. *The Mathematical Theory of Communication*.
    Urbana: University of Illinois Press.
Sperber, Dan. 1995. "How Do We Communicate?" In *How Things Are: A Science Toolkit for the
    Mind*, edited by John Brockman and Katinka Matson, 191–199. New York: Morrow.
Sperber, Dan, and Deirdre Wilson. 1986. *Relevance: Communication and Cognition*. Cambridge,
    MA: Blackwell.
Stewart, Ian, and Jack Cohen. 1999. *Figments of Reality*. Cambridge, UK: Cambridge University
    Press.
Thompson, Evan. 2001. "Empathy and Consciousness." *Journal of Consciousness Studies*
    8:1–32.
Tomasello, Michael. 2008. *Origins of Human Communication*. Cambridge, MA: MIT Press.
Tooby, John, and Leda Cosmides. 1992. "The Psychological Foundations of Culture." In *The
    Adapted Mind: Evolutionary Psychology and the Generation of Culture*, edited by Jerome
    Barkow, Leda Cosmides, and John Tooby, 19–136. New York: Oxford University Press.
Vakoch, Douglas. 1999. "The View from a Distant Star: Challenges of Interstellar Message
    Making." *Mercury* 28(2):26–32.
Wittgenstein, Ludwig. 1953. *Philosophical Investigations*. Oxford: Basil Blackwell.

Interspecies Altruism: Learning from Species on Earth

Denise L. Herzing

Abstract Many examples of altruism exist in the animal kingdom, both within species and between humans and non-human species. What can we learn about interspecies altruism from sentient species on Earth? Is it only social species that display altruism? Marine mammals, and other terrestrial species, have been observed in incredible displays of altruism with humans and non-human species. Some common features of interspecific altruism include complex social structure, potent individuality, and specific behavioral contexts. The competition versus cooperation argument made by evolution has recently gained new discussion points involving empathy, at least for social species. Mirror neurons also provide a potential mechanism for empathy. Altruism is communicated within and between species through a variety of sensory systems and social relationships. Do we share enough in common with certain species to recognize cues of distress, unhappiness, or danger to allow such interaction? For example, dolphins have been observed to help humans, and other dolphin species, in distress. Likewise humans often come to the rescue of dolphins in distress during both natural and human-made events. Although species might be able to recognize intention through real-time communication signals, the challenge of recognizing altruism from distant contact, or in one communication modality, remains. Given the difficulties in interpreting possible radio signal information, is it realistic to try to determine intention in extraterrestrial signals? We will explore all these questions relative to SETI and future potential contact with other beings in the universe.

Keywords Dolphins · Reciprocal altruism · Interspecies communication · Communication · Sociality · Primates · SETI

D. L. Herzing (✉)
Research Department, Wild Dolphin Project, PO Box 8436 Jupiter, FL 33468, USA
e-mail: wdpdenise@wilddolphinproject.orgdherzing@wilddolphinproject.org

D. L. Herzing
Department of Biological Sciences, Florida Atlantic University, 777 Glades Road, Boca Raton 33431, USA

D. A. Vakoch (ed.), *Extraterrestrial Altruism*, The Frontiers Collection,
DOI: 10.1007/978-3-642-37750-1_12, © Springer-Verlag Berlin Heidelberg 2014

1 Introduction

What can we learn about altruism from sentient species on Earth? Does inter-species altruism provide a non-terrestrial exercise in recognizing cross-species altruism? Altruism is usually defined in two ways. Kinship-based altruism suggests that relatedness drives such acts based on the evolutionary drive of the selfish gene and survival of kin. Reciprocal altruism suggests that acts are based on reaping benefits in the future from unrelated individuals. Historically, humans were thought to be the only species capable of reciprocal altruism, and kinship based altruism was thought to predominate over reciprocal altruism. Reciprocal altruism (Trivers 1971) was considered to be less common because the conditions for it to operate were believed to be rare (Hammerstein 2003; Stevens et al. 2005). Recent studies suggest that in tightly knit related kin, reciprocal altruism can supersede kinship. Many non-human species have shown advanced cognitive and commu-nicative abilities which enable reciprocal altruism. Early on, Darwin in 1871 (1982) speculated that animals that lived in stable social groups and had the intelligence and mental qualities necessary to form a system of personal friend-ships might be able to transcend the limits of familial relationships for altruistic acts. Recent work by de Waal (2009) and Bekoff (2008) suggest that at least for social species, cooperation wins out over competitive behavior, potentially driving the need for altruism. A degree of mutual dependence and long-lived proximity in social groups may increase the probability and evolutionary advantages of reci-procal altruism. Thus it is in long-lived social mammals, such as primates, ele-phants, and dolphins, that we see the emergence of reciprocal altruism.

Marine mammals, and other terrestrial species, have been observed in incred-ible displays of altruism with humans and non-human species. Are there factors in their social structure, individuality, or behavioral context/circumstances that might give us insight into recognizing altruism and intent from another species? Does the evolution of the brain and its associated infrastructure predict such altruism? The competition versus cooperation argument of evolution has recently gained new discussion points involving empathy, at least for social species.

Although our roots and similarity with primates is good basis for comparison, a better comparison for the challenges facing the search for extraterrestrial intelli-gence (SETI) probably comes from more alien species like dolphins or elephants. Since the ancient Greeks, dolphins said to save companions by biting through harpoon lines or by hauling them out of nets in which they were entangled. Dolphins also support sick companions near the surface to keep them from drowning, and stay close to females in labor. Whales tend to interpose themselves between a hunter's boat and an injured conspecific, or capsize the boat (Connor and Norris 1982).

In 1982 (p. 372) Connor and Norris described the potential for altruism in whales and dolphins:

> The evidence of epimeletic behavior, though based wholly on anecdotes, is so common as
> to be overwhelming in its broad detail. The data on school structure, based on reliable

marking studies at sea, clearly shows much fluidity of relationship, except in the larger dimorphic, polygynous species; but in both, intergeneric and interspecific cooperative behavior is clear. Reciprocal altruism carries with it the opportunity for the development of complicated social relationships involving, in addition to altruism, deceit, punishment of those who violate social rules, and complicated communication systems between school members.

Elephants, another intelligent social mammal, are also known to reassure distressed companions (Payne 1998; Poole 1996) and support or lift up others too weak to stand (Douglas-Hamilton et al. 2006; Joubert 1991).

Could empathic abilities drive cooperative tendencies both within and between species? If so, is it individually driven or society driven? Does such interaction depend on a relationship between individuals or species, or do existing social systems and exposure to other species as nonthreatening allow cross-species altruism? Conditions conducive to altruism may include relatedness, similarity or familiarity, social closeness, positive experience with the "Other," and friendship and fondness of an individual. Not all altruistic behaviors require empathy, although empathy, which is phylogenetically ancient, is a likely mechanism in that it allows recognition of another's pain, need, or distress. Recently "mirror neurons" that allow a being to "feel" what another being is experiencing has been proposed as a mechanism for empathy (di Pellegrino et al. 1992). Do we need similar sensory systems to engage in altruism? Perception of the emotional state of another being increases empathy and is consistent with both kin selection and reciprocal altruism theory. Mirror Self Recognition, recognizing that one's own body is reflected in a mirror as measured in primates (de Waal et al. 2005), elephants (Plotnik et al. 2006), and dolphins (Reiss and Marino 2001) is another mechanism that reinforces the perspective-taking tools available to social mammals.

Given the continuity of elements of life in the known universe and the convergence of both physical and mental features of the evolution of life on Earth, it is likely that altruism has evolved in at least social beings yet undiscovered in the greater universe.

How does the recognition of need, distress, vulnerability, danger, and gratitude manifest itself within and between species? Do species passively observe and act, or is altruism modeled in complex societies? We know that acts of human kindness create positive endorphins both in the actor and the observer. Are the contexts for altruistic acts situational and/or physiological, such as sensitive vulnerable states during reproduction and bonding?

Of course, species need to be able to communicate their needs, distress, and vulnerabilities to another being. How is altruism communicated within and between species? Are there examples in the natural world of potential communication signals used in these contexts? We will look to the history of non-human communication signals both within and between species to answer these questions.

2 Communication Within and Between Species

How might we begin to recognize cross-species signals of altruism? Do we share enough in common with certain species to recognize cues of distress, unhappiness, and danger, to allow such interaction? How much cross-species communication already takes place between species?

There are different models in human communication theory (Hockett 1960). One is the *Action model*. A sender projects an encoded message, via a channel, and a receiver decodes the message. In this system there can be information sent without intention (such as a yawn, a physiological reaction to low oxygen, potentially interpreted as boredom by a receiver). In the *Interactive model*, the message is modulated by feedback from each communicating party (after inquiry by the receiver, the sender explains a lack of sleep the previous night, which avoids misinterpretation). Thirdly, the *Transactional model* allows simultaneous messages across multiple modalities with constant feedback through interaction (the sender offers the lack-of-sleep explanation while yawning, while the receiver empathizes and stretches). In addition, levels of meaning may vary between individuals depending on their experience or history. To interpret meaning, there must be an understanding of the relationship, history, and the modalities available both to the receiver and the sender. For example, in humans eye contact is used to create an emotional linkage between people, but in other primates it is considered a threat to stare. In humans, gestures also serve to accentuate and mark periods of conversation. To look at either interpersonal (humans), or interspecific (between species) communication, we need to look at interaction of the organism and what potential information is available. Can we find forms of intraspecific communication within non-human animals that give us insight into possible parallels and continuity in communication signals between species?

3 Frameworks for Decoding

Humans and other species use multiple ways of communicating across a divide. Do we decode someone else's language, encode our own, or develop something mutual when interacting with the Other?

3.1 Decoding Through Observation

One way to decode is to listen passively, learn to understand some basic cues/signals, and tap into them for one's own survival. Researchers have successfully decoded the natural communication behavior of non-human animals with some species, including the vervet monkey and its abilities to refer to specific types of predators (Cheney and Seyfarth 1982), chimpanzees in their abilities to communication aggression through vocalizations and behaviors (de Waal 1982, Goodall 1971), and other similar studies such as in gorillas (Fossey 1983), ground squirrels (Leger et al. 1980), birds (Beer 1982; Kroodsma 1981), and honeybees (Gould and Gould 1982). Studies like Cheney and Seyfarth's work with vervet monkeys, which show that they respond to specific alarm calls that refer to distinct types of predators, dispel the idea that animal communication is only an emotional expression. Recently work on prairie dogs (Slobodchikoff et al. 2009) and domestic dogs (Miklósi 2009) suggest even great capacities among non-human animals. Signals that humans fail to discriminate between carry specific meaning for non-human animals and versatile forms of animal communication may elude us because of our limited sensory perception, abilities, or biases toward human uniqueness. Learning how particular species' communicate thoughts and feelings to each other might be a way of accessing other animal minds (Griffin 1981). The repertoire and use of communication signals in non-human species may be more complicated then we have imagined. Many animal systems rely on a small repertoire of individual signals, but use them in combinations and in circumstances that are context dependent or "situation bound" (Smith 1986). The widespread assumption that animals always express their motivational states immediately has been contradicted by many cases in which animals clearly retain the memory of some relationship about which they communicate only when circumstances are appropriate. We know little about the extent to which animals think about what they are doing and about the behavioral choices they make; only when behavioral science overcomes its antipathy to the concept of animal consciousness can we make any real progress in understanding it.

Co-opting signals from each other is also known in the natural world. Birds use alarm calls of various types to communicate (Sommer 2011) and those alarms are subsequently used by other species as information. Researchers have also reported on mixed-species associations and shared or co-opted signal use among fur seals (Page et al. 2002), birds (Dolby and Grubb 1998), dolphins (Herzing and Johnson 1997), tamarins (Windfelder 2001), and Diana monkeys (Zuberbuhler 2002), as well as between humans and harbor seals (Ralls et al. 1985), and a lion and baby Oryx (Munene 2002). Animals of different species may converse with each other to some degree and understand each other's signals. In a study of interspecific signal used between humans and domesticated dogs around the world, McConnell found that the dog trainers and handlers all used similar acoustic signals to communicate with their dogs (McConnell and Baylis 1985; McConnell 1990). Other studies between humans and dogs or horses (McKinley and Sambrook 2000; Rooney et al. 2001) have also shown understanding between species. Similarly, white crown sparrows learn interspecific songs of neighboring birds in order to blend in with the environment (Baptista and Morton 1981).

Humans have been known to immerse themselves when interacting with other human cultures and during enforced situations such as "wolf" children. Immersion takes on a more extreme meaning in the case of human beings raised and socially adapted to other species and known as "feral" or wolf children. Some fifty cases of animal adoptions of humans have been recorded. Children adopted by wolves, bears, leopards, monkeys and gazelles learned to walk on all fours, use animal communication signals, and had no verbal language. One such story is recounted by Jean Claude Armen who, in his travel across the Spanish Sahara, ran across a child adopted by gazelles (Armen 1976). Armen posed the question of how the two species learned to communicate and understand each other. He did not make an effort to "save" the child and bring him back to humans. Instead, Armen tried to blend in and gain the trust of the gazelle mother as well as dominant males in the gazelle group to gain access to the human child. He found that the boy had developed gazelle awareness: he would sniff the air and attend to subtle cues pertinent to the gazelles. Adapting to this cultural milieu, these human children have been adept at acquiring knowledge and interpreting and imitating animal behavior. As members of their new society, they had to adhere to its societal rules and obligations. By participating in animal societies other than our own, we may come to understand the meaning they give to their perceptual world. This experience may be as illuminating as those children who accidentally met the Other and became successfully incorporated into a non-human society.

3.2 Decoding Through Sharing Signals

Secondly, one can try to encode or share information from your own communication system/conceptual signals and teach another culture or species. In the natural world, such transmission of information seems to occur primarily within species or cultures. Teaching in dolphins (Bender et al. 2008) and primates (Boesch 1991) are examples of non-human information transition, in these cases between mothers and their offspring and potential with peers and non-related individuals.

This is the primary model that has been used between humans and primates. Since the beginning of culture, humans have been fascinated by the possibility of communicating with non-human animals, yet interspecies interaction is a newly evolving field of scientific inquiry. Our search for other intelligent minds has led us to primates, elephants, and cetaceans as well as other large-brained, social mammals. To date, the most serious efforts at interspecies interaction have occurred in laboratory settings, possibly because they enable researchers to control their studies and have access to the animals.

3.2.1 Laboratory Studies of Interspecies Communication

Although animal communication and behavioral studies span all taxonomic lines, the study of long-lived social animals has been the focus of cognitive ethology, most likely because of their closeness to humans, laboratory accessibility, and the complexity of their social structure and communication. Included are the great apes (Fouts 1973; Gardner and Gardner 1978; Miles 1978; Patterson 1978;

Premack 1976; Savage-Rumbaugh et al. 1980), marine mammals (Herman 1980; Herman et al. 1984; Lilly 1978; Marten and Psarakos 1995; Reiss and McCowan 1993; Schusterman and Krieger 1984), and birds (Pepperberg 1986).

Laboratory interspecies research has enabled scientists to ask direct questions and allowed animals to respond within a social context sometimes not entirely unlike their lives in the wild. This can be a powerful means of studying animal cognition and may include an open, arbitrary system that makes use of subtle variations to create an enormous variety of signals. Researchers have often focused on training their non-human subjects with artificially derived codes, rather than ones that are species-specific, because of the difficulty of knowing the natural communication signals of a species. This mutual approach, or two-way communication methodology, has often been chosen because of the difficulty of deciphering the complexities of non-human natural systems.

In the history of work with captive primates there has been a continuum of methodologies that researchers have used, from very objective and strict, to more interactive, involving bonding with the animals under study. One of the first studies was on a chimpanzee (*Pan troglodytes*) named Sarah who was acquired from the wild (Premack 1976). She was trained with the use of a keyboard, testing her ability to discriminate "same and different" objects. Premack used an interrogative concept and remained separate and "objective" during his work with Sarah. His work could be considered on the far end of the continuum.

The Rumbaughs were among researchers who used fewer restrictions in their experiments with Lana, a chimpanzee that learned "Yerkish," an artificial language developed for use on a keyboard through a computer interface (Savage-Rumbaugh and Rumbaugh 1978). Lana learned to push a sequence of buttons on the computer that would result in sentences, such as "give Lana apple" and so forth. Although some would argue that Lana simply learned to discriminate between simple associations, it seemed that she was able to make novel connections on her keyboard.

Over the years, the Rumbaughs moved to a more interactive approach with two other chimpanzees, Sherman and Austin. They spent time developing rapport with Sherman and Austin, and also tried to develop experiments that included referents or objects that were important to the chimpanzees. Allowing Sherman and Austin to work as a team, possibly mimicking the way communication systems evolved under natural conditions, was very successful. Subsequently they worked with Kanzi, a pygmy chimpanzee (*Pan bonobo*) of a species more closely related to humans than other chimpanzees (Savage-Rumbaugh 1994). Using a portable keyboard and a semi-natural environment, Kanzi is skilled at understanding complex symbolic associations.

Recognizing that the vocal capacity of chimpanzees was low, other researchers tried a bold experiment; they would try to teach American Sign Language (ASL) to a chimpanzee named Washoe (Gardner and Gardner 1978). Other researchers had been unsuccessful at teaching chimpanzees English vocally since chimpanzees do not have the necessary vocal anatomy. Finding a modality through which Washoe could express herself proved the key to communicating with chimpanzees.

Herbert Terrace was another researcher who tried to test whether chimpanzees could understand sentences through the use of ASL. Raising a young chimp in a human home, Terrace put "Nim" through strict classroom training, where he did learn to sign (Terrace 1979). Terrace's conclusion was that Nim could use sign language with the appropriate "cues" from his teachers but that he did not functionally understand syntax. Terrace was careful to avoid extraneous "cueing" of Nim, in part because he was concerned with the "Clever Hans" phenomenon.

Clever Hans was a horse that was taught to "count," among other skills, by his trainer (Prescott 1981). When a psychologist observed this interaction, he noticed that the trainer was giving Hans subtle cues, everything from face nods to slight movements. Without these "cues," Hans could not perform properly. The "double blind" method, not allowing the trainer or experimenter the knowledge of the answer, was subsequently used to avoid the exchange of subtle cues from trainer to animal. But what does the "Clever Hans" phenomenon suggest about biological communication in general? Could it be that these "cues" are the transactional signs or signals that create complex communication in the first place? If so, then we are asking non-human animals to do what we cannot; to be restricted to one modality, such as gestural signs or acoustic cues, and learn a multi-modal language that involves these exact subtle cues. This is what some researchers later concluded, which resulted in the creation of a more interactive and functionally expressive methodology for interspecies communication.

Roger Fouts, who originally worked with Washoe, continued this work on an even more interactive level (Fouts 1973). He argued that Terrace's study failed because it did not recognize the needs of the chimpanzees for social and interactive contact. Fouts reasoned that the chimpanzees would be much more interested in working with humans if they liked them and found them interesting. This would provide the social motivation for expressing themselves and communicating with someone they liked about something they desired or wanted to express. His study documented the spread of sign language from chimp to chimp, without human intervention, as an example of cultural transmission within chimpanzee society (Fouts and Couch 1976). Similar to the interactive approach and rapport development taken by Miles with orangutans (Miles 1978), Fouts tried to provide the chimpanzees with a natural environment rich with relationships and communication.

The most controversial interspecies study is that of Francine Patterson with the gorilla Koko (Patterson 1978). She used a highly interactive method with Koko, providing maximum bonding in all the channels through which they could interact. Critics say that because of this extreme interaction, her interpretations are skewed and she cannot remain objective. Patterson argues that this degree of bonding gives her insight into complexities and nuances of Koko's behavior, obtainable in no other way.

Despite the continuum of methodologies in all primate interspecies work, most researchers agree that creating rapport with non-human subjects is critical to the motivation of the animal during interaction.

The recognition of the social needs of many non-human animals has paved the way for new research, and is nowhere more evident than in the work of Irene Pepperberg and her African Gray parrot, Alex, a highly social yet unlikely subject to many. Pepperberg taught Alex to use English functionally (Pepperberg 1986). Using progressive methods that include social modeling and rivalry, which puts a human model in the role of a student competing for attention with Alex, she acknowledged Alex's need for social attention. A species so different from humans needs modeling and instruction in order to comprehend what a human researcher requests. Pepperberg's success has come from both intense social interaction and contextually relevant modeling. This functional acquisition of a non-species-specific communication code by a parrot is an instance of exceptional learning.

3.3 Decoding by Creating Mutual Signals

A third method is to develop a mutually understandable language, or use parts or a subset of signals that are used in "mutual situations/contexts" which are accessible through the sensory systems of both species. Humans of course have used this technique when human cultural language barriers exist. Such examples exist in the natural world, but this can present problems depending on the degree of conceptual similarity and overlap. For example, Orcas (*Orcinus orca*), or killer whales, live in close family pods. Orcas living in adjacent areas have distinct vocal dialects. When their areas overlap, each pod also has a subset of "shared" calls that they use only when interacting with the other pods (Weiß et al. 2011). Recently this overlap of shared vocal signals has also been documented in two species of dolphin in Costa Rica (May-Collado 2010). Again, shared signals are used when these two species meet, but not when they are interacting with their own species.

The fact that this seems to be a preferred method of interspecies communication might suggest that it is most easily adaptable between two disparate, but similar, beings. This might suggest a best-fit model for humans trying to understand and interact with another advanced being. Most illuminating to this process may be the existing examples of scientists studying non-human societies in the wild. These studies might suggest a three-step process involving immersion, observation, and exploration of communication signals.

Unlike the psychological and behavioral studies begun in the 1930s with Konrad Lorenz, and followed in an experimental laboratory approach by Niko Tinbergen (Lorenz 1952; Tinbergen 1972), ethological studies strongly emphasize spontaneous behavior in the natural environment. It is paramount for the ethologists to wait to see what the animals do of their own accord, instead of encouraging a particular kind of behavior for experimental purposes. The patient observation that distinguishes ethology is evident even in captive chimpanzee studies (de Waal 1982). In his study de Waal found long-term observation productive, and recording changes that took place over the years the most fascinating aspect of his work.

Like other students of natural behavior, de Waal recognized human limits in interpreting animal behavior.

Jane Goodall pioneered looking at the animal's own "Umwelt" in her work with wild chimpanzees. Among the critical aspects of her study were: accessing the chimpanzees and doing close-range observation; seeing enough behavior to make dependable observations; recognizing individuals to follow their long-term changes and life histories; and studying them in the context of their own community, without habitat or behavioral disruption (Goodall 1971). Her work helped us illuminate the chimpanzee in nature, as a highly intelligent, intensely social being, capable of close and enduring attachments and rich communication through sounds, gestures, and postures. In her writings, she describes the richness of chimpanzee life in the wild, including politics, families, and struggles. She has, in fact, become a cultural observer in the chimpanzee world. Even in her own fieldwork, Goodall immediately recognized that the chimpanzees she was studying were individuals with different personalities. She also wondered, when she returned after having been absent for a period of time, if the chimpanzees would have forgotten her. But when she returned to them, they were even more tolerant of her presence than before. Human researchers in the wild have also used the signals of their subjects, such as Dian Fossey's knuckle walking with gorillas, and Jane Goodall's pretense of nibbling roots with her chimpanzees.

An early attempt to teach English to dolphins was unsuccessful (Lilly 1978). However subsequent communication research that applied Pepperberg's techniques and framework and created mutual signals has been semi-successful (Xitco et al. 2004). In my own work with dolphins in the wild, I have used these frameworks to both develop a relationship and a level of comfort with an aquatic mobile culture of wild dolphins to tolerate human researchers in the water. This has afforded countless observations of natural behavior (Herzing 1996; 1997; Herzing and Johnson 1997). In unique circumstances dolphins were reported to use dolphin-appropriate interchanges with humans (Herzing and White 1998). To explore these human/dolphin signals, the deployment of a two-way interface in an attempt to "bridge the gap" between our two disparate species has been attempted (Herzing et al. 2012).

Speaking of the new aspects of work in the wild, Donald Griffin (1981, 157) stated that Jane Goodall

> ... approached a state of acceptance by wild chimpanzees without any attempt at morphological disguise, and she exchanged a few simple communication gestures, as others have done with captive apes. A new generation of ambitiously pioneering ethologists might open up an enormously powerful new science of participatory research in interspecies communication.

The late Dian Fossey followed Goodall's footsteps and used mimicry in her study of the highland mountain gorilla (Fossey 1983). From the Kabara group of gorillas, she learned to accept the animals on their own terms and never pushed them beyond the varying levels of tolerance they were willing to give. This gave her a strong understanding of the etiquette of her wild subjects: "Any observer is an intruder in the domain of a wild animal and must remember that the rights of the animal supersede human interests. An observer must also keep in mind that an animal's memories of one day's contact might well be reflected in the following day's behavior" (Fossey 1983, 14). Throughout her work, Fossey always remained concerned about human interaction with these wild animals, poaching, and habitat protection for this endangered species.

Cynthia Moss has also engaged in a decades-long study of another long-lived social mammal, wild elephants (Moss 1989). Elephants are very special animals: intelligent, complicated, and intense. In their natural social groupings, Moss has watched them give birth, mate for the first time, leave the security of their families, and defend their young. Like other researchers, Moss takes pride in knowing individuals, having a trustful relationship with them, and knowing about them as members of their society. And like humans, some elephants are curious and tolerant of humans, while others are not.

4 How Is Altruism Communicated?

How we might decipher intention through communication signals during real-time contact versus delayed and distant contact? Given the difficulties in interpreting anticipated radio-signal information, is it realistic to try to determine intention in distant signals? Is altruism always one-way or does it require the recognition of mutual or altruistic potential?

We know, for example, that as a species some humans fight and work for the preservation for whales, dolphins, and a myriad of other species on the verge of extinction or exposed to harmful human activities. This is typically driven by multiple reasons, including acknowledging the empathic abilities and complex social lives of dolphins, or recognizing that species should not be allowed to go extinct for moral or ecological reasons. So at least for our species, mutual recognition of altruism is not necessary for its one-way application in the real world. Could this be the case with beings on other planets? If so, how would we know? Of course this perspective must be overcome as we venture into space and potentially contact another advanced being, either in real time or distantly. During such contact, humans might find themselves facing the opposite problem, being on the other side, and with the need to communicate a benign and helpful intention to a more advanced being who would look upon us as we look upon non-human beings on Earth.

Anthropomorphizing, or giving non-humans the traits of humans, has long been considered inappropriate in describing their behavior, but some theorists have advanced the idea of critical anthropomorphism as a technique for "thinking about" non-human animal thinking (Griffin 1981; Ristau 1991). Even though direct experience of non-human minds may be closed to us, this method might enable researchers to observe data to develop inferences in objects or events (such as thoughts and minds of non-human animals) and gather evidence that mental states and processes in animals do exist independently of our observational and operational procedures. Aspects of critical anthropomorphism, including analogies, intuitions, intentions and holistic ideas, allow us to combine our human abilities with various kinds of knowledge, and may help us generate much-needed creative questions.

How do you encode and decode intention or altruism? There are many directions (vertical, horizontal, oblique) and mechanisms (trial and error, imitation, etc.) of information transmission and social learning (Herzing 2005). Many species employ a wide range of techniques, including observation and teaching, and these mechanisms need not be mutually exclusive. For instance, the fact that we can observe the specific facial expression of a fellow human being and intuitively understand it and share the emotional experience has arisen in the evolution of our species within our social environment. The subtle facial alterations that make up the repertoire of human facial expression have been documented (Eckman 1982), and are recognized by humans across social groups. Research on subliminal perception also provides evidence that humans often absorb important stimuli without conscious awareness (Shevrin 1973). We do not know how this occurs, but in the complex interaction of mind and body, interaction is essential (Pelletier 1978).

Dolphins might have different physical forms and live in an aquatic environment, but they share with humans strong family bonds, great investment in their young, and complex communication signals and learning abilities. It is here where we might expect to see the overlap of types of altruistic acts. Furthermore, overlapping sensory areas, such as vision and sound between our two species, may be critical for encoding and decoding. In a dolphin society, one of the most basic needs is for animals to breathe air at the surface. Sharks are also a major predator of dolphins. Altruistic acts by dolphins usually focus on these areas of need, such as holding each other up at the surface when one is distressed or injured, or dolphins circling humans, or herding humans to safety, when large sharks are nearby. In this case, the dolphins recognize both the need to breathe and the vulnerability of the Other, and they act as they would in their own society for their own survival. That such acts of altruism are often enacted without a close relationship with the individual (such as a strange human) suggests that dolphins are able to perceive the needs of an individual within a society, without knowing that individual. Perhaps perceptions of relationship extend to groups as well as individuals. This of course has tremendous implications for SETI, assuming that an advanced non-terrestrial being might live in a complex society. So perhaps when looking across species we should seek examples of similarity versus dissimilarity. Humans, for example, might not understand the dance of the honeybee if it was trying to communicate the location of danger to us. Every species will have both its assets and its limitations regarding sensory and communicative information. Probably the single most important piece of information we have learned from our work on Earth is that a species will more easily use what is available in its own system and should not be expected to learn another species' language.

The use of interspecies communication as an investigative tool for altruism may be a good exercise. There are, however, aspects of communication that might have some universal features. For example, prosodic features of communication, such as rhythm and intensity, might be used in similar ways across species to communicate emotions or motivation. Morton (1977) describes motivational features of birds and mammals based on a matrix of vocalizations varying in frequency and bandwidth that translate into fear, aggression, and distress across species. Whether we call it Gestalt or intuition, evidence for the importance of primary perceptions is supported by the phenomena of "synesthesia" (Marks 1978). This phenomenon of our nervous systems allows overlap and interaction between primary senses. For example, one may smell the color orange, or feel certain sounds. Information reaching our nervous systems may translate across senses, providing a multi-channeled and sensitive perceiving device to the human, or non-human, being. This defines common mechanisms and pathways for perception within the human body and is the synthesis of perception that may explain complex sensory inter-actions of our ability to perceive the world.

The study of deception (Griffin 1981) is especially relevant for societies that engage in altruistic acts. Deception can take place by the withholding or con-cealment of information critical to other animals. Therefore, non-human animals must monitor and predict the mental states and behaviors of members of their society and the consequences of their own behavior. And there must be a cost imposed on "cheaters," and such information must be decipherable. But to what extent do they use communication signals to manipulate each other? For deceptive behavior to occur, animals must anticipate and predict the behavior of other ani-mals. Deception also includes the use of individuals as social tools through triadic interactions or deceptions through alliance loyalty as seen with primates (de Waal 1982). This includes cooperative behavior and group action based on past and sustained relationships. Especially interesting is how an individual acknowledges an altruistic intervention or expresses gratitude. In one case, a humpback whale entangled in net was released by a group of human divers. The whale swam in circles and one by one touched or nudged each diver in apparent thanks (Fimrite 2005). In this case the animal used what it had available, touch, and did what whales and dolphins do to reconcile with each other. So we should expect a species to use what is available to it to express itself in unfamiliar or cross-species encounters.

5 Implications for SETI

Of course most of the examples in our natural world revolve around real-time interaction, where observers have the distinct advantages of direct observation of another species across many sensory modalities and for long periods of time. Recent observations may suggest that the building blocks of life are universal, and convergent evolution may suggest that social aspects, including communicative and sensory aspects, may also be universal. Insight may be gained from these methods for real-time contact. However, how might SETI gain from these examples for a distant contact with another being is yet undetermined? Without some universal feature of communication, as has been suggested by mathematical constructs for radio signal decipherment, or perhaps in the prosodic universals of languages on Earth, long-distance communication of empathy and altruism may be difficult to discover.

References

Armen, Jean Claude. 1976. *Gazelle Boy: A Child Brought up by Gazelles in the Sahara Desert*. London: Picador.

Baptista, Luis F., and Martin L. Morton. 1981. "Interspecific Song Acquisition by a White-crowned Sparrow." *Auk* 98:383–385.

Beer, Colin G. 1982. "Conceptual Issues in the Study of Communication." In *Acoustic Communication in Birds*, edited by Donald E. Kroodsma and Edward H. Miller, 279–310. New York: Academic Press.

Bekoff, Marc. 2008. *The Emotional Lives of Animals: A Leading Scientist Explores Animal Joy, Sorrow, and Empathy—and Why They Matter*. Novato, CA: New World Library.

Bender, Courtney E., Denise L. Herzing, and David F. Bjorklund. 2008. "Evidence of Teaching in Atlantic Spotted Dolphins (*Stenella frontalis*) by Mother Dolphins Foraging in the Presence of Their Calves." *Animal Cognition* 12(1):43–53.

Boesch, Christophe. 1991. "Teaching among Wild Chimpanzees." *Animal Behaviour* 41(3):530–532.

Cheney, Dorothy L., and Robert M. Seyfarth. 1982. "How Vervet Monkeys Perceive Their Grunts: Field Playback Experiments." *Animal Behaviour* 30:739–751.

Connor, Richard C., and Kenneth S. Norris. 1982. "Are Dolphins Reciprocal Altruists?" *American Naturalist* 119(3):358–374.

Darwin, Charles. 1982 [1871]. *The Descent of Man, and Selection in Relation to Sex*. Princeton, NJ: Princeton University Press.

de Waal, Frans B. M. 1982. *Chimpanzee Politics: Power and Sex among Apes*. San Francisco: Harper and Row.

de Waal, Frans B. M. 2009. *The Age of Empathy: Nature's Lessons for a Kinder Society*. New York: Random House.

de Waal, Frans B. M., Marietta Dindo, Cassiopeia A. Freeman, and Marisa Hall. 2005. "The Monkey in the Mirror: Hardly a Stranger." *Proceedings National Academy of Science USA* 102:11140–11147.

di Pellegrino, Giuseppe, Luciano Fadiga, Leonardo Fogassi, Vittorio Gallese, and Giacomo Rizzolatti. 1992. "Understanding Motor Events: A Neurophysiological Study." *Experimental Brain Research* 91:176–180.

Dolby, Andrew S., and Thomas C. Grubb, Jr. 1998. "Benefits to Satellite Members in Mixed-species Foraging Groups: An Experimental Analysis." *Animal Behaviour* 56:501–509.

Douglas-Hamilton, Iain, Shivani Bhalla, George Wittemyer, and Fritz Vollrath. 2006. "Behavioral Reactions of Elephants Towards a Dying Matriarch." *Applied Animal Behaviour Science* 100:87–102.

Eckman, Paul. 1982. "Methods for Measuring Facial Action." In *Handbook of Methods in Nonverbal Behavior Research*, edited by Klaus. R. Scherer and Paul Ekman, 45–90. Cambridge, UK: Cambridge University Press.

Fimrite, Peter. 2005. "Daring Rescue of Whale off Farallones." *San Francisco Chronicle*, December 14.

Fossey, Dian. 1983. *Gorillas in the Mist*. Boston: Houghton Mifflin Company.

Fouts, Roger S. 1973. "Acquisition and Testing of Gestural Signs in Four Young Chimpanzees." *Science* 180:978–980.

Fouts, Roger S., and T. B. Couch. 1976. "Cultural Evolution of Learned Language in Chimpanzees." In *Communicative Behavior and Evolution*, edited by Martin E. Hahn and Edward C. Simmel, 141–161. New York: Academic Press.

Gardner, Beatrice T., and Allen R. Gardner. 1978. "Comparative Psychology and Language Acquisition." *Annals of the New York Academy of Sciences* 309:37–76.

Goodall, Jane 1971. *In The Shadow of Man*. Boston: Houghton Mifflin Company.

Gould, James L., and Carol Grant Gould. 1982. "The Insect Mind: Physics or Metaphysics?" In *Animal Mind–Human Mind*, edited by Donald R. Griffin, 269–298. Berlin: Springer-Verlag.

Griffin, Donald R. 1981. *The Question of Animal Awareness: Evolutionary Continuity of Mental Experience*. Los Altos, CA: William Kaufman.

Hammerstein, Peter. 2003. "Why Is Reciprocity So Rare in Social Animals?: A Protestant Appeal." In *Genetic and Cultural Evolution of Cooperation*, edited by Peter Hammerstein. 83–94. Cambridge, MA: MIT Press.

Herman, Louis M., ed. 1980. *Cetacean Behavior: Mechanisms and Functions*. New York: John Wiley and Sons.

Herman, Louis M., Douglas G. Richards, and James P. Wolz. 1984. "Comprehension of Sentences by Bottlenosed Dolphins." *Cognition* 16:129–219.

Herzing, Denise L. 1996. "Vocalizations and Associated Underwater Behavior of Free-ranging Atlantic Spotted Dolphins, *Stenella frontalis*, and Bottlenose Dolphins, *Tursiops truncatus*." *Aquatic Mammals* 22(2):61–79.

Herzing, Denise L. 1997. "The Natural History of Free-ranging Atlantic Spotted Dolphins (*Stenella frontalis*): Age Classes, Color Phases, and Female Reproduction." *Marine Mammal Science* 13(4):576–595.

Herzing, Denise L. 2005. "Transmission Mechanisms of Social Learning in Dolphins: Underwater Observations of Free-ranging Dolphins in the Bahamas." In *Autour de L'Ethologie et de la Cognition Animale*, 185–193. Lyon: Presses Universitaires de Lyon.

Herzing, Denise L., and Christine M. Johnson. 1997. "Interspecific Interactions Between Atlantic Spotted Dolphins (*Stenella frontalis*) and Bottlenose Dolphins (*Tursiops truncatus*) in the Bahamas, 1985-1995." Aquatic Mammals 23(2):85–99.

Herzing, Denise L., and Thomas I. White. 1998. "Dolphins and the Question of Personhood." *Etica and Animalia* 9:64–84.

Herzing, Denise L., Fabienne Delfour, and Adam A. Pack. 2012. "Responses of Human-habituated Wild Atlantic Spotted Dolphins to Play Behaviors Using a Two-way Human/Dolphin Interface." *International Journal of Comparative Psychology* 25:137–165.

Hockett, Charles F. 1960. "The Origin of Speech." *Scientific American* 203:88–96.

Joubert, Dereck. 1991. "Elephant Wake." *National Geographic* 179:39–42.

Kroodsma, Donald E. 1981. "Geographical Variations and Functions of Song Types in Warblers (*Parulidae*)." *Auk* 98(4):743–751.

Leger, Daniel W., Donald H. Owings, and Deborah L. Gelfand. 1980. "Single Note Vocalizations of California Ground Squirrels: Graded Signals and Situation-specificity of Predator and Socially Evoked Calls." *Zeitschrift für Tierpsychologie* 49:142–155.

Lilly, John C. 1978. *Communication Between Man and Dolphin*. New York: Crown Publishers.

Lorenz, Konrad. 1952. *King Solomon's Ring: New Light on Animal Ways*. New York: Thomas Y. Crowell.

Marks, Lawrence E. 1978. *The Unity of the Senses: Interrelations Among the Modalities*. New York: Academic Press.

Marten, Ken, and Suchi Psarakos. 1995. "Using Self-view Television to Distinguish Between Self-examination and Social Behavior in the Bottlenose Dolphin (*Tursiops truncatus*)." *Consciousness and Cognition* 4:205–224.

May-Collado, Laura J. 2010. "Changes in Whistle Structure of Two Dolphin Species During Interspecies Interaction." *Ethology* 116:1065–1074.

McConnell, Patricia B. 1990. "Acoustic Structure and Receiver Response in Domestic Dogs, *Canis familiaris.*" *Animal Behaviour* 39:897–904.

McConnell, Patricia B., and Jeffrey R. Baylis. 1985. "Interspecific Communication in Cooperative Herding: Acoustic and Visual Signals from Human Shepherds and Herding Dogs." *Zeitschrift für Tierpsychologie* 67:302–328.

McKinley J., and T. D. Sambrook. 2000. "Use of Human-given Cues by Domestic Dogs (*Canis familiaris*) and Horses (*Equus caballus*). *Animal Cognition* 3:13–22.

Miklósi, Adam. 2009. "Evolutionary Approach to Communication Between Humans and Dogs." *Veterinary Research Communications* 33:S53–S59.

Miles, H. Lyn White. 1978. "Language Acquisition in Apes and Children." In *Sign Language Acquisition in Man and Ape*, edited by Fred C. C. Peng, 103–120. Boulder, CO: Westview Press.

Morton, Eugene S. 1977. "On the Occurrence and Significance of Motivation-Structural Rules in Some Bird and Mammal Sounds." *American Naturalist* 111:855–869.

Moss, Cynthia J. 1989. *Elephant Memories: Thirteen Years in the Life of an Elephant Family*. New York: Fawcett Columbine.

Munene, Mugumo. 2002. "Surprise in the Kenyan Wild as Lioness Adopts Oryx." *Daily Nation News*, January 7.

Page, Brad, Simon David Goldsworthy, Mark Andrew Hindell, and Jane McKenzie. 2002. "Interspecific Differences in Male Vocalizations of Three Sympatric Fur Seals (*Arctocephalus spp.*)." *Journal of Zoology London* 258:49–56.

Patterson, Francine G. 1978. "The Gestures of a Gorilla: Language Acquisition in Another Pongid." *Brain and Language* 5:72–97.

Payne, Katy. 1998. *Silent Thunder: In the Presence of Elephants*. New York: Penguin.

Pelletier, Kenneth R. 1978. *Towards a Science of Consciousness*. New York: Dell.

Pepperberg, Irene M. 1986. "Acquisition of Anomalous Communicatory Systems: Implications for Studies on Interspecies Communication." In *Dolphin Cognition and Behavior: A Comparative Approach*, edited by Ronald J. Schusterman, Jeanette A. Thomas, and Forrest G. Wood, 289–302. Hillsdale, NJ: Lawrence Erlbaum Associates.

Plotnik, Joshua M., Frans B. M. de Waal, and Diana Reiss. 2006. "Self-recognition in an Asian Elephant." *Proceeding National Academy of Sciences USA* 103(45):17053–17057.

Poole, Joyce. 1996. *Coming of Age with Elephants: A Memoir*. New York: Hyperion.

Premack, David. 1976. *Intelligence in Ape and Man*. Hillsdale, NJ: Lawrence Erlbaum Associates.

Prescott, John H. 1981. "Clever Hans: Training the Trainers, or the Potential for Misinterpreting the Results of Dolphin Research." *Annals New York Academy of Science* 364:130–136.

Ralls, Katherine, Patricia Fiorelli and Sheri Gish. 1985. "Vocalizations and Vocal Mimicry in Captive Harbor Seals, *Phoca vitulina.*" *Canadian Journal of Zoology* 63:1050–1056.

Reiss, Diana, and Brenda McCowan. 1993. "Spontaneous Vocal Mimicry and Production by Bottlenose Dolphins (*Tursiops truncatus*): Evidence for Vocal Learning." *Journal of Comparative Psychology* 107(3):301–312.

Reiss, Diana, and Lori Marino. 2001. "Mirror Self-recognition in the Bottlenose Dolphin: A Case of Cognitive Convergence." *Proceedings National Academy of Science USA* 98:5937–5942.

Ristau, Carolyn A. 1991. *Cognitive Ethology: The Minds of Other Animals*. Hillsdale, NJ: Lawrence Erlbaum Associates.

Rooney, Nicola J., John W. S. Bradshaw, and Ian H. Robinson. 2001. "Do Dogs Respond to Play Signals Given by Humans?" *Animal Behaviour* 61:715–722.

Savage-Rumbaugh, Sue. 1994. *Kanzi*. New York: Wiley.

Savage-Rumbaugh, E. Sue, and Duane M. Rumbaugh. 1978. "Symbolization, Language, and Chimpanzees: A Theoretical Reevaluation Based on Initial Language Acquisition Processes in Four Young *Pan troglodytes.*" *Brain and Language* 6:265–300.

Savage-Rumbaugh, E. Sue, Duane M. Rumbaugh, and Sarah T. Boysen. 1980. "Do Apes Use Language?" *American Scientist* 68:49–61.

Schusterman, Ronald J., and Kathy Krieger. 1984. "California Sea Lions Are Capable of Semantic Comprehension." *Psychological Records* 34:3–23.

Shevrin, Howard. 1973. "Brain Wave Correlates of Subliminal Stimulation, Unconscious Attention, Primary and Secondary Process Thinking, and Repressiveness." *Psychological Issues* 8:56–87.

Slobodchikoff, Constantine N., Bianca S. Perla, and Jennifer L. Verdolin. 2009. *Prairie Dogs: Communication and Community in an Animal Society.* Cambridge, MA: Harvard University Press.

Smith, W. John. 1986. "Signaling Behavior: Contributions of Different Repertoires." In *Dolphin Cognition and Behavior: A Comparative Approach,* edited by Ronald J. Schusterman, Jeanette A. Thomas and Forrest G. Wood, 315–333. Hillsdale, NJ: Lawrence Erlbaum Associates.

Sommer, Christina. 2011. "Alarm Calling and Sentinel Behaviour in Arabian Babblers." *Bioacoustics* 20(3):357–368.

Stevens, Jeffrey R., Fiery A. Cushman, and Marc D. Hauser. 2005. "Evolving the Psychological Mechanisms for Cooperation." *Annual Review Ecological Evolution Systems* 36:499–518.

Terrace, Herbert S. 1979. *Nim.* New York: Academic Press.

Tinbergen, Nikolaas. 1972. *The Animal in Its World.* Cambridge, MA: Harvard University Press.

Trivers, Robert L. 1971. "The Evolution of Reciprocal Altruism." *Quarterly Review Biology* 46:35–57.

Weiß, Birgitta M., Helena Symonds, Paul Spong, and Friedrich Ladich. 2011. "Call Sharing Across Vocal Clans of Killer Whales: Evidence for Vocal Imitation?" *Marine Mammal Science* 27(2):E1–E13.

Windfelder, Tammy L. 2001. "Interspecific Communication in Mixed-species Groups of Tamarins: Evidence from Playback Experiments." *Animal Behaviour* 61:1193–1201.

Xitco, Mark J., Jr., John D. Gory, and Stan A. Kuczaj, II. 2004. "Dolphin Pointing Is Linked to the Attentional Behavior of a Receiver." *Animal Cognition* 7:231–238.

Zuberbuhler, Klaus. 2002. "A Syntactic Rule in Forest Monkey Communication." *Animal Behaviour* 63(2):293–299.

Part IV
Universal Ethics and Law

Terrestrial and Extraterrestrial Altruism

Holmes Rolston III

Abstract How Earthbound are values and ethics? We humans enjoy a surprising transcendence of localized body and place. We are always situated somewhere, but it does not follow that all our knowledge is situational. True, science is a human enterprise. True, ethics is a human activity—even biologically-based. But can we expect to share some of our science and ethics with extraterrestrials? Perhaps in the search for extraterrestrial intelligence, the question to ask is not about the value of π, or the atomic number of carbon. A more revealing test might be to ask whether one should tell the truth, keep promises, or be just. The Golden Rule may be as universally true as is the theory of relativity.

Keywords Altruism · Ethics · Extraterrestrials · Golden Rule · Promises · Justice · Truth-telling

1 Introduction

We humans live in place, locally on landscapes and globally on this planet. We have powers of displacement too, of getting out of our places and taking up, whether empathetically or objectively, the situations of others, other humans, sometimes others than humans. There is a human genius for transcending location. Critics will at once object that whatever knowledge we gain has to "come through" at our native range. Have I forgotten that philosophers have perennially found themselves in an epistemic prison? There is no human knowing that is not looking out from where we are, using our senses and our brains, from an anthropocentric perspective.

H. Rolston III (✉)
Department of Philosophy, Colorado State University, Campus Delivery 1781, Fort Collins, CO 80523-1781, USA
e-mail: rolston@lamar.colostate.edu

D. A. Vakoch (ed.), *Extraterrestrial Altruism*, The Frontiers Collection, DOI: 10.1007/978-3-642-37750-1_13, © Springer-Verlag Berlin Heidelberg 2014

That seems uncontroversial, but does this imply that we can see and know no others as they are in themselves, nothing non-anthropocentric, nothing true apart from our looking, nothing external? Yes—so at least runs the currently fashionable answer. Knowledge is relative to our location, our embodiment, our size, our terrestrial habitat. We do not have a "view from nowhere," no "God's eye view." Everybody has a "body," everybody has a "standpoint," a "viewpoint." We are incarnate, in the flesh, and our particular form of embodiment poses "epistemological" problems—for which, in Western philosophy, a classic metaphor is Plato's men in the cave, or, in the East, the blind men and the elephant. In contemporary idiom, postmodernists insist that all our knowledge is "socially constructed." Knowledge is relative to our worldviews; realism, even a critical realism, is thought to be quite naive. Our knowledge wears a human face.

True, we never experience a dis-embodied, un-placed, or other-worldly mind. But we do gain views that look out from our bodies and places and see what is out-of-my-body, out-of-my-place. Had we been microbes or octopuses, our native ranges of perception and conception would be different. But we humans also enjoy a surprising transcendence of localized body and place. We are always situated somewhere, but it does not follow that all our knowledge is situational.

My progression of argument here first explores this search for objectivity and transcendence in science. Then we press a parallel question in ethics, pushing toward universality. If you prefer another way of phrasing this, avoiding such disfavored terms as "objective" and "universal," we are asking how "inclusive" and "comprehensive" can science and ethics be? This question is relevant even if our interests stay Earthbound. The question will further figure into the search for intelligent life elsewhere. Can we expect to share some of our science and ethics with extraterrestrials?

2 Terrestrial and Extraterrestrial Science

Once upon a time natural history produced humans, and some of these humans have become scientists. Some of these scientists have given us an account of natural history, out of which they (and all other living beings) evolved. This science is done with native endowments, brains, eyes, hands; this science has built up over centuries of cumulative transmissible cultures. We are not born scientists; we are educated to become chemists or biologists. Such science is a dominant fact of modern culture.

The natural sciences certainly seem to open up a world that is objectively there; they contact, catch, describe it more and less accurately: evolutionary history with its trilobites and dinosaurs, ecology with its phenotypes, genotypes, and trophic pyramids. True, the scientists have a network of theories they have constructed, but this network hits up against a hard world of experience. Some things have been "found" out. The human mind sometimes seems to be able to reach truths about realms that it does not inhabit, extrapolating and reasoning from the realms it does.

We learn this even at native ranges in biology. Humans reach outside their own sector to study warblers, viruses, and dinosaurs. We learn much that underlies our native range of experience but is not self-evident: life is carbon-based and photosynthesis is the primary biochemistry for capturing solar energy.

Humans become still more universal in their physics and inorganic chemistry, learning about the microworlds of elementary waves and particles, about the astronomical worlds of outer space. Martians too, when doing chemistry, will have to figure the conservation of charge, energy, and mass, and balance their equations. Science employs biologically evolved perceptual and conceptual faculties, it is a social construct; but, for all that, it sometimes flowers to discover objective truths—such as relativity theory or the atomic table, which are true all over our universe.

How far can we see out of our native range? Microscopes and telescopes come immediately to mind, making visible the microbes and galaxies. Today neither the microscopes nor telescopes have to be light-based. Such instrumentation expands to Geiger counters, mass spectrometers, and on and on. Alone among the species on Earth, all embodied, *Homo sapiens* is cognitively remarkable. With our instrumented intelligences and constructed theories, we now know of phenomena at structural levels from quarks to quasars. We measure distances from picometers to the extent of the visible universe in light years, across 40 orders of magnitude. We measure the strengths of the four major binding forces in nature (gravity, electromagnetism, the strong and weak nuclear forces), again across 40 orders of magnitude. We measure time at ranges across 30 orders of magnitude, from picoseconds to the billions-of-years age of the universe. Nature gave scientists their mind-sponsoring brains; nature gave scientists their hands. Nature did not give scientists radiotelescopes with which to "see" pulsars, or relativity theory with which to compute time dilation. These come from human genius.

These extremes are quite beyond our embodied experience. No one experiences a light year or a picosecond. But they are not beyond our comprehension entirely, else we could not use such concepts so effectively in science. The instrumentation is a construction (microscopes and mathematics), a "social construct," if you must. Certainly the data on these instruments comes through at native range, as we read the meters or computer output. But precisely this construction enables us dramatically to extend our native ranges of perception. The construction distances us from our embodiment.

No one has an everyday "picture" of a quark or a pulsar. So if one wants to lampoon the perfect "mirror" image at these extremes of range, this is easy to do. But we have good theory why nothing can be "seen" at such ranges in ordinary senses of "see," which requires light in the wavelength range of 400–700 nm, with quarks and pulsars far outside that range. We can ask whether a molecule is too small to be colored, or whether an electron, in its superposition states, is so radically different as to have no position, no "place" in the native range sense, but only a probabilistic location.

"Absolute" is a forbidding word; it seems too metaphysically transcendent. "Universal," also a term currently disliked by philosophers, is still welcomed by scientists; the term is at least this-worldly, oriented toward "our" universe. Scientists have what Michael Polanyi (1962, 65) called "universal intent"; they hope for theories true at all times and places, true for all peoples everywhere. If this is true in science, we can also ask whether the human mind can reach ethical principles that may transcend our somatic embodiment. It might be that in the search for extraterrestrial intelligence, the question to ask is about the value of *pi*, or the atomic number of carbon. Another, more revealing test of their intelligence might be to ask whether one should tell the truth, keep promises, or be just.

3 Earthbound Ethics?

Millennia before scientists, natural history produced moral agents, incrementally perhaps, but still appearing where there were none before. Ethics, like science, is distinctive to humans. There is no ethics in wild nature. To be ethical is to reflect on considered principles of right and wrong and to act accordingly, in the face of temptation. This is a possibility only human life, so that we expect and demand that persons behave morally and hold them so responsible.

Peter Singer's *Ethics* has a section "Common Themes in Primate Ethics," including a section on "Chimpanzee Justice," and he wants to "abandon the assumption that ethics is uniquely human" (Singer 1994, 6). But many of the behaviors examined (helping behavior; dominance structures) are more pre-ethical than ethical; he has little or no sense of holding chimpanzees morally culpable or praiseworthy.

Frans de Waal (1996, 209) finds precursors of morality, but concludes:

> Even if animals other than ourselves act in ways tantamount to moral behavior, their behavior does not necessarily rest on deliberations of the kind we engage in. It is hard to believe that animals weigh their own interests against the rights of others, that they develop a vision of the greater good of society, or that they feel lifelong guilt about something they should not have done. Members of some species may reach tacit consensus about what kind of behavior to tolerate or inhibit in their midst, but without language the principles behind such decisions cannot be conceptualized, let alone debated.

Jerome Kagan (1998, 91) puts it this way: "What is biologically special about our species is a constant attention to what is good and beautiful and a dislike of all that is bad and ugly. These biologically prepared biases rend the human experience incommensurable with that of any other species." Completing a careful survey of behavior, Helmut Kummer (1980, 45) concludes, "It seems at present that morality has no specific functional equivalents among our animal relatives."

After her years of experience with chimpanzees, and although she finds pair bonding, grooming, and the pleasure of the company of others, Jane Goodall writes

> I cannot conceive of chimpanzees developing emotions, one for the other, comparable in any way to the tenderness, the protectiveness, tolerance, and spiritual exhilaration that are the hallmarks of human love in its truest and deepest sense. Chimpanzees usually show a

lack of consideration for each other's feelings which in some ways may represent the deepest part of the gulf between them and us (van Lawick-Goodall, 1971, 194).

In social animals, where reciprocal aid develops, as when monkeys give alarm calls, one animal can and does aid another, and the result (averaged over the population) is increased conservation of somatic and genetic value. (This is often called "reciprocal altruism," a misnomer, better to call it "reciprocity.") The result is maximal protection of values held in common by the animals involved. It would be a mistake in evolutionary strategy for one animal to lose where this did not bring high enough gains in kindred lives bound with it in community. Natural selection selects against such behavior.

Although we might not want to call it altruism, biologists are increasingly inclined to see that natural selection can promote not only competition but also co-operation. Martin Nowak claims, both on the basis of observation in the wild and of computer simulation, "Competition does not tell the whole story of biology. Something profound is missing. Creatures of every persuasion and level of complexity cooperate to live.... This is the bright side of biology" (Nowak 2011, xii–xiv). "Cooperation is the master architect of evolution," Nowak (2011, xviii) continues. "I have accumulated a wide range of evidence that competition can sometimes lead to cooperation," he writes (Nowak 2011, 14). "By cooperation, I mean that would-be competitors decide to aid each other instead" (Nowak 2011, xiv). Organisms are quite interrelated, living in communities, ecosystems, with myriad co-actions, co-operations, interdependencies.

Yes, but even if animals learn to cooperate as well as to compete, they are still seeking their self-interests, and this continues in human social animals. We next meet a claim often made by Darwinians, especially the evolutionary geneticists. Ethics is a survival tool; individual humans construct their ethics to keep their genes in the next generation. The explanation—so this claim runs—lies in a naturalized ethics, a Darwinized morality. That, of course, will be Earthbound, local, and likely individualized, or family, or tribal.

Michael Ruse, a philosopher, joins E. O. Wilson, a biologist:

> Morality, or more strictly, our belief in morality, is merely an adaptation put in place to further our reproductive ends. Hence the basis of ethics does not lie in God's will... or any other part of the framework of the Universe. In an important sense, ethics... is an illusion fobbed off on us by our genes to get us to cooperate (Ruse and Wilson 1985, 51–52).

"Morality is a biological adaptation no less than are hands and feet and teeth," Ruse (1994, 15).

A Darwinized ethic will be Earthbound: "A conclusion of central importance to philosophy [is] that there can be no genuinely objective external ethical premises.... No such extrasomatic guides exist." Ethics is "idiosyncratic" to the biology of a species (Ruse and Wilson 1986, 186, 173). Ruse and Wilson are especially anxious to conclude that there is nothing absolute or permanent, or extraterrestrial, about these ethical commandments; they fit our human biology and nothing more.

Although a morality that conserves human genetic material is welcome enough, these claims bring deeper trouble. Evolution produces this fertility through a radical selfishness incompatible with ethics. George Williams (1988, 385) claims "Natural selection… can honestly be described as a process for maximizing short-sighted selfishness." Robert L. Trivers (1971, 35) claims that these are "models designed to take the altruism out of altruism." After these biological reinterpretations, we will not find altruism even on Earth. If so, we might not have sense enough to search for it among extraterrestrials in space.

Meanwhile at least genes have to be spread around; that is the only way they can be conserved. Fitness is not measured by an individual's own survival, long life, or welfare. Fitness is measured by what any individual can contribute to the next generation in its environment, fitness in the flow of life to pass life on. Survival of the fittest turns out to be survival of the better senders of whatever is of adaptive value in self into others in the next generation.

Genetic evolutionary natural history has generated "felt caring," when the sentient organism is united with or torn from its loves. The Earthen story is not merely of goings on, but of "going concerns." Animals hunt and howl, find shelter, seek out their habitats and mates, feed their young, flee from threats, grow hungry, thirsty, hot, tired, excited, sleepy. They suffer injury and lick their wounds. In physics and astronomy, we meet a causal puzzle, one of creation *ex nihilo*. Biology adds creation *ex nisu*, creation *per laborem*. To mechanical cause, there is added proactive care. There is death, but, with labor and regeneration, life ongoing. Selfish genes, once again, need to be cooperators in keeping life surviving in the midst of its perpetual perishing.

Listening to such biologists, extrapolating from the behavior of genes to the behavior of persons, we are puzzled by several facts: (1) Natural history produced persons with capacities with which they sometimes become scientists. (2) Some, but only some, of these scientists think ethics is an earth-bound illusion. (3) Natural history produced persons with capacities with which they sometimes (at least seem to) become altruists. (4) Ethicists claim that the human self achieves the novel possibility of acting as a moral agent. (5) The individual self can recognize and make its own concern questions about the worth of others outside its own local sector: contributing aid for the Ethiopians or for conservation of whales. The person can and ought defend values more comprehensive than those of self-love; a person can love the other as well as oneself. And such persons can flourish.

A possible line of resolution is to add culture to biology. In culture, one can gain enlarged interests and so an enlarged sense of identity. The first cultural unit is the nuclear family, where there is also genetic identity, kinship identity. But there is also tribal identity; here theories of "group selection" have been returning in ethics. In *Unto Others*, Elliott Sober and David Sloan Wilson find both self-interest and at least in-group altruism: "Natural selection is unlikely to have given us purely egoistic motives," but rather a "motivational pluralism," enabling the tribe to survive, although they also insist that there is no "universal benevolence" (Sober and Wilson 1998, 9, 12, 324). So, within the community, we find the patriots in battle, the Rotarians building their public spirit, even the Christians loving both self and neighbors.

The cultural self, like the biological self, lives the life of myriads of interconnections, extending far beyond the family. One works for a business firm, serves on a town council, is a volunteer at the hospital, spends time in military service, makes a donation to the college of which he or she is an alumnus or alumna, supports a scientific research project, teaches a class of students with kindred interests, joins a conservation society, leaves a will with a bequest not only to children but to those institutions he or she wishes to see continue after death.

Almost everything that the self cares about has to be cared about in concert with others, in reciprocity. The cultural self comes to transcend, even to replace, in part, the biological self. But neither does this stay tribal or patriotic. What one wishes to survive is one's ideas, one's values, or, more accurately, those ideas and values into which one comes to be educated and in which one meaningfully and critically participates. That is true with the scientists; the scientists do not claim that by promoting their science they are simply attempting to leave more genes in the next generation. Nor are they simply defending the survival of their tribes, either racial or professional. So why should we think it true of the ethicists? In science, scientists believe that, although they operate within their biologies and their cultures, they sometimes know truths external to their interests, external to human affairs. Why deny that ethicists might do the same?

There are twin truths: Ethics does have to be "naturalized," to fit human biology, including human reproductive needs. One does not want an ethic of no Earthly use. A species-blind moral system would be inadequate. Ethics needs to be situated where the moral agents live, internal to the agents' conditions—physical, environmental, social, cultural. We can expect that human ethics will have evolved to suit human biology and to defend the sorts of values that persons can instantiate, enjoy, lose, and protect.

If resources are in short supply, there will need to be an ethic about stealing. If killing is possible, there will need to be an ethic about murder. If the moral agent is sexed, there will need to be an ethics of sexuality. If humans have lusts, there will need to be a command not to covet. Ethical principles will need to fit human sociology and psychology, as this is superimposed on our biology. Ethical principles will also need to fit human cultural institutions—such as contracts of marriage, or business dealings, or citizenship in states—many of which have little precedent in nature. Ethics will need to protect, both by defending and sharing, the multiple capacities that humans have for enjoying values. In this sense, information about behaving ethically where one does not reside is irrelevant.

Yet there is an exodus out of wild, spontaneous nature into the freedom of spirit in cultural life, superimposed on biological life. We never become free from nature, but we do become free within nature. Humans receive a biological legacy, but are elevated into a cultural process superimposed on the evolutionary one. Ethics is social and informs us how to make a way through both the social and the natural worlds. Perhaps humans do not need any ethics for larger realms than those in which they actually reside. But we do have sciences, as we were noticing, that in the course of discovering useful theories and laws for human affairs, also discover what is true for extraterrestrials. We are still wondering whether that might be also true in ethics.

4 Inclusive Global Ethics

Humans have expanded their territories all over the globe. They are also capable of asking how ought humans to live as they reside on this Earth we have so much occupied. Should each maximize his or her own self-interest, reciprocating as needs be to accomplish this? Should the rich grow richer, the privileged more privileged? Should each person maximize the number of his or her offspring? Each family? Each tribe? Each nation state? Should *Homo sapiens* as a species maximize the high intrinsic value of our kind? The most convincing answers urge a more global, more generous defense of value.

This better ethics, these cosmopolitans will argue, has to be universally shared; it generates concern for other humans near and far, relating to them with the moral values of justice, love, and respect. The commitment that one has to make transcends one's genetics and one's society. Any account of in-group altruism to achieve out-group competitive success is powerless to explain the universalism in the major world faiths. Humans can enter into an ethical contract with other humans, the principles of which are oblivious to the specific circumstances of time and place, genome or culture. Ethics—at least in ideal, if not in real—has "universal intent."

Immanuel Kant claimed that the fundamental moral precepts must be "universalizable," that is, what he called "categorical imperatives" applicable to all moral agents. He thought the Golden Rule was one such formulation of a universal morality (Kant [1785] 1964). The Golden Rule is indeed widespread in one form or another, especially in the classic world religions (Wattles 1996). Similarly with respect for life, or the greatest good for the greatest number. The Ten Commandments—at least the moral table—are for all humans, pan-culturally, pan-genetically. "Do not murder." "Do not steal." "Do not covet."

Ethics, according to John Rawls' influential account, takes up an "original position," based on judgments about what one would do if one did not know the particulars of one's birth and local circumstances (Rawls 1971). Ethics becomes globally inclusive—without denying that there may be differing duties to family, friends, community, business, heritage, ethnic or interest group, nation state, and humanity at large. Human values are widely distributed and shared, as witnessed, for example, in the *Universal Declaration of Human Rights.*

Such hope for an inclusive global ethics is not without its critics. To the claims of a Darwinized ethic, we must add a second main area of challenge, cultural relativism, also claiming that ethics is quite parochial. Ethics, these critics reply, is quite relative; do not the myriad behaviors of the myriad peoples amply illustrate that? Indeed, we have no one agreed-upon ethics even within the Western traditions, as differences between the utilitarians and the deontologists illustrates. Agreement on broad principles ("the greatest good for the greatest number") quickly splinters into fragmented actual decisions in practice (how far there should be public welfare, or whether to employ capital punishment). Agreement on the Golden Rule is more apparent than real; differing cultures and centuries put it into practice quite differently. Noble global principles are lofty but never functional. Not one person in ten million ever acts in specific circumstances after deliberating globally the greatest good for the greatest number. If one does so deliberate, in most of the contested decisions, a person seldom knows which of the immediate alternatives faced will result globally in the most good.

Some argue that ethics is set in the context of one's heritage (MacIntyre 1981). Ethics always goes into some larger interpretive framework, some worldview. In reaching ethical conclusions, differing peoples give different weights to the importance of individuality versus family versus community, or rights versus duties versus virtues, or justice versus fairness versus benevolence, or freedom versus determinism, or forgiveness versus reparations versus retaliation, or past versus present versus future generations, or reason versus emotion, or censure versus praise, grace versus merit. Ethics is continually evolving, as with decisions about brain death, or homosexuality; we next face decisions about genetically modifying ourselves. Only the historically naive will claim that humans have reached any established cultural ethic, much less some transglobal ethic.

Nevertheless, ethical systems do reach around the globe. Differing contexts and evolving traditions do not preclude some commonality in ethics. The major world faiths have escaped parochial tribalism, not only in ideal but also in the real proportionately to their success. It seems impossible to explain their ecumenical and inclusive global missions on the basis of either genetics or local traditions. Somehow, somewhere these missionary ethicists reached insight into a better standard of what is right. One's culture is subject to on-going critical re-evaluation.

These religions criss-cross races, nations, and centuries; they operate in diverse times and cultures, and involve some logic of the mind that is tracking what is transgenetically and transculturally right or of value, no matter whether one has this or that set of genes or was born in this or that culture. Genetic success in each local community is necessary but not sufficient to explain this universalism. Nor are these widespread and persistent concerns so easy to "deconstruct" as nothing but local "social constructions." It is more plausible to conclude that such religions have been discovering what is trans-tribally, trans-culturally valuable.

In interhuman ethics there is already a striking novelty, unprecedented in prior natural history. In environmental ethics, rather than using mind and morals as survival tools for defending the human form of life, or my tribal heritage, mind generates an intelligible view of the whole and defends the varieties of life in all their forms, not just other peoples but nonhumans too. Humans transcend their own concerns. Humans can get "let in on" more value than any other kind of life. The novelty is class altruism emerging to coexist with class self-interest, sentiments directed not simply at one's own species but at other species fitted into biological communities. Humans can and ought to think from an ecological analogue of the original position, a global position that sees Earth objectively as an evolutionary ecosystem. Interhuman ethics has spent several millennia waking up to human dignity. Environmental ethics invites awakening to the greater story of which humans are a consummate part.

So both in interhuman ethics and in interspecific ethics, humans do seem to be able to reach more inclusive and comprehensive standards. Humans can aspire to global dimensions in their ethics. But perhaps ethics is still Earthbound?

5 Universal Ethics

We know only one ethical species and it is difficult to universalize from only one known case. Perhaps humans do not need any ethics for larger realms than those in which they actually reside; a global ethic is aspiration enough. Even if we contact extraterrestrial intelligence, we are quite unlikely to have any direct, interactive dealings with civilizations many light years away, certainly not moral responsibilities resulting from our actions affecting them for better or worse. Nor they with us.

Still it may be worth inquiring whether to expect that extraterrestrial intelligence will be moral. One pointer in that direction will involve conclusions about whether and how far we humans already know what such intelligences will know. The human mind sometimes seems to be able to reach truths about realms that it does not inhabit, extrapolating and reasoning from the realms that it does (as we were claiming with scientists). So we can ask whether the human mind can reach ethical principles that may transcend our somatic embodiment, that are "objective" to our humanity, and therefore likely to be shared with extraterrestrials.

In the ultra-sophisticated circles of postmodern epistemology, if describing universal nature as if it objectively exists is an illusion, finding universal moral values is still more illusion, illusion-on-stilts. That would require us further still to do what we cannot do, get out of our skins, get a "view from nowhere," get (God forbid!) the "God's eye view." "Absolute" is a forbidding word; epistemologists will be reticent about using it. "Universal" is not so "absolute"; perhaps we can claim that some of our moral insights, if perhaps not true in all possible worlds, are at least true elsewhere in our universe.

Extrapolating from our progress toward an inclusive global ethics, some insights in our human moral systems may be transhuman. Keep promises. Tell the truth. Do not steal. Respect property. Love your enemies; do good to those who hate you. Such commandments may be imperatives on other planets where alien species of moral agents inhabit inertial reference frames that have no contact with ours. Wherever there are moral agents living in a culture that has been elevated above natural selection, one can hope that there is love, justice, and freedom, although we cannot specify what content these activities will take in their forms of life.

In these worries about how comprehensive humans can be in their ethical systems, the focus on altruism is central, because that is particularly hard to explain. But perhaps we should recall that neither in deontological ethics nor in utilitarianism, the two main Western traditions, is altruism the sole ethical principle. The moral agent does what is just, giving to each his or her due; and whether this due is to self or other is secondary. The agent does the greatest good for the greatest number, which might mean benefits to self and/or to other, depending upon options available. The Golden Rule urges one to love neighbor as one does oneself, but this is not other love instead of self-love. "Do to others as you would have them do to you" seeks parallels in the self doing for others with others doing for the self, suggesting reciprocity as much as antithesis between self and other. Doing the right, the good is a matter of optimizing values, which often indeed means sharing them, but this is never simply a question of benefiting others instead of oneself. Ethics is about optimizing and distributing moral and other values. This is a more comprehensive question than whether the self is preferred over others or vice versa.

G. K. Chesterton states our conclusion forcefully:

> Reason and justice grip the remotest and the loneliest star. Look at those stars. Don't they look as if they were single diamonds and sapphires? Well, you can imagine any mad botany or geology you please. Think of forests of adamant with leaves of brilliants. Think the moon is a blue moon, a single elephantine sapphire. But don't fancy that all that frantic astronomy would make the smallest difference to the reason and justice of conduct. On plains of opal, under cliffs cut out of pearl, you would still find a notice-board, "Thou shalt not steal" (Chesterton [1911] 1976, 23).

It certainly does not follow that nothing generally true can appear in human morality because it emerges while humans are in residence on Earth. There is nothing particularly Earthbound about: "Do to others as you would have them do to you." That could be true whether or not the moral agents are *Homo sapiens*. If there are moral agents with values at stake in other worlds, this could be universal truth.

If visitors from outer space were to arrive and wish to set up a space station that required capturing humans as slaves to build their station, would we not urge "universal human rights" upon these extraterrestrials. If they proceeded, would we not censure them? If they could not understand our urging, we would think their intelligence inferior to ours. If they tried to justify their action by claiming their advanced intelligence, we would judge them morally retarded. If they also set up their space station in ways that required destroying a rich tropical forest ecosystem, filled with endangered species, would we not urge this non-Earthen species to respect Earth's biodiversity? If they did not, again, we would censure them and think them morally inferior.

The ethics that humans have reached, although certainly one appropriate to our Earthen residence, discovers values that are objective enough to urge on moral agents from whatever extraterrestrial origin. There would be something censurable about moral agents anywhere, anytime who lied, cheated, stole, hated, or were unjust. Or caused unwarranted pain in animals. Or destroyed endangered species without adequate justification. One can plausibly venture the claim that if there are moral agents anywhere, anytime, they have not matured until they have reached the capacity for altruism—indeed, if we listen to sensitive religious ethicists, until they have acquired the capacity for suffering love.

The surprising thing is that we humans have enough wit to do this, whether by revelation and prophecy, or by creative human inspirations and breakthroughs that rise to new levels of respect, caring, and compassion. As ethics matures, we are left wondering if it does not move beyond both nature and culture, glimpsing universals come to expression point on Earth. If there is life elsewhere, we have this logical warrant for expecting that even superintellects will be moral agents who have reached parallel convictions.

References

Chesterton, G. K. [1911] 1976. *The Innocence of Father Brown*. New York: Garland Publishing.
de Waal, Frans. 1996. *Good Natured: The Origins of Right and Wrong in Humans and Other Animals*. Cambridge, MA: Harvard University Press.
Kagan, Jerome. 1998. *Three Seductive Ideas*. Cambridge: Harvard University Press.
Kant, Immanuel. [1785] 1964. *Groundwork of the Metaphysic of Morals*. New York: Harper and Row.
Kummer, Helmut. 1980. "Analogs of Morality among Nonhuman Primates." In *Morality as a Biological Phenomenon*, edited by Gunter Stent, 31–47. Berkeley: University of California Press.
MacIntyre, Alasdair C. 1981. *After Virtue: A Study in Moral Theory*. Notre Dame, IN: University of Notre Dame Press.
Nowak, Martin A., with Roger Highfield. 2011. *Supercooperators: Altruism, Evolution, and Why We Need Each Other to Succeed*. New York: Free Press.
Polanyi, Michael. 1962. *Personal Knowledge: Towards a Post-Critical Philosophy*. New York: Harper and Row.
Rawls, John. 1971. *A Theory of Justice*. Cambridge, MA: Harvard University Press.
Ruse, Michael. 1994. "Evolutionary Theory and Christian Ethics: Are They in Harmony?" *Zygon: Journal of Religion and Science* 29:5–35.
Ruse, Michael, and Edward O. Wilson. 1985. "The Evolution of Ethics." *New Scientist* 108 (17 October):50–52.
Ruse, Michael, and Edward O. Wilson. 1986. "Moral Philosophy as Applied Science." *Philosophy* 61:173–192.
Singer, Peter. 1994. *Ethics*. New York: Oxford University Press.
Sober, Elliott, and David Sloan Wilson. 1998. *Unto Others: The Evolution and Psychology of Unselfish Behavior*. Cambridge, MA: Harvard University Press.
Trivers, Robert L. 1971. "The Evolution of Reciprocal Altruism." *The Quarterly Review of Biology* 46:35–57.
van Lawick-Goodall, Jane. 1971. *In the Shadow of Man*. Boston: Houghton Mifflin.
Wattles, Jeffrey. 1996. *The Golden Rule*. New York: Oxford University Press.
Williams, George C. 1988. "Huxley's Evolution and Ethics in Sociobiological Perspective." *Zygon: Journal of Religion and Science* 23:383–407.

Kenotic Ethics and SETI: A Present-Day View

George F. R. Ellis

Abstract Contact with extraterrestrials will raise acute issues regarding ethics—both theirs and ours. The evolutionary theory that ethics is solely about preserving one's genes is not the whole story, indeed it does not reflect the true nature of ethics. This chapter makes the case that the deep nature of ethics is kenotic, and that it is likely an advanced alien intelligence will recognize this—because long-term survival is dependent on such a stance. We should approach any contact with them on this basis.

Keywords Extraterrestrials · Ethics · Evolution · Moral realism · Kenosis · Human survival

As Doug Vakoch has pointed out, the Search for Extraterrestrial Intelligence (SETI) is crucially impacted by the ethical stance of any extraterrestrials we might some day encounter. This is to some degree a truism: H. G. Wells' *The War of the Worlds*, C S Lewis' trilogy (*Out of the Silent Planet, Perelandra*, and *That Hideous Strength*), and innumerable other science fiction novels, together with films such as *Independence Day, Alien, E.T., Contact,* and many others, all implicitly raise the issue, and more often than not assume hostility rather than friendliness (with Lewis' novels in particular being a notable exception). The pervasiveness of this theme in science fiction can perhaps dull us to the fact that this could indeed become a very real issue for humankind—literally a matter of life and death. If we did encounter them, which ethics would they follow? Perhaps we can approach the question now in a more scientific way than before, because of recent writing on the nature and origin of morality. Many of the points one makes in any systematic analysis of the issue are obvious, nevertheless it may be useful to have it set out in a systematic way.

G. F. R. Ellis (✉)
Mathematics Department, University of Cape Town, Rondebosch 7701,
Cape Town, South Africa
e-mail: George.Ellis@uct.ac.za

D. A. Vakoch (ed.), *Extraterrestrial Altruism*, The Frontiers Collection,
DOI: 10.1007/978-3-642-37750-1_14, © Springer-Verlag Berlin Heidelberg 2014

The first point is that any extraterrestrials we may encounter in the foreseeable future will either be communicating with us or will have travelled near to us, and so will be at least as advanced as we are technologically, or (more probably) will be technologically well in advance of us. There is thus good reason to believe that if they wished to destroy us, they would possess the technological means to do so, for almost every advanced technology can be applied towards constructive or destructive ends. As has always been the case, superior technology can usually easily win wars if turned to that end in a context where there are only enemies on the ground (one is not trying to keep the support of some of the affected population, so one can destroy the lot—unlike the situation in Vietnam and Iraq). Furthermore, it would be prudent for even a peace-loving species that had entered into the adventure of space travel to be prepared with some suitable weaponry in case they need to defend themselves, for they—like us—could not necessarily assume everyone they meet will be friendly.

The second point however is that if they seriously engage in space travel, they must have developed technological civilization for a very substantial time, during which time the race between ethics and technology will have been run and presumably effectively won. The issue is that as beings evolve consciousness and then become aware of other consciousnesses and cultures around them, it seems— at least on Earth—that the natural first reaction is to destroy such alien consciousnesses when one meets them, or at least to bring them into servitude, because they threaten a world view that one has only recently gained, and hence which is on precarious ground. This tendency is strengthened by any struggle for resources that may occur, where a zero sum game is envisaged by one or both sides: you have to get it, rather than allowing them to have it. The advance of technology then opens the possibility that one faction can completely annihilate the other; and as fundamental physics and biotechnology research proceeds, fundamentalists who believe in an Armageddon will probably eventually have the possibility of destroying all life on Earth.

However as experience is gained, many learn that not only can one survive the encounter with alternative worldviews, one can even gain from the associated widened understanding, indeed it can provide crucial new ideas and perspectives that greatly enhance one's own capabilities and understanding. Thus the innate fear of the other can evolve to an appreciation of their natures and capacities: an appreciation of difference, as characterized by Jonathan Sacks (2002) in his book *The Dignity of Difference*. Furthermore the struggle for resources need not be a zero sum game, in many cases all can gain in a symbiotic way. Additionally technology—if appropriately harnessed, and in particular if based on careful use of information rather than straightforward use of power—can greatly reduce the resource base we need to survive, and so remove the need to capture resources from others. The question, though, is which happens first: the development of sufficient technological capacity to destroy the others completely, or the ethical transition that enables one to relate to them and enjoy them?

This question is still open on Earth: we may easily destroy ourselves in the next thousand years or so, as the means of destruction becomes ever more readily available to fundamentalists of all persuasions, while ethical progress, if any, is slow—certainly very much slower than technological progress, as Sir John Templeton has famously remarked. However we on Earth are nowhere near being ready for serious interstellar travel: our technology has a long way to go before reaching that peak. In our case one can make the guess that if we survive long enough to have that technical ability, precisely because it will also bring with it ever enhanced possibilities of destructive capacities but we will have survived that long, we will by then have learned to live peacefully with each other. Perhaps this will involve some kind of uniformly accepted world government in association with generally accepted conflict resolution mechanisms that finally remove the need to go to war, substituting an ethic of sharing and compromise for that of possession and confrontation.

If this argument is accepted, we may then attempt to extrapolate to other civilizations that may evolve elsewhere in the galaxy. It seems likely that they too will face the need for an ethical transition in the same way, as their consciousness and their technology both evolve. If they fail this crucial test it is unlikely they will ever attain serious space travel—*inter alia* because the energy and resources that would have gone into that enterprise would in fact be taken up with deadly rivalries at home instead. This is of course not a logical proof of impossibility—maybe the imperialist armies envisaged in so many science fiction films could indeed manage to attain the necessary technology while also destroying their enemies and additionally holding their subjects in oppression. The question is whether they could hold a sufficient monopoly on destructive power; and on Earth that is very difficult because of the spread of communication abilities across the world that has arisen as a result of technology. One might think this would be a common feature of reasonably advanced technologies, partly because the progress of technology depends on innovative and individualistic people who do not lightly take to shackles being imposed on their thoughts and communication.

The third point, however, is that to be truly effective in terms of promoting tolerance and peace in a long-term situation, the ethic that develops must be *kenotic* (Murphy and Ellis 1996; Polkinghorne 2001): it must involve the capacity and willingness for forgiveness and reconciliation (Helminck and Petersen 2001; Worthington 2003), which in turn means giving up the need for revenge in order to attain the greater good. On occasion it can involve the ultimate of sacrifice on behalf of the other, even the enemy, because that is the way to turn the hardened heart, to convert an enemy to a friend, and so to create the true security that comes from being surrounded by friends rather than enemies. The implication is not that one is always sacrificing on behalf of others, but rather that one is prepared to do so when the context is such that a move of this kind can have a transformational quality. This moves the nature of ethics to a totally new regime: the arena of deep ethics that can indeed transform context and situation, in a way that is paradoxical because what was impossible in the old context becomes possible in the new. The hardened heart can indeed on occasion be touched and transformed. The suggestion I am making is that the ethical transition is only secure when it has this kenotic element—which is recognized as an aspect of the highest good by all the major religions on Earth (Templeton 1999). Thus if a civilization is both technologically and ethically advanced, this is the ethics one would expect to eventually dominate, both because it is the deep ethics that is eventually recognized by all truly deep spiritual strands in society, and because it is the only ethics that can create true security by fundamentally transforming the situation.

This is a considerable assumption here. The underlying issue is whether there are universal mechanisms that would underlie such transitions on planets other than our own. Would there be some reason why the nature of ethics would be the same there as here? Would they reasonably be expected to evolve to a kenotic stage? If we can make that case, we can back up the above argument.

This is where we encounter the claims of sociobiology and evolutionary psychology [see, e.g., the summary in Pinker (2003) and the relation to altruism in Post et al. (2003)]. They suggest that Darwinian evolution applied to behavioral patterns would not only lead to deadly competition between rivals, but would also lead to kin altruism that would encourage the start of an ethic of sharing and even in some cases sacrifice on behalf of kith and kin. It is the group gene that must be preserved, and individuals may give up on behalf of closely related others to let this happen. If that is true here on Earth, the logic must apply elsewhere: it is not based specifically on the nature of life on Earth, but on general tendencies in the evolutionary process that we might expect to hold on any planet where life evolves through a Darwinian process (and we believe that is the only process that can lead to complex life).

However this in fact is far from what is needed for the ethical transition noted above, because the very idea of saving the group's genes in the face of competition with others defines in-groups and out-groups, and gives no credence to the idea that one should not be hostile to out-groups. It sets the very stage on which intertribe and interracial conflict thrives (Pinker 2003), let alone interspecies conflict either on Earth or in an interplanetary setting. The selfish gene is basically selfish, if one buys into that viewpoint. If you run into competing genes, do them in before they do you in.

Luckily this bottom-up reductionist viewpoint is not the whole story. Consciousness and culture can overcome what genetically based tendencies there may be to destroy the other in order to preserve one's own genes—whether that aim is conscious or sub-conscious (Pinker 2003). We have choice, and that can override both genetic tendencies and culture if we work hard enough at it— attaining the transcendent state of "ecstasy" characterized by Peter Berger (1963) in his book *Invitation to Sociology*, whereby social and intellectual innovators can attain new insights that transform society. One of those insights is the extension of ethical understanding: from selfishness to kin altruism to ethical standards such as the Golden Rule, and on to the deep patterns of transformational ethics, where forgiveness and kenosis enter (Helminck and Petersen 2001; Polkinghorne 2001).

Several points are relevant. First, it is highly doubtful that sociobiology or evolutionary psychology can ever aspire to explaining development of this level of morality for a number of reasons, but in particular both because they do not take culture seriously (Pinker 2003), and more fundamentally because they in fact explain away morality rather than truly explaining it (Murphy 1998). Second, one can make a good case—as mentioned already above—that spiritual humans of whatever religious persuasion agree on this deep nature of ethics (Templeton 1999); it can be claimed to have the status of a cultural universal amongst those who take spiritual and moral issues seriously.

Those who understand this issue then believe that the same would be true also for beings on other planets: that is, the development of understanding of ethics would proceed along broadly similar lines everywhere, in the case of beings that had developed sufficient sensitivity to their environment and to each other over a sufficiently long time. So the final point is an expectation that conscious life on other planets, if it exists, would have comparable notions of kenosis.

What is the underlying basis for such an idea? It is the belief this is the true deep layer of ethics, which is universal and indeed imbedded in the nature of reality, as suggested by Murphy and Ellis (1996) in *On the Moral Nature of the Universe*. This is a realist view of ethics, which is what is also proposed by Stephen Pinker (2003) in *The Blank Slate*, because this is what is needed if ethics is to have a truly normative nature, independent of culture: that is, if ethics is to truly have the nature of a morality that makes claims on behavior of a universal nature for all humans. In that case, the development of ethical understanding is of the nature of a discovery of what is, rather than having the nature of the construction of a human invention.

So we can believe this ethics is universal: it should be found by all deeply developed civilizations near Alpha Centauri and in the Andromeda galaxy because it is a characteristic of the universe, in-built into its very structure in some way, just as much as are the laws of logic, energy conservation, and the number π (we don't know how this happens in any of these cases, but we do know this is the way things are). Understanding of the true nature of ethics would then be built up in any society on the basis of a foundation laid by sociobiological processes, but completed through social forces and individual understanding as the result of a process of discovery of the way things are, just as in the case of science and mathematics.

One can suggest a kind of moral realism without linking it into a religious foundation, as Pinker (2003) does. However to many it seems to make more sense to see deep ethics as having a religious foundation, if it is real; indeed that the profound kenotic nature of morality derives from the *Telos* or purpose of the universe, which in turn is an expression of the nature of God. This is the argument for a kenotic ethics laid out in depth in my book with Nancey Murphy entitled *On the Moral Nature of the Universe* (Murphy and Ellis 1996). The extra point then is that in this case, this underlying purpose would be discovered by religiously sensitive people partly through their sensitive responses to their experiences but partly assisted by some kind of process of revelation as envisaged in most spiritual traditions. The sense of what is the true nature of things underlying ethical action would be the same everywhere because it would be an expression of the universal nature of God, which is unchanging and perfect. This revelation would be conveyed to people through their religious experience and tradition, transforming them and enabling them in turn to act in a transformational way by being channels for the power of God [on this view, a central aspect of kenosis is letting go of one's own wishes and being open to the inner voice or "inner light" (Gorman 1973; Hubbard 1964)]. Thus a key issue would be what kind of religious views any aliens might have. On the view put here they would come to views that might be very different in intellectual terms but would have a deep ethical grounding that we could recognize and respond to, as for example in the life of Mahatma Gandhi and Martin Luther King.

If we accept the argument above, we can be hopeful about our meeting with aliens if it ever takes place: there may be reasonable grounds to hope they will have reached an advanced ethical state which will then perforce be kenotic. However as pointed out above, that does not mean they will be likely to be kenotic to the point of being suicidal: they probably will be armed and willing to use those armaments if necessary for their survival (sacrifice on behalf of the enemy is sometimes called for, but not always; whether it is appropriate or not depends on the circumstances). The final point then is that we will have a serious need to persuade them that we too are kenotic, and not locked in some dangerous ethical state where we will be a danger to their survival because we are too primitive to respond appropriately to a kenotic approach from their side.

Now as so ably pointed out by Carl Sagan (1985) in the book *Contact* (and the resulting film of the same name), we have made a bad start in this regard: in the sphere of TV signals spreading outwards from the Earth at the speed of light, the vanguard is the film of Hitler addressing the Olympic Games in Munich in 1935. Furthermore if aliens can decode the present-day stream of violence in the TV programs emanating from Hollywood and elsewhere, also spreading out into the space around us at the speed of light, they will almost certainly conclude that we are a race of homicidal maniacs that should either be strictly confined to Earth where we can kill ourselves to our hearts content without affecting others, or we should be destroyed to prevent our homicidal madness from spreading elsewhere. In other words, our own lack of ethical transition at the present time, profoundly manifest in the sickening stream of violence pouring out from our TV stations 24 h a day every day, could be a great danger to our future survival.

It is in this context that Vakoch's proposal to send out kenotic messages to interstellar space makes sound sense. It is the one hope we have of convincing anyone else out there that we are not totally demonic. It is then crucial to note that as well as sending out those more positive messages, we have to work in all ways possible to help the Earth as a whole to make the required ethical transition, so that those positive signals begin to correspond to reality on Earth. Indeed that is probably crucial to our survival, whether or not there is anyone else out there in position to decode our signals. For the reasons given above, our continued existence in a hundred, a thousand, or ten thousand years from now probably hinges on our ethical progress. Kenosis may well be the key to our long term survival. If we make that transition, we will be suitably ready for aliens when we eventually contact them: we will each recognize that kenotic state of enlightenment in each other (Sacks 2002).

References

Berger, Peter. 1963. *Invitation to Sociology*. New York: Doubleday.

Gorman, George H. 1973. *The Amazing Fact of Quaker Worship*. London: Friends Home Service Committee.

Helminck, Raymond G., and Rodney L. Petersen. 2001. *Forgiveness and Reconciliation: Religion, Public Policy, and Conflict Transformation*. Philadelphia: Templeton Foundation Press.

Hubbard, Geoffrey. 1964. *Quaker by Convincement*. London: Penguin.

Murphy, Nancey. 1998. "Supervenience and the Nonreducibility of Ethics to Biology." In *Evolutionary and Molecular Biology: Scientific Perspectives on Divine Action*, edited by Robert John Russell, William R. Stoeger, and Francisco J. Ayala, 463–489. Vatican City State: Vatican Observatory Publications, and Berkeley, CA: Center for Theology and the Natural Sciences.

Murphy, Nancey, and George Ellis. 1996. *On the Moral Nature of the Universe: Theology, Cosmology, and Ethics*. Minneapolis: Fortress Press.

Pinker, Stephen. 2003. *The Blank Slate*. London: Penguin Books.

Polkinghorne, John, ed. 2001. *The Work of Love: Creation as Kenosis*. Grand Rapids, MI: Eerdmans.

Post, Stephen G., Byron Johnson, Michael E. McCullough, and Jeffrey P. Schloss. 2003. *Research on Altruism and Love: An Annotated Bibliography of Major Studies in Psychology, Evolutionary Biology, and Theology*. Philadelphia: Templeton Foundation Press.

Sacks, Jonathan. 2002. *The Dignity of Difference: How to Avoid the Clash of Civilisations*. London: Continuum.

Sagan, Carl. 1985. *Contact*. New York: Simon and Schuster.

Templeton, Sir John. 1999. *Agape Love: A Tradition Found in Eight World Religions*. Philadelphia: Templeton Foundation Press.

Worthington, Everett L. 2003. *Forgiving and Reconciling: Bridges to Wholeness and Hope*. Downers Grove, IL: InterVarsity Press.

Altruism, Metalaw, and Celegistics: An Extraterrestrial Perspective on Universal Law-Making

Adam Korbitz

Abstract The existence of extraterrestrial intelligence (ETI) contemporaneous with the existence of the human species remains an unproven hypothesis. It is likely that until we achieve contact with ETI, we will know with certainty little or nothing about the ethics, morals, philosophy or laws and legal systems of advanced ETI civilizations. This lack of knowledge extends to fundamental questions regarding ETI behavior, such as whether an ETI civilization that is likely to be vastly older and more advanced will adopt an altruistic and benevolent disposition toward the human race or an egoistic, selfish and possibly destructive one. Efforts to speculate about the ethics, morals and legal precepts of ETI civilizations predate the theoretical foundations of the scientific effort now known as SETI. *Celegistics* is proposed as a reference term for explicating theories of ETI legal systems. Over 50 years ago, Andrew Haley first suggested the concept of Metalaw to refer to fundamental legal precepts of universal application to all intelligences, including ETI. In the decades since, several other writers have pursued this celegistic exercise and further developed Haley's ideas regarding Metalaw. All of these efforts have incorporated various assumptions about the altruism of ETI civilizations; the reliability of these assumptions is unknown because of the dearth of empirical evidence regarding the existence and nature of ETI. Various avenues (including celegistics), while limited due to their anthropocentric bias, may exist to allow us to further empirically explore the validity of our ideas regarding ETI's ethics, morals, and legal precepts, including Metalaw and its celegistic assumptions regarding extraterrestrial altruism.

Keywords Altruism · Andrew G. Haley · Celegistics · Ernst Fasan · Extraterrestrial intelligence · Metalaw · SETI · Thermoethics

A. Korbitz (✉)
CeleJure, 410 Midland Lane, Monona, WI 53716-3827, USA
e-mail: a.korbitz@att.net

D. A. Vakoch (ed.), *Extraterrestrial Altruism*, The Frontiers Collection,
DOI: 10.1007/978-3-642-37750-1_15, © Springer-Verlag Berlin Heidelberg 2014

1 The Unknown Ethics of Extraterrestrial Intelligence

If extraterrestrial intelligence (ETI) currently exists in our galactic neighborhood or elsewhere in the universe, its morals, ethics and laws are currently unknown to us, including whether or not it recognizes a concept similar to what humans have labeled altruism. (This chapter explores extraterrestrial altruism as a philosophical or ethical concept relevant to the human concepts of law and morality, as opposed to a biological, psychological or evolutionary concept.) In this discussion, the terms "morals" and "ethics" are used interchangeably and should be understood in a normative legal sense, describing laws or rules that compel or restrict behavior or otherwise regulate behavior between intelligent beings or between intelligent beings and non-intelligent entities, living or not. This chapter focuses on the history and the current state of human speculation regarding the laws or ethics of extraterrestrial intelligence (ETI) in the context of ethical altruism.

Few scientists or legal writers have published papers discussing the potential legal implications of human contact with ETI. The reasons for this are probably varied. One likely reason is that, more than fifty years into the SETI experiment, we have not yet achieved contact with ETI. If and when we do detect a confirmed radio or other signal from an advanced extraterrestrial civilization, we will know at least one thing: ETI exists—or at least did exist at the time the signal was sent (unless the signal was sent by a robotic beacon or sentry transmitting long after its creators became extinct). However, we may know more. We may have a good idea of the location of the signal's origin (its direction and approximate distance from Earth), telling us at least where ETI was located when the signal was first broadcast. If the signal is more than a carrier wave and contains intelligible information we are able to decode and understand, we may learn a great deal more. We may have some idea of the civilization's age, level of scientific and technological development, physical constitution, habitat, culture, law and ethics, just to name a few potential properties that may be illuminated by such a message. These would all be facts the human race could consider in debating and deciding whether or not to respond to such a message and how to respond.

Contrast such a hypothetical scenario with the reality facing us now in the context of discussing extraterrestrial laws, ethics and morality—including the (human) concept of ethical altruism. Anything we think we know about ETI is, at best, speculation, even if it is speculation informed by analogy from the only advanced technological civilization we know, the human one (Denning 2011). Stating the obvious and speaking of ETI specifically, we really have no empirical evidence (drawn from extraterrestrial examples) on which to base our ideas. The stark reality is that we do not even know if ETI exists anywhere in the universe, let alone in our own Milky Way galaxy. Therefore, we really know nothing about the morals, ethics, laws, motivation or capabilities of ETI, if ETI exists at all contemporaneously with our own civilization. At present, the very best we can do is engage in conjecture and thought experiments based on the human example, as well as non-human terrestrial examples of animal behavior. While valuable, the inherent limitations of such speculation should be kept in mind. Such speculation may or may not be predictive of extraterrestrial societies; we simply do not know.

However, the discussion of these issues is important because the questions asked and their potential answers have a direct bearing on many significant public policy questions, such as whether or not we should respond to an ETI signal if one is received, whether activities such as Active SETI or the deliberate broadcast of transmissions into space should be allowed, and principles of risk communication and perception relevant to discussing these issues with the public. Central to all of these questions is whether or not ETI would regard altruism toward the human race as an ethical or legal virtue.

2 Altruism and Extraterrestrial Law: Haley, Fasan, and Metalaw

The dearth of evidence regarding the ethical, moral or legal nature of ETI does not mean that we are utterly unprepared to explore this subject. Several hardy adventurers have set forth ahead of us to pioneer this largely undiscovered country. However, the trail does become more sparse when we narrow our focus to what legal or ethical principles, including altruism, may guide ETI in its dealings with other intelligent societies such as our own.

Three years before Cocconi and Morrison (1959) proposed the theoretical foundations for the 50-year experiment now known to the world as SETI, pioneering space lawyer Andrew G. Haley began exploring the potential moral and (more specifically) legal ramifications of what the human race might face if and when it succeeds in establishing contact with ETI. In an article published in the student newspaper of the Harvard Law School, Haley (1956) coined the term *Metalaw*, referring to fundamental legal precepts of theoretically universal application to all intelligences, human and extraterrestrial.

As his basic metalegal concept, Haley (1956) proposed that only one principle of human law could be projected onto our relations with ETI: the stark concept of absolute equity. In this same article, Haley first proposed his Interstellar Golden Rule: Do unto others as they would have you do unto them. As a metalegal principle, Haley rejected traditional formulations of the Golden Rule ("do to others as you would have done to you") as articulated by various religions and philosophers throughout the ages. Haley (1956) envisioned Metalaw as dealing "with all frames of existence—with sapient beings different in kind. We must do unto others in different frames of reference... To treat others as we would desire to be treated might well mean their destruction. We must treat them as they desire to be treated."

Haley (1963) developed these concepts somewhat further before his death in 1966, positing various derivative clauses from his Interstellar Golden Rule on a more or less *ad hoc* basis, such as one stating that no force of any kind may be used in establishing spatial relationships between civilizations. In general, however, Haley did not pursue further systematic explication of Metalaw after 1963.

Haley (1963, 413) deduced his metalegal principles primarily from Kant's Categorical Imperative and natural law theory in an effort (which others would find inadequate, as discussed below) to prepare the human race for dealing "with intelligent beings who are by nature different in kind and who live in environments which are different in kind. Although these propositions open great areas of juridical speculation, it is sufficient at this time to establish the simple proposition that we must forgo any thought of enforcing our legal concepts on other intelligent beings."

2.1 Ernst Fasan

Austrian space law pioneer Ernst Fasan (1970) significantly elaborated Haley's ideas in later decades. Several authors (Freitas 1977; Korbitz 2010; Sterns 2004; Vakoch 2007) have given extensive exposition to the work of both Haley and Fasan, which is briefly summarized here only to frame the discussion regarding the relationship of their ideas to extraterrestrial altruism. The reader is encouraged to consult the references for a deeper understanding of their concepts.

Fasan (1970, 42) defined Metalaw as being the entire sum of legal rules regulating relationships between different races in the universe. Fasan (1970, 41) also said Metalaw provides the ground rules for a relationship if and when we establish communication with or encounter another intelligent race in the universe. Like Haley, Fasan (1970, 41) envisioned these metalegal rules as governing both our conduct and that of an alien race in order to avoid mutually harmful activities; Metalaw is the first and basic "law" between interstellar societies.

Extrapolating from what he believed would be the essential characteristics of ETI, and building on Haley's initial work, Fasan (1970, 71–72) proposed the following rank order of eleven metalegal principles, starting with the strongest first.

1. No partner of Metalaw may demand an impossibility of another.
2. No rule of Metalaw must be complied with when compliance would result in the practical suicide of the obligated race.
3. All intelligent races of the universe have in principle equal rights and values.
4. Every partner of Metalaw has the right of self-determination.
5. Any act which causes harm to another race must be avoided.
6. Every race is entitled to its own living space.
7. Every race has the right to defend itself against any harmful act performed by another race.
8. The principle of preserving one race has priority over the development of another race.
9. In case of damage, the damager must restore the integrity of the damaged party.

10. Metalegal agreements and treaties must be kept.
11. To help the other race by one's own activities is not a legal but a basic ethical principle.

Fasan (1998, 678–679) later distilled a much simpler three-part formula from the above principles. That formula involves:

1. A prohibition on damaging another race.
2. The right of a race to self-defense.
3. The right to adequate living space.

3 Metalaw: Extraterrestrial Altruism in Practice?

Andrew Haley did not explicitly address the issue of altruism in his discussions of Metalaw. However, an obvious thread of ethical altruism runs throughout Haley's metalegal theories. Haley's basic metalegal principle, absolute equity between species, seems at first to lie squarely between ethical altruism (where the interests of the other are paramount to the interests of the actor) and ethical egoism (which elevates the interests of the actor over those of all others). Under Haley's Metalaw, the starting point is that both the interests of the other and the actor are equal. However, from the perspective of the actor (whether human or ETI), the balance seems to shift quickly in favor of altruism toward the other, because under Haley's Interstellar Golden Rule, interstellar actors are to conform their behavior to the norms and desires of those acted upon (the other intelligent race). The reasons for this appear to be explicitly altruistic: because an alien race may be different from us in ways we can only imagine, treating them the way we wish to be treated may actually be harmful to the alien race. The unstated assumption that it would be ethically wrong to harm the alien race (at least without good reason) lies behind other rules proposed by Haley, such as the general prohibition on the use of force.

This unstated assumption becomes more explicit in Fasan's rank-order list of eleven metalegal principles. Most of the principles in Fasan's list are founded directly in at least a modified form of ethical altruism, or at least an avoidance of harm to the other. On the other hand, some of the principles, namely the third, fourth, sixth, seventh and tenth, seem to be based squarely on Haley's stark concept of absolute equity between interstellar societies, balanced somewhere evenly between altruism and egoism. Fasan's eleventh principle is the most obviously altruistic (and is also given the lowest or weakest ranking in the list), stating explicitly that altruistic behavior toward another interstellar society is not a legal but a basic ethical principle.

Fasan's distinction between ethical altruism and legal duty has strong parallels in the legal system of the United States of America. While some seemingly altruistic behavior is legally enforced in American law (for example, parents must provide adequate care for their children, or at least must not neglect them), in general American law does not compel altruism. (Some people who pay taxes that support a variety of governmental initiatives might disagree and regard those taxes as enforced altruism.) Notable exceptions would be (in a negative sense) prohibitions on compensation beyond legitimate expenses for organ donations, adoption, or surrogate motherhood, in which case American law generally compels these activities to be altruistic if they are to be pursued. In other words, the activity itself (e.g., organ donation) is not compelled, but it generally cannot be engaged in for monetary profit by those donating.

An interesting illustration of the relationship between ethical altruism and legal duty was set forth by law professors William Landes and Richard Posner (1978). (Posner is now a judge on the U.S. Court of Appeals for the Seventh Circuit in Chicago.) They examined the role of altruism in the rescue of people or the property of strangers in high-transaction-cost situations. Employing an economic definition of altruism as any transfer of wealth that is not compensated, they suggested that society (at least American society of the 1970s) recognizes altruism as an economizing force that functions as a low-cost method of internalizing (for the rescuer) the external benefits of rescue (to the rescued). As such, it is more efficient as a motivator than legal measures, such as laws that impose liability for the failure to rescue strangers. At the time the paper was published, only one U.S. state, Vermont, had such a law. Altruism is also more efficient than laws that mandate a reward to the rescuer, which are also generally lacking. Exceptions noted by Landes and Posner involve unusually high costs to the rescuer, such as the salvage of property at sea, where the salvager is legally due a reward, and the payment of wages to medical personnel and other professionals whose careers involve the rescue or aid of strangers.

Landes and Posner contrasted situations such as the rescue of strangers, where altruism appears to internalize external *benefits*, with situations where altruism appears to be inadequate to internalize external *costs*, and legal duties are therefore imposed. They cite auto accidents as an example; the law does not rely on human altruism to avoid them. Rather, a driver is generally liable for any accidents he or she causes. The point of this contrast is that society has recognized that altruism alone is generally sufficient to motivate humans to aid strangers, especially when the benefit to the stranger is substantial compared to the cost to the rescuer, and that neither liability nor reward are generally required to so motivate people— subject to narrow exceptions for situations with a high cost to the rescuer. On the other hand, altruism is insufficient to motivate people to avoid accidents if they feel that the cost (to the driver) is potentially substantial. (Think of someone who recklessly speeds to avoid being late to work and possibly losing their job, disregarding the hazard they are creating for other people.) Therefore, Landes and Posner argue, the law imposes liability in such situations as a financial (and sometimes penal) premium that motivates accident avoidance; it does not rely upon altruism to do so.

In many ways, Metalaw's vision of the ethics and laws of ETI is optimistically altruistic (perhaps unreasonably so), as summed up in Haley's Interstellar Golden Rule and, to a lesser extent, in Fasan's various formulations. In light of Landes' and Posner's work on altruism, it is interesting that Fasan (1970) ranked his most explicitly altruistic principle as eleventh and therefore weakest. That principle states that altruistic behavior toward another race is not a legal but a basic ethical principle, just as Landes and Posner point out that American law does not generally impose a duty to aid strangers but relies upon them to be Good Samaritans out of altruistic motivations. Fasan ranks as the ninth (and therefore stronger) principle one that actually imposes something resembling legal liability; this principle states that a society that damages another must restore the integrity of the damaged party, something closely akin to make-whole remedies recognized under law. This would be analogous to Landes' and Posner's example of laws that impose liability on people who cause auto accidents, rather than relying upon altruism to deter such accidents.

Economists (like lawyers) have generally not discussed extraterrestrial societies, which is unfortunate because economic theory would appear to have much to offer the debate. One troubling observation that Landes and Posner make is that altruism is not a trait with a positive survival value in a competitive market; in an economic situation, competition will weed out altruistic and other high-cost actors. Seth Shostak (2010, 1028), who suggests that any extant ETI is likely to be of a machine nature, has posited that machine ETI will employ a "winner take all" survival strategy. In other words, the first machine intelligence to arise will dominate, at least in a given area of space in which communication is possible, such that a second genesis of machine intelligence would not be able to catch up. At least within a local area, the oldest machine culture may dominate.

Shostak's proposition that a "winner take all" survival strategy will attain among machine extraterrestrial cultures may raise the most troubling questions about Haley's and Fasan's visions of Metalaw, steeped as they are in the concept of "absolute equity" among intelligent races in the universe. If Shostak (2010, 1028) is correct in also suggesting that machine intelligence would have no obvious limit to its temporal existence, and might even undertake interstellar travel because of its immortality (thus having no obvious limit to its spatial existence either), the implications are troubling indeed for the metalegal foundation laid by Haley and Fasan. A kind of cut-throat competition, not altruism or even Haley's "absolute equity," may be the norm among ETI, and any society encountering ETI may put itself at a competitive disadvantage by engaging in altruistic behavior, per the suggestions of Landes and Posner (1978). Shostak's thesis regarding the likely machine nature of ETI points to a need to rethink the entire subject of ground rules that would govern our interaction with alien intelligences and their potential ethical attitudes regarding altruism (Korbitz 2010).

While Metalaw appears to put humanity's best foot forward (at least from the perspective of ethical altruism), we have no reason to conclude that ETI would or would not share such ethical views. Both Haley and Fasan relied on the Categorical Imperative and natural law theory in making this assumption. Whether their assumption is justified, we simply do not know. At the very least, Haley's suggestion that we cannot expect to enforce our legal concepts on other intelligent beings begs the question of whether that would ever even be possible, given interstellar distances and the current state of scientific thought regarding the likelihood that any ETI currently extant in our galactic neighborhood would be of much greater antiquity and technological development (Shostak 2010, 1026). In the end, all this speculation may simply be too anthropocentric to be descriptive or predictive of ETI's legal or ethical mores, including altruism.

4 Early Criticism of Metalaw

Several commentators have noted that Haley's formulation of Metalaw depends on subjective or relative (and therefore inadequate) concepts of "good" and "bad" (Lyall and Larsen 2009, 557). Others have noted that there is no guarantee another civilization would abide by Haley's observation that to treat others as we would desire to be treated might well mean the destruction of the other (the alien) (Reynolds 1992).

However, the sharpest criticisms of Metalaw have been leveled by those critical of its reliance on an approach to legal science and jurisprudence known as "natural law theory." An extensive discussion of natural law theory and alternative legal theories (such as legal positivism, legal interpretivism or legal realism) is beyond the scope of this chapter. At the risk of oversimplifying, in jurisprudence "natural law theory" refers to the view that links universal law directly to morality and proposes that just laws are immanent in nature and independent of the lawgiver, waiting to be discovered or found (as opposed to created by humans), usually by means of reason alone (as opposed to empirical data) (Robinson 1969). As Fasan (1970, 31) quotes Kant, "... moral principles are not based upon that which is typical of human nature, but must exist a priori of themselves, namely on principles from which can be deduced practical rules for every intelligent nature, and thus for the human one as well." Fasan (1970, 52) explicitly asserted that when discussing legal rules valid for every intelligent race in the universe, the starting point must be those principles which are deducible by and from reason alone.

The metalegal formulations of Haley and Fasan are squarely rooted in natural law theory and flow from Kant's Categorical Imperative in a largely deductive manner rather than being drawn from actual human legal institutions in an inductive fashion. Haley appeared to recognize the obvious anthropocentric limits of natural law theory but could not ultimately divorce Metalaw from this intellectual construct. This failure led one commentator to note that the cultural concept of rules or law is itself anthropocentric (Robinson 1969).

How these various shortcomings of Metalaw might be addressed in the future are discussed in the concluding section of this chapter.

5 Celegistics: Later Elaboration of Metalaw

5.1 Celegistics

Haley (1956) coined the term Metalaw to refer to fundamental legal precepts of theoretically universal application to all intelligences, human and extraterrestrial; Fasan (1970) later proposed that Metalaw is the first and basic "law" regulating interactions between interstellar societies. In the decades since, only a few writers have elaborated significantly on the metalegal ideas of Haley and Fasan. The work of two of these writers is summarized below, particularly as it bears on the issue of extraterrestrial altruism. But what do we call this activity of proposing or explicating universal legal rules that apply, not only to human society, but to extraterrestrial societies we now can only imagine in our science and our science fiction? What do we call speculation about ETI legal systems? There seems to be no term for it. For the reasons outlined below, the present writer proposes the term *celegistics*.

A *legist* is one who is expert in or studies the law, in particular ancient law. This seems like an appropriate place to start, since as Shostak (2010, 1026) points out, it is now a well-established truism of SETI that if SETI succeeds in detecting a signal from ETI, it will likely come from a civilization vastly older and more advanced than the human race—possibly millennia or more beyond our current level of technological development. It may make sense, then, to consider that other, non-technological aspects of an extant ETI, such as its legal system, would also be vastly older and more developed than human legal systems. In studying potential legal schema of extraterrestrial (and possibly universal) application, we may be talking about legal systems and rules that are actually quite ancient compared to human legal systems and rules. In fact, the antiquity of extraterrestrial legal schema (assuming they exist at all) may be the one characteristic of Metalaw we can predict with any degree of confidence.

Taking our etymological journey a step further, *legistics* is a term sometimes used to refer to collections of writings regarding the drafting or development of legislative texts or law. Married with the prefix *cele-* (as in celestial), which is derived from the Latin source of the words meaning "sky" in most of the Romance languages, we arrive at *celegistics*, or writings about "celestial" law of probably ancient origin (and possibly universal application).

5.2 Robert Freitas

The scientist and inventor Robert A. Freitas, Jr. was an early practitioner of what we may call celegistics. Like other critics of Metalaw, Freitas (1979) (by then also a newly minted lawyer) expressed his skepticism of Metalaw's reliance on Kant's Categorical Imperative and, by implication, Metalaw's reliance on the natural law theory of jurisprudence. Freitas (1979, Sect. 25.1.3) points out that Kant ignores the possibility of sentient beings of a qualitatively higher order than humanity, as posited later by Shostak (2010) and others. He suggests that Fasan falls into a similar anthropocentric trap by regarding human-level intelligence as the highest possible. Pointing out that multiple orders of higher sentience are possible (and quite likely, given the probable antiquity of extraterrestrial intelligence compared to humans), Freitas (1979, Sect. 25.1.3) succinctly asserts that Kant's Categorical Imperative cannot be valid for interactions among beings of qualitatively different orders of sentience, any more than it can be used to guide human relations with insects.

To overcome this shortcoming, Freitas (1979, Sect. 25.1.3) recasts Fasan's work using a concept Freitas dubs thermodynamic ethics or *Thermoethics*. In one sense, Thermoethics is a recasting of natural law theory that attempts to divorce natural law from its anchor forged from human moral sensibilities and give it something resembling an empirical basis. Freitas' basic thesis is that all living beings seek to reduce the degree of entropy or disorder in the universe. From this basic idea, Freitas (1979, Sect. 25.1.3) draws the Principal Thermoethic, which states that intelligent beings should act so as to minimize the total entropy of the universe; to put it another way, intelligent beings should seek to maximize the total negentropy, or negative entropy.

In Freitas' scheme, societies that most reduce entropy are the most ethical. This gives rise to his Corollary of Negentropic Equality, which states that entities of equal negentropy have equal rights and responsibilities; the more negentropic an entity, the greater are its rights and the deeper are its responsibilities (Freitas 1979, Sect. 25.1.3). This appears to be a significant departure from Haley's stark concept of absolute equity, but rather links a society's rights and duties and as well as its level of ethical development to its level of negentropy.

From these two ideas—the Principal Thermoethic and the Corollary of Negentropic Equality—Freitas (1979, Sect. 25.1.3) formulates three thermoethical Canons, which he believes should apply to interactions between all intelligent beings, a priori:

Canon I: Actions that increase the entropy or disorder of another society should be avoided.
Canon II: Every society holds its negentropy (or information) in trust for all intelligent beings, and should do everything possible to preserve it.
Canon III: Actions that increase the negentropy or order of another race should be carried out.

In a manner similar to Fasan, Freitas (1979, Sect. 25.1.3) then helpfully simplifies these as follows:

Canon I. *Destroy not.* (Avoid harming, if it is at all possible.)
Canon II. *Preserve*, if in preserving you do not destroy.
Canon III. *Create*, if without harm and the creation may be preserved.

Freitas (1979, Sect. 25.1.3) then uses each canon to generate a number of specific metalegal laws governing particular situations, most of which roughly parallel if not mirror Fasan's earlier rank-order listing of metalegal principles discussed above but recast them in Freitas' Thermoethics language. Several of Freitas' metalegal laws are explicitly altruistic. For example, Freitas' first metalegal law, the Entropic Censorship Rule, prohibits a society that comes into contact with another for the first time from sharing any information that may cause harm to the other. This is roughly parallel to Fasan's fifth metalegal principle.

5.3 G. Harry Stine

Another early practitioner of celegistics was the science fiction author G. Harry Stine. Stine (1980, 44–45) renamed Haley's Interstellar Golden Rule (or "First Principle of Metalaw") and dubbed it, simply, Haley's Rule; he also proposed (in a manner not unlike Freitas) collapsing Haley's Rule and Fasan's eleven principles of Metalaw into six general precepts, which he viewed as a system rather than a set of distinct rules.

Stine (1980, 43) thought Haley's Rule is not just the only way to treat ETI, but that it is also the only way to treat other human beings; doing unto others as they would have you do unto them demands a respect for others as individuals and a respect for their cultural background. Stine (1980, 45–47) further suggested the metalegal concepts first proposed by Haley and Fasan might be models not only for some possible future legal relationship with an advanced extraterrestrial civilization, but could also illuminate human interpersonal relationships and international diplomacy on Earth as well. Stine (1980, 42) suggested that the human invention of law (and lawyers) may be humanity's single most important advance in avoiding deadly and bloody conflict. Rather than relying routinely upon force or violence to address conflict at the interpersonal level, human societies have developed specialists in human conflict: lawyers, attorneys, barristers, solicitors, advocates and other legists. These professionals write the laws our society lives by, and also interpret and enforce them (sometimes by using other professionals, such as courts and the police).

Stine later revisited Metalaw in a science fiction novel, which he published under his science fiction pen name, Lee Correy (1986). Stine had obviously given some additional thought to Metalaw in the years since he published his original 1980 article on Metalaw. His later novel begins with a preface stating his revised metalegal principles, which he also renamed "Canons," adding two to the original six. Below are Stine's Canons from his 1986 novel (his sixth principle from the 1980 version became his Seventh Canon in the 1986 rework, and he added the new Sixth and Eight Canons in the list below) (Correy 1986, 7–8):

First Canon (Haley's Rule): Do unto others as they would have you do unto them.
Second Canon: The First Canon of Metalaw must not be applied if it might result in the destruction of an intelligent being.
Third Canon: Any intelligent being may suspend adherence to the first two Canons of Metalaw in his own self-defense to prevent others from restricting his freedom of choice or destroying him.

Fourth Canon: An intelligent being must not affect the freedom of choice or the survival of another intelligent being and must not, by inaction, permit the destruction of another intelligent being.

Fifth Canon: Any intelligent being has the right of freedom of choice in life style, living location, and socio-economic-cultural system consistent with the preceding canons of Metalaw.

Sixth Canon: Sustained communication among intelligent beings must always be established and maintained with bilateral consent.

Seventh Canon: Any intelligent being may move about in a fashion unrestricted by other intelligent beings provided that the Zone of Sensitivity[1] of another intelligent being is not thereby violated without permission.

Eighth Canon: In the event of canonical conflict in any relationship among intelligent beings, the involved beings shall settle said conflict by non-violent concordance.

Stine's elaboration of Metalaw in both his original 1980 article and his 1986 novel is creative and imaginative, and his suggestion that metalegal principles could inform not only interpersonal relations but international relations is intriguing if possibly a tad too optimistic. Stine owes an obvious debt to several metalegal thinkers who preceded him. Stine (1980, 43–44) credits not only Haley and Fasan but also Isaac Asimov's (1950) Three Laws of Robotics, which clearly have echoes in Stine's formulations of Metalaw. Stine (1980, 44) also acknowledges the work of Robert Freitas, whose expansion and systematization of Fasan's eleven metalegal precepts enjoys an obvious priority in time over Stine's.

6 The Relevance of Altruism in Extraterrestrial "Law": Why Metalaw and Celegistics Matter

6.1 Reminding Ourselves of What We Do Not Know

Metalaw, as first proposed by Haley and Fasan and later elaborated upon by early pioneers of celegistics such as Freitas and Stine, makes many assumptions regarding the presumably altruistic behavior of any ETI that may exist. For Haley and Fasan, these assumptions were drawn from Kant's Categorical Imperative and natural law theory, a philosophical foundation that has been vigorously criticized by others for both its anthropocentrism and its lack of an empirical basis. Freitas

---

[1] For differing definitions of this term as used by Stine/Correy in 1980 and 1986, see (Stine 1980, 44–45; Correy 1986, 7). Stine's concept (although not its precise definition) of a Zone of Sensitivity expressed in both 1980 and modified in 1986 actually originated in Haley's work. Haley (1963, 418) intended the term to refer to "the closest distance of approach outside of which no possible effect can be exerted upon a hypothetical being." Haley considered areas outside such zones to be free space open to all, subject to an interstellar "freedom of the seas."

(1979) attempted to address this weakness by founding Metalaw in his concept of Thermoethics, which elevated negentropy as the pre-eminent ethical precept over both equality and altruism. Stine restored both equality and altruism as central tenets of Metalaw in his formulations, but the basis for his doing so seems unclear.

Korbitz (2010) has criticized existing formulations of Metalaw for failing to anticipate the metalegal implications of the likely machine nature of any ETI the human race may one day communicate with or otherwise encounter, as proposed by Shostak (2010). A similar critique could be made regarding Metalaw's uncritical assumptions, in its various formulations to date, that ETI shares either altruism or equality as ethical virtues. (To be sure, making the opposite assumption, that ETI would be egoistic, or selfish, or predatory, is equally uncritical in the absence of empirical data about ETI.) The bottom line is that we simply do not know. Virtually all arguments, for or against altruism, or equality, or selfishness, are based on human or at least Earth-bound examples. Unfortunately, such examples are all we have, but their predictive or explanatory power is currently unknown when it comes to ETI. It is worth taking stock of how little we actually know about life in the universe, let alone intelligent life. Standing as a backdrop to this review, of course, is the stark reality that we do not even know if ETI exists at all. Fasan (1990) correctly noted that we can only speculate regarding the physical and intellectual characteristics of ETI. As Fasan (1970, 55–56) earlier stated, "Without knowing anything about the different intelligent races of the universe, we cannot therefore definitely establish rules of Metalaw."

6.2 How We Might Discover More Through Celegistics

If SETI does achieve success in this century or later, that success may come as predicted under the established and accepted SETI paradigm, whereby we detect a radio or perhaps laser signal sent in our direction by a distant civilization, probably hundreds or thousands of years ago. Metalegal questions will not likely be very pressing under this scenario until someone on Earth decides to respond (which is probably inevitable). As Douglas Vakoch (2011) points out, it generally requires two actors to initiate a legal relationship. Patricia Sterns (2004, 125) has stated the transmission of a response will formally engage the Earth and ETI in a form of legal relationship. Even so, under this scenario it will be a long-distance and long-term relationship likely taking place on a temporal scale of many generations. In the context of considering regulations regarding possible success of the established SETI paradigm, Vladimir Kopal (1990) has suggested that such a legislative process can only begin when the boundary between mere possibilities and well-established realities has been crossed. Space lawyers will have to depend on the advice of scientists as to when and to what degree the emerging problems posed by contact with ETI should be considered real and urgent, Kopal (1990, 125) concluded. We may not know how urgent, however, until a signal from ETI is actually detected (Billingham 1998).

The established SETI paradigm is not the only possibility for contact, however. Paul Davies (2010) has urged that SETI re-evaluate and consider expanding its methodologies to include possibilities such as searching the L4 and L5 Lagrange points for signs of extraterrestrial probes of probable ancient origin; he has also proposed that ETI might utilize communication methods that include nanoprobes of microscopic scale, viruses, or even coded messages in genetic material. The metalegal implications of such possibilities may be mundane unless communication with ETI on a timescale that fits at least within a human generation or so is somehow possible. Fasan (1970, 30), like Davies (2010), speculated that "contact" may take the form of the discovery, on Earth or elsewhere in our solar system, of some ancient technology left long ago by a transiting civilization. The metalegal questions in such a situation would also not be very pressing unless there was reason to think that ancient ETI or its descendents still inhabited our galactic neighborhood.

The potential deficiencies of current metalegal concepts stemming from the failure to realistically assess ETI's potential ethical constitution, including its disposition toward altruism or egoism, are not just an abstraction. As evidenced by our failure to respond with effective strategies to deal with climate change, the human race has proven it can be very poor at long-term planning. We may be even worse at preparing for eventualities that, while uncertain or unlikely, are still within the realm of possibility and have a high potential impact if they do happen. Contact with ETI would appear to fit within this category (Almár and Tarter 2011). However, failure to prepare virtually guarantees that we will be unprepared if the unlikely occurs. While a great deal of thought has been put into how various human institutions would react to the discovery of, or contact with, ETI (Michaud 2007), little thought since Haley and Fasan has been put into what impact such an event would have on human legal institutions or what rules would govern such a relationship. Davies (2010, 194) notes that, if we ever do make contact with an advanced extraterrestrial civilization, the aliens will approach godlike status in our eyes. We should at least contemplate the legal ramifications of that possibility. Lawyers and scientists should both begin to explore this subject, not only to be prepared for the unlikely, but to learn what that process can tell us about ourselves and our legal culture. Echoing similar thoughts by Harry Stine, Kopal (1990, 125) has noted that the task of elaborating rules to govern relations between humans and ETI (what this chapter calls *celegistics*) would add a new dimension to space law that could also have a beneficial influence on relations between the nations and peoples of Earth.

If we know little about the very existence of extraterrestrial civilizations, we know even less about the ethics, morals, laws, motivations or intentions of ETI or what value they may place on principles similar to the human concepts of altruism or equality. Andrew Haley and Ernst Fasan were pioneers in their field, but their celegistic work remains unfinished. Unfortunately, in recent decades few lawyers or other writers beside Robert Freitas and Harry Stine have shown interest in continuing these labors or venturing into the intellectual terrain of celegistics that Haley and Fasan first courageously explored.

If space lawyers and other jurisprudents wish to work side by side with scientists engaged in SETI, they should adopt the methods and approaches of science. This may mean abandoning natural law theory as the intellectual foundation of metalegal concepts and precepts, along with its reliance on a priori and deductive reasoning. If law is a science (as jurisprudents like to claim), it is a social science. Indeed, A. A. Cocca (1998) has argued that law is the oldest social science, dating from when humans first interacted, those interactions ultimately giving rise to the fields of criminal law, family law, civil and commercial law (among others). Those studying law and its application to SETI and related scientific endeavors should at least adopt the inductive and empirical methods of other social scientists.

Fortunately, several schools of legal thinking exist that rely, at least in part, on observation of what lawyers, judges, and lawmakers do in actual practice. These include the school of legal interpretivism (Dworkin 1986) as well as the critical legal studies movement (Unger 1986) along with its progenitor, American legal realism (Nourse and Shaffer 2009). All three schools share, at least in part, an empirical approach to law that supports the notion that "law" ultimately is what lawyers and legal systems actually *do* in practice, not a system of ethereal moral precepts waiting to be grasped, a priori, by legal philosophers.

The human race is the only law-based civilization currently known to us. As such, it is our only empirical source of knowledge and insight about how legal relationships and legal institutions work (Reijnen 1990). The lessons drawn from human examples are obviously anthropocentric, but they are the best we can do with an example of one. Generalizing or analogizing from the human example to the extraterrestrial may be our only empirical avenue under these circumstances. This limitation has not stopped other fields, such as the study of human religions (Dick 2000), from utilizing the lessons of human experience to illuminate our study of what ETI may be like. As another example, see Albert Harrison's (1997) extensive application of James Grier Miller's Living Systems Theory to extraterrestrial civilizations.

George Robinson (1969, 270) vigorously criticized Haley and Fasan for relying on natural law theory as the intellectual foundation of Metalaw and urged that space lawyers, when examining the possibilities of our legal relationships with ETI, adopt an empirical approach similar to that used, for example, by cultural anthropologists; he argued for an empirical analysis of Metalaw by studying human values formed with respect to totally alien concepts in all ecological and cultural situations.

Possibilities for new directions in empirical studies related to Metalaw are easily imagined. Principles illuminating celegistic and metalegal concepts might be elaborated further by analogizing from legal concepts established or under development in fields involving relationships with the alien "other," such as international and diplomatic relations (Michaud 1972; Goodman 1990), artificial intelligence (Hall 2007; Kurzweil 2005), robotics (Allen and Wallach 2009), transhumanism (Launius and McCurdy 2008) and even animal rights (Garner 2005). It might also be instructive to examine ethical principles and guidelines (in particular those related to altruism) that have been suggested to apply to our potential discovery of non-intelligent exobiological life forms in our solar system (Race and Randolph 2002). However, the severe predictive and descriptive limitations of such investigations must also be remembered when it comes to speculation regarding the ethical constitution of ETI civilizations.

References

Allen, Colin, and Wendell Wallach. 2009. *Moral Machines: Teaching Robots Right from Wrong.* New York: Oxford University Press.
Almár, Iván, and Jill Tarter. 2011. "The Discovery of ETI as a High-consequence, Low-probability Event." *Acta Astronautica* 68(3–4):358–361.
Asimov, Isaac. 1950. *I, Robot.* New York: Doubleday & Company.
Billingham, John. 1998. "Cultural Aspects of the Search for Extraterrestrial Intelligence." *Acta Astronautica* 42(10–12):711–719.
Cocca, A. A. 1998. "Legal Science as Catalyzer of SETI Science, Engineering and Operations." *Acta Astronautica* 42(10–12):671–675.
Cocconi, Giuseppe, and Philip Morrison. 1959. "Searching for Interstellar Communications." *Nature* 184(4690):844–846.
Correy, Lee. 1986. *A Matter of Metalaw.* New York: Daw Books.
Davies, Paul. 2010. *The Eerie Silence.* New York: Houghton Mifflin Harcourt.
Denning, Kathryn. 2011. "Ten Thousand Revolutions: Conjectures about Civilizations." *Acta Astronautica* 68(3–4):381–388.
Dick, Steven, ed. 2000. *Many Worlds: The New Universe, Extraterrestrial Life & the Theological Implications.* Philadelphia: Templeton Foundation Press.
Dworkin, Ronald. 1986. *Law's Empire.* Cambridge, MA: Belknap/Harvard University Press.
Fasan, Ernst. 1970. *Relations with Alien Intelligences: The Scientific Basis of Metalaw.* Berlin: Berlin Verlag, pp 71–72.
Fasan, Ernst. 1990. "Discovery of ETI: Terrestrial and Extraterrestrial Legal Implications." *Acta Astronautica* 21(2):131–135.
Fasan, Ernst. 1998. "Legal Consequences of a SETI Detection." *Acta Astronautica* 42(10–12):677–679.
Freitas, Robert A. 1977. "Metalaw and Interstellar Relations." *Mercury* 6(March–April):15–17.
Freitas, Robert A. 1979. *Xenology: An Introduction to the Scientific Study of Extraterrestrial Life, Intelligence, and Civilization.* Sacramento, CA: Xenology Research Institute. Accessed August 29, 2012. http://www.xenology.info/Xeno.htm .
Garner, Robert. 2005. *The Political Theory of Animal Rights.* Manchester: Manchester University Press.
Goodman, Allan E. 1990. "Diplomacy and the Search for Extraterrestrial Intelligence (SETI)." *Acta Astronautica.* 21(2):137–141.
Haley, Andrew G. 1956. "Space Law and Metalaw—A Synoptic View." *Harvard Law Record,* November 8.
Haley, Andrew G. 1963. *Space Law and Government.* New York: Appleton Century Crofts.
Hall, John S. 2007. *Beyond AI: Creating the Conscience of the Machine.* Amherst, NY: Prometheus.
Harrison, Albert A. 1997. *After Contact: The Human Response to Extraterrestrial Contact.* New York: Plenum Trade.
Kopal, Vladimir. 1990. "International Law Implications of the Detection of Extraterrestrial Intelligent Signals." *Acta Astronautica* 21(2):123–126.

Korbitz, Adam C. 2010. "The Limits of Metalaw and the Need for Further Elaboration." Paper presented at the 39th Symposium on the Search for Extraterrestrial Intelligence, 61st International Astronautical Congress, Prague, Czech Republic, September 27–October 1.

Kurzweil, Ray. 2005. *The Singularity Is Near*. New York: Penguin.

Landes, William M., and Richard A. Posner. 1978. "Altruism in Law and Economics." *The American Economic Review* 68(2):417–421.

Launius, Roger D., and Howard E. McCurdy. 2008. *Robots in Space*. Baltimore, MD: Johns Hopkins University Press.

Lyall, Francis, and Paul B. Larsen. 2009. *Space Law: A Treatise*. Burlington, VT: Ashgate Publishing Company.

Michaud, Michael A.G. 1972. "Interstellar Negotiation." *Foreign Service Journal* 49(December):10–14, 29–30.

Michaud, Michael A.G. 2007. *Contact with Alien Civilizations: Our Hopes and Fears about Encountering Extraterrestrials*. New York: Copernicus Books.

Nourse, Victoria F., and Gregory C. Shaffer. 2009. "Varieties of a New Legal Realism: Can a New World Order Prompt a New Legal Theory?" *Cornell Law Review* 95:61–138.

Race, Margaret S., and Richard O. Randolph. 2002. "The Need for Operating Guidelines and a Decision Making Framework Applicable to the Discovery of Non-intelligent Extraterrestrial Life." *Advances in Space Research* 30(6):1583–1591.

Reijnen, G. C. M. 1990. "Basic Elements of an International Terrestrial Reply Following the Detection of a Signal from Extraterrestrial Intelligence." *Acta Astronautica* 21(2):143–148.

Reynolds, George H. 1992. "International Space Law: Into the Twenty-First Century." *Vanderbilt Journal of Transnational Law* 25:225–255.

Robinson, George S. 1969. "Ecological Foundations of Haley's Metalaw." *Journal of the British Interplanetary Society* 22:266–274.

Shostak, Seth. 2010. "What ET Will Look Like and Why Should We Care." *Acta Astronautica* 67:1025–1029.

Sterns, Patricia M. 2004. "Metalaw and Relations with Intelligent Beings Revisited." *Space Policy* 20:123–130.

Stine, G. Harry 1980. "How to Get Along with an Extraterrestrial... or Your Neighbor." *Analog Science Fiction/Science Fact* 2(February):39–47.

Unger, Roberto M. 1986. *The Critical Legal Studies Movement*. Cambridge, MA: Harvard University Press.

Vakoch, Douglas A. 2007. "Metalaw as a Foundation for Active SETI." Paper presented at the International Institute of Space Law Session on the 40th Anniversary of the Outer Space Treaty, and Other Legal Matters, 58th International Astronautical Congress, Hyderabad, India, September 24–28.

Vakoch, Douglas A. 2011. "Responsibility, Capability, and Active SETI: Policy, Law, Ethics, and Communication with Extraterrestrial Intelligence." *Acta Astronautica* 68(3–4):512–519.

A Logic-Based Approach
to Characterizing Altruism
in Interstellar Messages

Alexander Ollongren

Abstract This chapter discusses the encoding (characterization) of the human notions of morality and altruism using a system for interstellar communication—in fact, some *Lingua Cosmica*. In order to show how that might be achieved, two existing, extensively documented linguistic systems of this kind, LINCOS and NEW LINCOS, are considered. The foundations of these systems, briefly reviewed, clearly show that they are completely different from one another—in conceptual setup and in use. In the LINCOS system, encoded conversations between humans are the means for representing information, worked out for the field of elementary mathematics, but also applicable for describing aspects of human behavior, such as altruism. This point of view implies that for understanding message content, a substantial amount of reflection on the basics of the linguistic system is mandatory at the receiving end of a communication line. The system is self-contained, but also flat because no recourse to external sources of information is required. In the astrolinguistic system NEW LINCOS, the encoding of information is by using constructive logic. Therefore, receivers of an encoded message on behavior in that system need first of all to recognize that logic is involved: this is a necessary condition for the interpretation of message content. In view of the simplicity of the logic employed, it can be assumed that receivers can achieve relatively quickly, without much guessing, at least some understanding of the basic conventions, and subsequently an appreciation of message structure and content. The system is self-contained, multi-level and not flat. In the second part of this chapter, characterizing altruism in some detail from a logic point of view is considered, using NEW LINCOS as a means for the encoding. In both systems many examples of encoded human behavior can be included in messages and redundancy need (should) not be avoided. Both LINCOS and NEW LINCOS are not unsuitable for expressing human *morality* and *altruism*—even if the necessary encoding might not be perspicuous in all respects.

A. Ollongren (✉)
Advanced Computer Science, Leiden University, Niels Bohrweg 1 2333 CA Leiden,
The Netherlands
e-mail: Gunvor.Ollongren@ziggo.nl

D. A. Vakoch (ed.), *Extraterrestrial Altruism*, The Frontiers Collection,
DOI: 10.1007/978-3-642-37750-1_16, © Springer-Verlag Berlin Heidelberg 2014

Keywords Astrolinguistics · Constructive logic · Intuitionistic logic · *Lingua Cosmica*

1 Introduction

The present chapter, a contribution to current discussions in the field of message design for interstellar communication, is concerned with the problem of describing some well-defined aspects of human behavior by means not strictly related to natural languages. Abstractions are the means preferred here. A strong motivation for such practice is the fact that an extraterrestrial party at the receiving end of a communication line can safely be assumed to be completely unacquainted with languages as they are spoken, written and understood by Earthlings. This applies evidently to concrete *surface* structures of our languages (sentences themselves, written or uttered in some natural, terrestrial language in the Chomskyan sense), but might very well also be relevant for the underlying abstract *deep* structures, i.e., the conceptual patterns in the human brain from which sentences are derived.

On the other hand, it can be argued that in case interstellar textual messages plainly written or encoded in some specific natural language would be emitted by us from Earth, receivers might possibly understand that indeed one of our languages is involved. Messages of such kind could then perhaps be helpful for recipients to gain an understanding of some aspects of the particular language used. How much they might be able to comprehend is another matter, and would depend on the universality of linguistic constructs as employed by Earthlings—an unknown quality. Textual messages of this kind should therefore somehow be supplemented with extra-linguistic information, providing an aid for intelligent extraterrestrials for the task of understanding the semantic message contents, even if only partly. In addition to this, designers of messages written in some natural language should keep in mind the existence of multitudes of subtleties in the grammars of our languages and avoid as much of them as possible in message construction.

In view of these considerations, it is a rather natural step in the area of message design to consider using some kind of *Lingua Cosmica* for the purpose of explaining for extraterrestrial intelligence aspects of our behavior and ourselves.

2 History of *Lingua Cosmica*

A *Lingua Cosmica* should be based on a "common ground," i.e., a body of knowledge assumed to have universal validity, and expressible using a formal system of notation. To date, two comprehensive linguistic systems of this kind have been proposed and worked out in much detail by Hans Freudenthal (1960) and Alexander Ollongren (2013).

Prof. Freudenthal [*1905–†1991], from Utrecht University in the Netherlands, was a renowned all-round mathematician of German origin who had, by the way, a special interest in education in mathematics at the high school and early college level. His interest in the development of a *Lingua Cosmica* was a challenging sideline. He, or at least his publisher and the editors Professors L. E. J. Brouwer, E. W. Beth and A. Heyting (all of them Dutch, well-known important mathematicians and logicians, most of them of the same generation as Freudenthal's), considered this project to belong to the field of *logic*. His book LINCOS was therefore published in the series Studies in Logic and the Foundations of Mathematics. In hindsight this is rather remarkable in view of the fact that formal logic is not prominently present in the project.

Alexander Ollongren [*1928–], the author of the present chapter, emeritus professor in Theoretical Computer Science and Dynamical Astronomy from Leiden University in the Netherlands, is of Swedish-Russian descent and was educated in a multilingual family setting. He has always been interested in languages and logic, and got deeply engaged in *constructive logic* (the base of his *Astrolinguistics*) after leaving his chair at the Institute of Advanced Computer Science at Leiden University in 1993. Constructive logic has its roots in constructive mathematics, more in particular in the mathematical philosophy of intuitionism, introduced by Prof. L. E. J. Brouwer in Amsterdam in his dissertation in 1907 and developed by him and his followers in the first half of the century. Notably, in intuitionism the law of the excluded middle is denied. That is also the case in constructive logic. The present author did get acquainted with the ideas of the intuitionists by attending lectures by Prof. Heyting from Amsterdam in the fifties, but realised later that they were not really relevant for his interest in Hamiltonian dynamics at the time. Important advances in constructive logic were achieved with the realization of computer implementations of (intuitionistic) type theories from the end the twentieth century onwards at INRIA in France (e.g., Coq Development Team 2010).

In the present chapter, Freudenthal's *Lingua Cosmica* is referred to as LINCOS and to the author's setup as NEW LINCOS in order to keep the linguistic systems apart from one another. The two systems are very different, because of the *common grounds* chosen but also because of the topics of human behavior treated and the way they are expressed. *Grosso Modo,* the mentioned common ground of the first system, is mathematics in a general sense and a restricted part of formal logic, while the second system is based on computer-implementable intuitionistic (or constructive) logic, a relatively recent development in formal and applied logic. LINCOS was published at the time when the first projects Search for Extraterrestrial Intelligence (SETI) got going, and set a landmark for the area of message construction for ETI. It has great merits, but the chosen setup implies that the instruments used for expressing human behavior are not easily interpretable for receivers of messages. NEW LINCOS, on the other hand, is relatively young and should be much more easily interpretable.

3 Human Behavior Expressed in Freudenthal's LINCOS

The book contains chapters on mathematics, time, behavior, space-motion-mass. In the third chapter (Freudenthal 1960, 90–167), about one-third of the book, human behavior is discussed *in extenso*. The chapter sets out with an interesting statement:

> For the time being it would be premature to try to describe human behaviour by a system of general rules.... Instead we shall *show* behaviour by quasi-general examples, from which the receiver may derive as many general behaviour rules as he pleases. The things to be shown are events. The program of showing events could be transformed into a program of reporting on events (Freudenthal 1960, 90).

The claim that specific (but rather general) behavior rules can be formulated in LINCOS, means that the linguistic system lays a claim for a kind of *universality*— but of course no proof of that has been given. The remark on programs can be interpreted as follows: a program showing events consists of a sequence of instructions that are or can be carried out (e.g., by an information processing machine), but a program can also be a recording a sequence of events that have happened. The recording can serve as a *memory*—to be used whenever necessary. In the present chapter, we will not distinguish between programs showing and those reporting on events.

Events, basic entities in LINCOS, are considered to be *acts of speech* carried out by *persons*, humans with names, e.g., *Ha*, *Hb*, *Hc*. .... Human behavior is modeled by events of various kinds. Let *P* be some proposition (i.e., in LINCOS something that can be said). An example of a basic (elementary) event is then

Ha Inq *Hb*. *P*:

meaning: *Ha* informs *Hb* about *P*. Inq is from Latin inquit—says. In this expression it is left open whether the proposition *P* (the thing said) is right or wrong in some sense, it can even be a question. However, LINCOS contains a mechanism that can be used to express approval or disapproval of something said previously in another event. For that purpose, two valuating words are available: Ben (from Latin bene—well) for approval and Mal (from Latin male—badly) for disapproval. Events are not labeled, so approval or disapproval must be expressed immediately after the relevant Inq, or with the help of points in and intervals of time. Representations of time-instants and time-intervals are indeed explicitly available in LINCOS. Ben and Mal are also acts of speech. In this manner, the logical constants Ver (from Latin verum—true) and Fal (from Latin false—falsum), are reserved for the values of expressions in classical propositional logic. Note that LINCOS does not use the concept of declarations.

Remark: delimiters like declarators and many others in computer programming languages date from the time of the first formal syntactic definition of an algorithmic programming language, i.e., the Algol 60 report published in 1960.

Once a person *Ha* is introduced as a constant, it is supposed to keep its identity throughout the program—all constants should have this property. Variables are introduced as needed, following usual practice in mathematics, omitting explicit declaration, and their identity is not global.

The chapter BEHAVIOUR contains many programs of events expressed in LINCOS, in fact exchanges of information (dialogues) between actors—but the author of the language "is neither inclined nor able to settle behaviour rules ... or pronounce valuating judgements otherwise than by the mouth of an acting person" (Freudenthal 1960, 91). As mentioned, the actors are supposed to be individuals, persons, human beings. From the point of view at the receiving end, they could be material robots, other artifacts of some kind, or immaterial spirits (computer programs?) as well. Freudenthal had realised this and proposed (unpublished) adding to messages coded in LINCOS schematic drawings of dolls depicting humans more or less realistically (note: modern Japanese manga's would not qualify!). As the language is not designed as a multi-level system, it is essentially flat.

Here is an example of behavior involving two persons Ha and Hb. The LIN-COS text is between the two #. The interpretation is evident.

Ha Inq Hb. ?x.100 $x = 1010$: Hb Inq Ha. 1010/100 : Ha Inq Hb. Mal :
  Hb Inq Ha. 1/10 : Ha Inq Hb. Mal : Hb Inq Ha. 101/10 : Ha Inq Hb. Ben #
Valid in any arithmetical system.

Humans occupied with information processing are on the stage of action. In the book, examples of actions chosen for descriptions show that the chapter is concerned with *human* behavior, over a wide range of topics. In a number of conversations, concepts of elementary mathematics are discussed to begin with: constants, free and bound variables, sets and functions appear in order "to assimilate LINCOS to colloquial mathematics," as the author remarks (see above example). Besides events of this kind, LINCOS admits expressions for an impressive set of linguistic phenomena, such as adversatives, oblique speech, and quotations. LINCOS words are as a rule abbreviations of Latin words. Words for "but," "though," and "whether" are provided. There is a word for the verb "to know" a solution to some question. That word, quoting the author, "does not aim at a (more or less mysterious) mental state of the person knowing, but simply at a certain behaviour that will be exhibited by means of a number of examples." If a person gained knowledge by witnessing an event, another word for "to know" is used. Yet another word is used for knowledge obtained by "finding it out" in some way. "Now," "here," "I," "this" and "that" are avoided. "I" is dispensed of by the convention that a person speaking about himself uses his proper name.

A remarkable discussion is presented on *modalities* of human behavior involving constraints of various kinds (Freudenthal 1960, 151–167). These constraints could be: legal or moral, those for decency's sake and those exercised by an actor's wish. Constraints can be abstracted and personified by a human. Hg for example could represent not just one good person, but all good people, those who wish only good things to happen. Hg then would be a personification of goodness.

The latter observation is significant in the context of the present chapter. Because personifications can be used, it appears that LINCOS possesses a strong starting point for expressing altruistic human behavior. In conversations between humans, relevant aspects of morality, such as benefits for self and for others, could be discussed more or less extensively. This is an important point. We shall, however, refrain here from going into this in more detail because many more ingredients of LINCOS would have to be brought into play in order to make messages containing aspects of this kind of behavior understandable for alien recipients. Instead we shall see how the notions of morality and altruism can be expressed in NEW LINCOS.

4 Outline of NEW LINCOS

It is appropriate to review now briefly the author's (astro)linguistic system NEW LINCOS, because the relevant basic ideas are rather different from those of LINCOS. The new *lingua cosmica* uses the branch of logic known as Calculus of Constructions with Induction (CCI). CCI is based on ideas from *intuitionistic logic* (van Atten 2012), *type theory* and the so-called *typed Lambda Calculus*. Gentle introductions to the latter calculi are in the author's recent book on NEW LINCOS, i.e., the monograph *Astrolinguistics* (Ollongren 2013). As we strive to achieve a measure of self-containment in this section, we collect here relevant information on the basics of NEW LINCOS.

The main basic assumption of the system is that all terms (expressions) are typed—so we use strong typing. Type is the bottom type, itself untyped. Global constants, represented by identifiers (names), themselves terms, are introduced by declaring their types. A term may have residents, i.e., entities typed by that term. If a term has a resident, it is justified. Locally-bound variables are introduced as abstractions giving their types, which specify their domains. An abstraction also specifies the lexicographic range of the variables. Mappings from domains to ranges are powerful expressive terms. These are considered to be *bona fide* functions, but they are at the same time logic implications. NEW LINCOS contains mechanisms for defining functions (inherited from CCI). Inductive definitions are admitted. *True* and *False*, as well as the abovementioned Ben and Mal of LINCOS, are absent. However, an implication can be *verified*; it is proved if and only if a justification for it, i.e., a resident of it, can be constructed. At the base we have the declaration of two fundamental constants

CONSTANTS Prop, Set : Type.

There are several admissible operations over the residents of Prop. We use the binary logical connective → (implication), but the binary ∧ (and), ∨ (or) and the unary ∼ (not) are also available. Therefore residents of Prop usually are (but need not be) propositions in the classical sense. We shall need in addition two global types

CONSTANTS D, R : Type.

Note the following conventions

- if D has a resident, say because CONSTANT d : D. was declared, then d justifies D, or alternatively D is verified by d
- [LAMBDA x : D] < b > introduces the local variable x with as scope the body < b >
- a mapping from domain D to range R is of type D → R. For example if r : R then [LAMBDA x : D]r : D → R (Note. The meaning of → depends on the context)
- [LAMBDA a : D]a : D → D is the identity function, noting for sake of clarity (ALL b : D) ([LAMBDA a : D]a b) : D.

Instruments used in verifications of facts, *an sich* elementary but fundamental, are also inherited from CCI. All expressions (i.e., not only introductory declarations and definitions, but also verifiable facts, logic terms in general in the present sections) are in the formalism of NEW LINCOS.

5 Moral Behavior Expressed in NEW LINCOS

We shall now show how aspects of moral behavior of humans can be modeled using abstractions rooted in logic. More particularly, we formalize aspects of morality and altruism associated with it using NEW LINCOS. The closely related concept of *empathy* is not considered. For points of view different from the exposé in the present section, we refer to Chap. 12 (Human Altruism) in Ollongren (2013) and the paper on communicating reciprocal altruism (Vakoch and Matessa 2011). Indeed, the choice of building stones strongly influences the formalizations achieved in the end.

The general notion of moral behavior (and so also of altruism) considered as a logic relation over some space is not reflexive. This means that, e.g., self-altruism (a kind of egoism) is excluded. In the treatment below, it is assumed that actors (in fact persons, humans) and actions (kinds of moral behavior) are involved. The (moral) acts of persons are concerned with (ethical) goods yielding in general benefits for others. There are three kinds of benefits represented by abstractions: none, moderate and large. Moral (altruistic) behavior is in the present view considered to be an act (of a person), involving ethical goods, under the assumption that there is a beneficiary (generally another person). So we have a stage consisting of actors and their actions, while there are beneficiaries and benefits involved (we shall use three types). The situation can be formalized in detail using concepts of NEW LINCOS. First the types of actors (a beneficiary is also an actor), ethical goods and benefits are introduced by

CONSTANTS Actor, Goods, Benefit : Set.

and then individual actors of type Actor, goods of type Goods, benefit of type Benefits, and the three distinct benefits are declared by

CONSTANTS Ha, Hb : Actor.
CONSTANT goods : Goods. CONSTANT benefit : Benefit.
CONSTANTS none, moderate, large : benefit.

In view of the previous section we use Ha and Hb for actors (not necessarily humans), Ha representing an altruistic actor and Hb a beneficiary (in general these entities are distinctive). Goods and Benefit are left unspecified and are considered to represent some kind of material or immaterial entity. An act of moral behavior (by a person Ha) will be represented by coupling goods, a beneficiary (a person Hb), and benefits to form the type

Actor → Goods → Actor → Benefit → Prop.

The symbol → means here that mathematical injective mappings are used. Prop is the mentioned type of propositions. The benefits of moral behavior are of three "sizes," large, moderate or none. An action of moral behavior is then typed by

Action-moral-behavior : Actor → Goods → Actor → Benefit → Prop.

This means that an action of moral behavior needs two actors (the first one performing an action of moral behavior involving goods, the second one being the beneficiary) while there is a benefit on the stage as well. Altruistic behavior of actor Ha with respect to actor Hb, is characterized by the following cases

(Action-moral-behavior Ha goods Hb moderate) : Prop.
(Action-moral-behavior Ha goods Hb large) : Prop.
(Action-moral-behavior Ha goods Ha none) : Prop.

Note that this characterization is seemingly symmetric: consider e.g.

(Action-moral-behavior Hb goods Ha moderate) : Prop.

Note, however, that if (Action-moral-behavior Ha goods Hb moderate) is the case, then there is no reason for (Action-moral-behavior Hb goods Ha moderate) to be the case as well.

Finally, we obtain a general definition of altruism as follows.

DEFINE Altruism1 : Actor → Goods → Actor → Benefit → Prop :=

  [Ha : Actor; goods : Goods; Hb : Actor; moderate : benefit]

  (Action-moral-behavior Ha goods Hb moderate).

DEFINE Altruism2 : Actor → Goods → Actor → Benefit → Prop :=

  [Ha : Actor; goods : Goods; Hb : Actor; large : benefit]

  (Action-moral-behavior Ha goods Hb large).

DEFINE Altruism3 : Actor → Goods → Actor → Benefit → Prop :=

  [Ha : Actor; goods : Goods; Ha: Actor; none : benefit]

  (Action-moral-behavior Ha goods Ha none).

DEFINE Altruism : Actor → Goods → Actor → Benefit → Prop :=

[Ha : Actor; goods : Goods; Hb : Actor; b : benefit]

(Altruism1 Ha goods Hb b) ∨
(Altruism2 Ha goods Hb b) ∨
(Altruism3 Ha goods Ha b).

Here the disjunction ∨ operator is used ranging over propositions. Given four arguments of the types indicated, the application of the function Altruism yields a proposition. So far the descriptions are conspicuous, at least for those of us who know the basics of the linguistic system, or the mathematical/formal logic at its base. This evidently does not apply immediately to aliens—even if they know the formal background. So we need to include supplementary information—a kind of aid function for the interpretation of an interstellar message on altruism.

NEW LINCOS is designed as a multi-level system; it is not flat. In message design we model ideas by writing a set of formal expressions. A designer has some aspect of reality in mind, and his/her views are reflected in the expressions, admitting the interpretation the designer has in mind. He/she (in fact we) should keep in mind that a receiver of the expressions need not necessarily give the same interpretation to them. It seems realistic to assume that a receiver will start off by trying an interpretation far beyond the intended meaning. Therefore the set of expressions should be accompanied by other information on the subject treated. So a *Lingua Cosmica* should not be flat. Illustrations, music, texts in some natural language, etc., should be part of messages. In that way an aid for (partial) clarification is provided and should be helpful. For describing and clarifying the topic of extraterrestrial altruism, we could very well use a kind of comics as supplementary information, simple drawn pictures depicting so to say altruism in operation.

6 Conclusion

In the field of message construction for interstellar communication evidently a comprehensive and consistent *Lingua Cosmica* system should be chosen. It cannot be expected that a receiver of a message will be able to understand the contents right away. An amount of necessary guessing at the receiving end on the structure and characteristics of the system employed is unavoidable, but should be done as little as possible. Therefore the introductory parts of a message (whether or not about ourselves) should be derived from some easily interpretable body of knowledge, such as elementary mathematics as advocated in LINCOS. In NEW LINCOS the underlying logic is conceptually simple, but should be supplemented with many examples easily understandable for aliens. Message designers should meet these necessary conditions. Conditions as outlined above being fulfilled, the present chapter shows that there are ample possibilities for explaining in cosmic messages our views on morality and altruism.

References

Coq Development Team. 2010. *The Coq Proof Assistant. Version 8.3*. Paris: INRIA.

Freudenthal, Hans. 1960. *LINCOS: Design of a Language for Cosmic Intercourse*. Amsterdam: North Holland Publishing Company.

Ollongren, Alexander. 2013. *Astrolinguistics: Design of a Linguistic System for Interstellar Communication Based on Logic*. New York: Springer.

Vakoch, Douglas A., and Michael Matessa. 2011. "An Algorithmic Approach to Communicating Reciprocal Altruism in Interstellar Messages: Drawing Analogies Between Social and Astrophysical Phenomena." *Acta Astronautca* 68(3–4):459–475.

van Atten, Mark. 2012. "The Development of Intuitionistic Logic." In *The Stanford Encyclopedia of Philosophy*, edited by Edward N. Zalta. Accessed December 28, 2012. http://plato.stanford.edu/archives/fall2012/entries/intuitionistic-logic-development/.

Equity and Democracy: Seeking the Common Good as a Common Ground for Interstellar Communication

Yvan Dutil

Abstract What cultural traits could we possibly share with any extraterrestrial civilization? It could be argued that every civilization has to face the same challenge: How to distribute resources between different tasks and/or individuals. Limitations of resources and conflicting interests are likely to be universal problems. In many cases, no perfect solution exists and cultural tradition plays a role in the choice of the allocation procedure. Being simultaneously universal and culturally oriented, the allocation problem is especially suitable as a topic of discussion between civilizations. Two theories attempt to solve the problem of fair resource distribution: equity theory and social choice theory. Both theories are described within a mathematical framework, which eases their translation into interstellar messages. Equitable sharing procedures and electoral procedures are intellectual tools developed to deal with conflicting individual interests for the best outcome for the group at large. Therefore, both equity and social choice theories are products of civilizations seeking to better manage interactions between individuals with selfish tendencies. In such circumstances, altruism should emerge as a prized quality that would be encouraged by extraterrestrial societies, but that is difficult to achieve at the level of individuals due to natural tendencies for selfishness. Therefore, social choice and equity theories are topics of discussion that should provide common ground for communication with any civilization that is struggling, like we are, to build a fairer society.

Keywords Equity theory · Social choice theory · Fairness · Resource distribution · Democracy · SETI

Y. Dutil (✉)
École de technologie supérieure, 2917 de Summerside, Québec, QC G1W 2E9, Canada
e-mail: yvan.dutil@sympatico.ca

D. A. Vakoch (ed.), *Extraterrestrial Altruism*, The Frontiers Collection,
DOI: 10.1007/978-3-642-37750-1_17, © Springer-Verlag Berlin Heidelberg 2014

1 Introduction

Basic mathematical and physical notions are generally accepted as some of the very first concepts that will be exchanged between human and extraterrestrial civilizations, since math and physics appear to be universal. However, such notions only reflect a small fraction of a much larger body of knowledge that could be shared in an interstellar conversation. Notoriously, no formal scheme for exchanging knowledge has been proposed for the social sciences, which unlike the physical sciences, are not yet mathematically formalized very much.

Within the infinite numbers of possible cultures, we might wonder what cultural traits we could possibly share with any extraterrestrial civilization. Nevertheless, it could be argued that every civilization has to face the same challenge: How to distribute resources between different tasks and/or individuals. Limitations of resources and conflicting interests are likely to be a universal problem. In many cases, no perfect solution exists and cultural tradition plays a role in the choice of the allocation procedure. Being simultaneously universal and culturally oriented renders the allocation problem especially suitable as a topic of discussion between civilizations.

2 Fair Resource Distribution

Two theories attempt to solve the problem of fair resource distribution. Equity theory, which treats of fair sharing methods of goods or chores, is a very powerful tool to explain the notion of distributive justice. Social choice theory, which analyzes the fundamental problem of finding the optimal "common good" from conflicting individual choices, offers a direct path to democracy. Moreover, these two theories are described within a mathematical framework, which eases their translation into interstellar messages. They also generate various paradoxes and impossibilities, both practical and theoretical, which are by themselves interesting topics of discussion.

2.1 Dividing Goods Between Two Actors

Let us consider the fundamental problem of dividing goods between two persons. The simplest solution is to split goods in two equal parts. The basis of this procedure is the principle of proportionality, which can be traced back to Aristotle's book on ethics. Due to the broad influence of Aristotle on Western culture, this principle is at the base of the system of justice in use in many countries. Still, other visions of justice exist on Earth. For example, the Babylonian Talmud poses the following question: *Two people hold a garment; one claims it all, the other claims half. What is an equitable division of the garment?* According to proportionality, the share should be 2/3 and 1/3, but the solution of the Talmud is 3/4 and 1/4. The difference comes from the fact that the Talmud only considers the *contested* parts (Brams and Taylor 1996).

This is a rather ideal case: claims are restricted and goods are divisible. In real life, claims tend to be unrestricted and often goods (e.g., houses or cars) are not divisible. In addition, we do not have always access to an impartial judge. A good sharing rule must not only be neutral for each actor (*equitable*), but must also be *efficient*, which means that all actors should be as happy as possible at the outcome of the process. In addition, it should *not create any envy* between participants at the end of the process. This last property is stronger than proportionality and supersedes it when there are more than two players.

For the construction of an interstellar message, any mathematical notation could be used (e.g., Dumas 2011; Dutil and Dumas 1998; Freudenthal 1960; McConnell 2001; Ollongren 2012). Here, for sake of simplicity, we will use as much as possible the standard mathematical notation, with pseudo-code describing relevant algorithms. In addition, due to the limited space available, we will skip some preliminary steps where these operators are defined. Indeed, discussion about altruism or any other advanced concepts would be deferred until after tens of pages of introduction to simpler notions. Freudenthal proposed to used "plays" as a way to define values like "good" and "bad." These plays would depict the interaction between humans on various topics. Indeed, the following algorithmic descriptions could be included in such "plays." Of course, descriptions in plays would be more extensive than "Two hold a garment...," "one claims...," and the "other claims." Within an interstellar message, we would write "Human A" and "Human B," "Objects," "Values A," "Values B," etc.

In this formalism, the three qualities of a fair share between two actors (a and b) can be described in terms of:

- Equitability: $\sum [U_a(S_a)] = \sum [U_b(S_b)]$
- Efficiency: $\sum [U_a(S_a)] + \sum [U_b(S_b)]$ is maximized
- Envy-freeness: $\sum [U_a(S_a)] \geq \sum [U_a(S_b)]$ and $\sum [U_b(S_b)] \geq \sum [U_b(S_a)]$,

where $\sum$ is the symbol for summation, S_a is the is the share of the player a, and U_a is the utility function of the player a. The utility can be described as the subjective value given to a good. This is a psychophysiological perception, which changes from person to person as well as varies over time for the same individual.

2.1.1 Alternation

One very simple rule for distributing non-divisible goods is alternation: each actor chose an item alternately. Every child learns this simple algorithm in kindergarten, where it is the usual procedure for constituting a sport team. If each player is acting sincerely, this method can be described by this simple algorithm:

Objects=[Object$_1$ Object$_2$ Object$_3$ Object$_4$]
S$_a$=[]
S$_b$=[]

Do until Size(Object)=0

  Move(max(U$_a$(Objects)),S$_a$)
  Move(max(U$_b$(Objects)),S$_b$)

End

Strict alternation is not a very good allocation procedure. First, envy-freeness cannot be guaranteed and it is not very efficient, since there is often a different allocation that pleases the actors more. Worst, even equitable sharing is not granted. It is possible to solve partially the problems of envy and inequity. Nevertheless, this method is fundamentally inefficient since it is based on an object-by-object comparison instead of a global comparison.

Alternation has the great advantage to be easily extendable to an arbitrary number of choosers. This is why the professional sport teams use it to allocate new players. It can also be adapted to handle unequal choosers. Such an adaptation was used to build the Gemini telescope observing schedule (Puxley 1997). In that peculiar case, telescope time should not only be distributed proportionally to each member, but should also take into account the scientific value and technical feasibility within each member's allocated time. Although alternation is frequently used, it is only one of a variety of allocation algorithms.

2.1.2 Divide and Choose Procedure

One of the oldest allocation methods is the divide and choose procedure: *One actor (the divider) divides the good(s) in two halves, then the second actor (the chooser) chooses his or her preferred half.* Since the divider doesn't know which part will be chosen, the divider has a strong incentive to divide the two halves as equally as possible from his own view point. The chooser is always guaranteed to get at least half of the total value of goods from his standpoint, as evaluated by the chooser.

The divide and choose method can be described by this simple algorithm:

Objects=[Object$_1$ Object$_2$ Object$_3$ ... Object$_n$]
S$_a$=[]
S$_b$=[]
S$_1$=[Object$_1$ ... Object$_{n/2}$]
S$_2$=[Object$_{n/2+1}$ ... Object$_n$]
Limit=x;

% A split the goods in two equal parts
Do until Sum(U$_a$(S$_1$))-Sum(U$_a$(S$_2$))<limit

      Diff=Sum(U$_a$(S$_1$))-Sum(U$_a$(S$_2$))

$\quad$ If Diff<0 & min($U_a(S_2)$)<Diff

$\qquad$ Move(min($U_a(S_2)$),S_1)

$\quad$ Else if Diff>0 & min($U_a(S_2)$)<Diff

$\qquad$ Move(min($U_a(S_1)$),S_2)

Else

$\qquad$ For I=1:Size(S_1)

$\qquad\qquad$ For J=1:size(S_2)

$\qquad\qquad\qquad$ Adjust(I,J)=$U_a(S_1(I))$-$U_a(S_2(J))$

$\qquad$ End

$\quad$ End

Swap(Min(Ajust-Diff))

End

% B choose the best share from its point of view
If Sum($U_b(S_1)$)>Sum($U_b(S_2)$)

$\quad$ $S_a=S_2$
$\quad$ $S_b=S_1$

Else

$\quad$ $S_a=S_1$
$\quad$ $S_b=S_2$

Table 1 provides an example of the divide and choose procedure.

Table 1 Numerical example of the divide and choose procedure

Divide and choose procedure				
	U_a	U_b	S_1	S_2
Object$_1$	6	14	X	
Object$_2$	6	14	X	
Object$_3$	17	2	X	
Object$_4$	17	1	X	
Object$_5$	4	4	X	
Object$_6$	6	2		X
Object$_7$	2	21		X
Object$_8$	8	14		X
Object$_9$	17	14		X
Object$_{10}$	17	14		X
$\sum U_a$	100		**50**	50
$\sum U_b$		100	35	**65**

Here we face a key problem common to almost all procedures: strategic manipulation by the divider. If the divider knows the value of each item (or some items) from the chooser's point of view, the divider could split the lot in such a way that the halves are worth, for example, 80 and 20 % of the total from the divider's standpoint, while they are worth 49 and 51 %, respectively, for the chooser. This trick would the lure chooser to take the second half (worth 51 % from the chooser's perspective), while leaving the divider with 80 % of the perceived value from the divider's perspective. However, there is some risk in adopting this strategy. For example, it could backfire badly if the chooser also knows the preference of the divider. The chooser could then decide to sacrifice a small fraction of wealth by choosing the second half! This would send a strong statement to the divider, warning never to try to fool someone again.

The opposite situation can also occur if the divider and the chooser have both very different needs and if each of them shares this information. In such circumstances, each actor may ends up with 80 % of the perceived value of the lot, if they collaborate. Unfortunately, this win-win solution is difficult to achieve with this simple procedure.

2.1.3 Adjusted Winner Procedure

To address this problem of achieving a win-win solution, the American mathematician Steven Brams developed the *adjusted winner procedure*. This method is simple. Each contestant first has to allocate a value to each disputed item; the total value is fixed. Then, the most desired items are temporally given to each contestant. Once all items are distributed, the total value of all items received by each contestant is calculated. Then to equilibrate gain, an adjustment is made. This is done by an appropriate splitting of the most contested goods when possible, or sharing the monetary value of this good after having sold it. This procedure achieves the remarkable feat of being simultaneously equitable, envy-free, and efficient. Like any other allocation procedure, the adjusted winner procedure is sensitive to strategic manipulation. However, since the adjusted winner is an efficient procedure and therefore the outcome is expected to be as good as possible, there is a strong incitative not to cheat but instead to play fair. This procedure can be expressed by the following algorithm:

Objects=[Object$_1$ Object$_2$ Object$_3$ Object$_4$]
S_a=[]
S_b=[]
%Favorite objects are distributed between actors
For I=1: Size(Objects)

      If U_a(Objects(I))>U_b(Objects(I))

          Move(Objects(I),Sa)

      Else If (Objects(I))<U_b(Objects(I))

End

%Transfert of the tied items
For I=1:Size(Objects)

      If Sum(Ua(Sa))>Sum(Ub(Sb))

          Move(Objects(I),Sa)

      Else Move(Objects(I),Sb)

End

Do Until Sum(Ua(Sa))=Sum(Ub(Sb))

      If Sum(Ua(Sa))>Sum(Ub(Sb))

          Mcont=Sa(min(Ua(Sa)/Ub(Sa)))
          F=(Sum(Ua(Sa))-Sum(Ub(Sb)))/(Ua(Mcont)-Ub(Mcont))
          If F≤1

              Move(Mcont*F, Sb)
              STOP

          Else Move(Mcont,Sb)

Else Sum(Ua(Sa))<Sum(Ub(Sb))

      Mcont=Sb(min(Ub(Sb)/Ua(Sb)))
      F=(Sum(Ub(Sb))-Sum(Ua(Sa)))/(Ub(Mcont)-Ua(Mcont))
      If F≤1

          Move(Mcont*F, Sa)
          STOP

      Else Move(Mcont,Sa)

End

An example of the adjusted winner procedure is shown in Table 2.

Unfortunately, it is impossible to extend this method to more than two actors. For larger numbers of actors, only two qualities of a fair deal can be satisfied at the same time. For example, proportional division is equitable and envy-free, but not efficient. Procedures that provide envy-freeness and efficiency (Reijnierse and Potter 1998) or efficiency and equitability (Wilson 1998) have been developed. We might wonder if some extraterrestrial civilizations have chosen one of these as their favorite allocation procedure.

Table 2 Numerical example of the adjusted winner procedure

Adjusted winner procedure							
	U_a	U_b	S_a''	S_b''	S_a'	S_b'	S_a'	S_b'
Object$_1$	6	14		X		X		X
Object$_2$	6	14		X		X		X
Object$_3$	17	2	X		X		X	
Object$_4$	17	1	X		X		X	
Object$_5$	4	4	–			X		X
Object$_6$	6	2	X		X		X	
Object$_7$	2	21		X		X		X
Object$_8$	8	14		X		X		X
Object$_9$	17	14	X				13.08	3.08
Object$_{10}$	17	14	X				X	
$\sum U_a$	100		74		74		70.08	
$\sum U_b$		100		63		67		70.08

2.2 Democratic Processes: Aggregating Individual Preferences

Since the application of fair attribution rules is difficult when the number of participants is large, other allocation mechanisms have been developed. These methods are based on the aggregation of individual preferences. Even if these procedures are much cruder at the level of the individual, and therefore more frustrating, they are much easier to apply than fair distribution procedures at the collective level. This trade-off between efficiency and applicability must be universal, as is the frustration about politics and democracy that this trade-off fuels.

At the most fundamental level, democratic processes are procedures of information collection. In this vision, democracy is only an extension of the motto: two heads are better than one. Therefore democracy generally produces better results in average than despotic governance, because decisions have a better chance of being more appropriate to the largest number of individuals.

2.2.1 Evolution of Democracy

Natural selection favors the emergence of democracy. Indeed, democracy-like processes are observed in animal species. For example, red deer decide to move when about 62% of the adults stand up. Gorillas move when 65 % of adults call. Swans fly when signal intensity indicated by head movements reaches a given threshold. Hamadryas baboons travel in the direction designated by the majority vote. African buffalos move in the average direction of the adult female gaze. Democratic decision-making appears to be beneficial mostly because it tends to produce less extreme decisions (Conradt and Roper 2003). In the context of discussing possible extraterrestrial societies, the American psychologist Albert

Harrison (1997; 2013, in this volume) concludes that democracy has been advantageous on Earth, where he notes democracy has made rapid progress. Generally, democratic governments are less war prone than totalitarian states, allowing democracies to reserve resources for other more important issues.

Despotism beats democracy only if the group is small and if the difference in information between the despot and the rest of the group is large (Conradt and Roper 2003). In his book *Democracy and Its Critics*, the American political scientist Robert Dahl (1991) has listed seven fundamental features of democratic institutions: elected officials, free and fair elections, inclusive suffrage, right to run for an office, freedom of expression, alternative information, and associational autonomy. All are related to the free access to and unbiased flow of information.

2.2.2 Condorcet's Method

Two French scientists laid the mathematical foundation for the analysis of democracy: Marie Jean Antoine Nicolas Caritat, marquis de Condorcet and Jean Charles, chevalier de Borda. The first was a preeminent mathematician and author of the 1791 French Constitution; the second was a physicist, hero of the American Revolution, and inventor of various scientific instruments. In addition, both distinguished themselves by putting in place the metric system.

The goal of Condorcet was to axiomatize the propositions made the Swiss philosopher Jean-Jacques Rousseau (1762) in his book *Du contrat social, ou Principes du droit politique*. Condorcet's big hope was that social sciences would follow a path of progress similar to the one that has been observed in physics since Newton's fundamental work, allowing his descendants to surpass those of his day in wisdom and in enlightenment (Condorcet 1847 [1782]). Condorcet's first work was to prove the value of the majoritorian rule by a probabilistic approach. He demonstrated that the error rate of a majoritorian vote is much lower than the individual error rate if the error rate of the individual voters is below 50%. This demonstration, known as the Condorcet Jury, has been since be proven over a wide set of initial conditions. Condorcet then tried to extend his conclusion to the problem of multicandidate elections, only to uncover one of the major paradoxes of democracy. In the Condorcet's vision, the true winner of an election is the candidate who beats all others candidates in a sequence of pairwise elections (Condorcet 1785).[1]

Even if this condition appears obvious, Condorcet soon discovered that this definition can lead to some apparent inconsistencies. For example, a situation can exists where candidate A beats candidate B, B beats C, and C beats A. Such cycles are inevitable and highlight a fundamental and universal limitation in any democratic process. They can happen even with a fully rational and sincere electorate.

---

[1] This definition is so compelling that the ability to select the Condorcet's winner has become the standard benchmark used to compare election methods.

Therefore, even the most advanced extraterrestrial civilization faces the same challenge as us in implementing an effective democracy.

2.2.3 Borda's Method

Borda recognized the value of Condorcet's approach but considered it much too complex to be practical. Instead, he proposed his own election method already in use at the French Académie des Sciences (de Borda 1781). In Borda's method, each voter gives his order of preference for each candidate. Each candidate receives a number of points equal to the number of opponents he beats. Much simpler than Condorcet's approach, this method was much easier to implement.

However, a weakness soon became apparent: Borda's method is highly sensitive to strategic manipulation. For example, the voter may choose to rate lower a candidate that is a menace for the voter's favorite, but who would be otherwise better evaluated. Borda himself accepted this weakness, claiming that his method was designed for an honest man. Even worse, the outcome of an election using Borda's voting method can be affected by the introduction of dummy candidates. Let's suppose there is two parties (A and B) with equal support in the population. Some supporters (10 %) of A split a create a new party a, which is less popular than either A or B in the rest of the population because is more extremist. The election result with 100 electors using Borda's method would be before the split:

$$50 : U_A > U_B \quad A = 100 \; B = 0$$

$$50 : U_B > U_A \quad A = 0 \; B = 100$$

and after the split

$$45 : U_A > U_a > U_B \quad A = 90 \; A- = 45 \; B = 0$$

$$5 : U_a > U_A > U_B \quad A = 5 \; A- = 10 \; B = 0$$

$$50 : U_B > U_A > U_a \quad A = 50 \; A- = 0 \; B = 100$$

After the split, A win by a large margin thanks to a marginal party. Even if B partisand try to manipulate strategically their vote by modifying their preference order to $B > A- > A$, this not change the result much: $A = 95 \; A- = 105$ and $B = 100$! only if in a perfect strategic manipulation, 10 % of the B supporters choose to vote $B > A > A-$, will they be able to force a draw!

2.2.4 Election by Plurality

What about plurality as an election method? It is one of the worst systems since it behaves very badly as soon there is more than two choices. It is not a very efficient way of picking the Condorcet winner, and it also strongly encourages strategic voting. Under plurality, giving a vote to a small party is almost equivalent to not voting at all. Therefore, most voters prefer to vote for the least worst of all parties that still has a chance of winning. The direct consequence of such a strategy is the emergence of a two party system, described by *Duverger's Law* (Duverger 1951). It will be difficult for extraterrestrials to understand why we still use a voting scheme that not only poorly measure voter intentions but also make them contort their own opinions. Nevertheless, a simple modification, allowing electors to vote for more than one candidate, can largely reduce strategic voting. This method, approval voting, is also much more likely to select the Condorcet winner (Merrill 1984).[2]

2.2.5 Comparing Election Methods

Does a perfect election method exist? Scientist have sought it for a long time and in the process have created a large number of voting methods: Baldwin, Black, Bucklin, Bullet (anti-plurality), Copeland, Coombs, cumulative voting, Dabagh, Dogson, Hare (Single transferable vote), Instant runoff, Nanson, Range Voting, Two-turn Runoff, Simpson, etc. Unfortunately for us, the answer is no! The American economist Kenneth Arrow (1963) demonstrated this negative conclusion in 1951 and received the Nobel Prize in Economics for his work. Still some methods are better, as we have learned from researchers. Condorcet's method is generally considered as the best one, but it is very difficult to apply in practice. Any method relying on the ordering of candidates is weakened by our inability to order effectively more than 4 or 5 candidates. And all methods involving vote transfer behave in such a way that voting for a candidate may actually hurt that candidate! Maybe extraterrestrials more advanced than us will use Condorcet's method without a problem, or maybe they will use approval voting, a less efficient method but good enough for most purposes.

3 Fair Civilizations: On Earth and Beyond

The existence of complex decision processes provides powerful insights into the species that uses them. For example, such processes cannot be put in place unless individuals are able to perceive and intellectualize the reality of other individuals. The ability to evaluate the fairness of a situation is likely to appear by natural evolution alone in social animals. For example, the capuchin monkey possesses a notion of fairness (Brosnan and de Waal 2003). Otherwise, only despotic decision

---

[2] Approval voting was used in the longest living democracy in history, the Republic of Venice, from 1268 to 1797. It is used now by the United Nations as well as by many scientific societies, including the Mathematical Association of America, the American Statistical Association, and the Institute of Electrical and Electronics Engineers.

processes are possible. Consequently, we should not expect to meet a democratic beehive-like civilization.

Equitable sharing procedures and electoral procedures are intellectual tools developed to deal with conflicting individual interests for the best outcome for the group at large. Therefore, both equity and social choice theories are products of civilizations seeking to manage better the interactions between individuals with selfish tendencies.

In such circumstances, altruism emerges as a prized quality that is encouraged by extraterrestrial societies, but which is difficult to achieve at the level of individuals due to natural tendencies for selfishness (Fehr and Fischbacher 2003). Therefore, we should expect social choice and equity theories to be topics of discussion common to any civilization that is struggling, like we are, to build a fairer society. We can hardly find a better topic to discuss with our interstellar interlocutor.

Another aspect underlying these procedures is their capacity to deal with another person's reaction. The capacity to put oneself in the other's person shoes is fundamental to the concept of altruism. (Altruism comes from the Latin *alter*, meaning "someone else"). Here we have presented altruism in relation to the concept of fairness, which is probably the most rational version of altruism. However, this is only one of many visions of altruism. Indeed, researchers studying the emergence of cooperation have constructed various other scenarios where altruism in its broader definition would act. As such, this chapter is only the tip of the iceberg, describing the version of altruism that is easiest to grasp.

Some problems and solutions born from allocating limited resources in the face of conflicting interests are likely to be civilization-specific, which make them especially interesting for interstellar communication. Since communication channel capacity and processing power are the ultimate factors limiting the efficiency of social interaction, we might discover that intelligence, wisdom, and altruism are closely tied together.

References

Arrow, Kenneth J. 1963. *Social Choice and Individual Values,* 2nd edition. New York: Wiley.

Brams, Steven J., and Alan D. Taylor. 1996. *Fair Division: From Cake-cutting to Dispute Resolution.* Cambridge, UK: Cambridge University Press.

Brosnan, Sarah F., and Frans B. M. de Waal. 2003. "Monkeys Reject Unequal Pay." *Nature* 425:297–299.

Condorcet, Marie Jean Antoine Nicolas Caritat. 1847 [1782]. "Discours prononcé dans l'Académie Française le jeudi 21 février 1782." In *Oeuvres de Condorcet*, Volume 1, edited by A. Condorcet O'Connor and M. F. Arago, 389–415. Paris: Firmin Didot.

Condorcet, Marie Jean Antoine Nicolas Caritat. 1785. *Éssai sur l'application de l'analyse à la probabilité des décisions rendues à la pluralité de voix.* Paris: De L'Imprimerie Royale.

Conradt, Larissa, and Timothy J. Roper. 2003. "Group Decision-making in Animals." *Nature* 421(6919):155–158.

Dahl, Robert A. 1991. *Democracy and Its Critics.* New Haven, CT: Yale University Press.

de Borda, Jean Charles. 1781 [1770]. *Mémoire sur les élections au scrutin*. Histoire de l'Académie Royale des Sciences. Accessed January 1, 2013. http://gerardgreco.free.fr/IMG/pdf/MA_c_moire-Borda-1781.pdf.

Dumas, Stéphane. 2011. "A Proposal for an Interstellar Rosetta Stone." In *Communication with Extraterrestrial Intelligence (CETI)*, edited by Douglas A. Vakoch, 403–411. Albany, NY: State University of New York Press.

Dutil, Yvan, and Stéphane Dumas. 1998. "Active SETI: Targets Selection and Message Conception." Paper presented at the annual meeting of the American Astronomical Society, Austin, Texas, January 9.

Duverger, Maurice. 1951. *Les partis politiques*. Paris: Colin.

Fehr, Ernst, and Urs Fischbacher. 2003. "The Nature of Human Altruism." *Nature* 425:785–791.

Freudenthal, Hans. 1960. *LINCOS: Design of a Language for Cosmic Intercourse*. Amsterdam: North Holland Publishing Company.

Harrison, Albert A. 1997. *After Contact: The Human Response to Extraterrestrial Life*. New York: Plenum.

Harrison, Albert A. 2013. "Cosmic Evolution, Reciprocity, and Interstellar Tit for Tat." In *Extraterrestrial Altruism*, edited by Douglas A. Vakoch, 3–22. Heidelberg: Springer.

McConnell, Brian. 2001. *Beyond Contact: A Guide to SETI and Communicating with Alien Civilizations*. Sebastopol, CA: O'Reilly Media.

Merrill, Samuel, III. 1984. "A Comparison of Efficiency of Multialternative Electoral Systems." *American Journal of Political Science* 28(1):23–48.

Ollongren, Alexander. 2012. *Astrolinguistics: Design of a Linguistic System for Interstellar Communication Based on Logic*. New York: Springer.

Puxley, Phil. 1997. "Execution of Queue-scheduled Observations with the Gemini 8 m Telescopes." In *SPIE Proceedings*, Vol. 3112, *Telescope Control Systems II*, edited by Hilton Lewis, 234–245.

Reijnierse, J. Hans, and Jos A. M. Potters. 1998. "On Finding an Envy-free Pareto Optimal-Division" *Mathematical Programming* 83(2):291–311.

Rousseau, Jean-Jacques. 1762. *Du contrat social, ou Principes du droit politique*. Amsterdam: M. M. Rey.

Wilson, Stephen J. 1998. *Fair Division Using Linear Programming*. Preprint. Ames, IA: Department of Mathematics, Iowa State University.

Patterns of Extraterrestrial Culture

William Sims Bainbridge

Abstract Much psychiatric literature concerns problems that affect an individual's ability to feel compassion toward others, and thus to behave at least occasionally in an altruistic manner. To the extent that extraterrestrial societies differ in their psychology, much of that variation ought to be comparable to the range of personalities exhibited by humans, including the psychopathologies. A classical perspective that combined anthropology with psychiatry was culture and personality studies, which now can be revived in the context of theorizing about the modal personalities of extraterrestrial cultures. This study explores these issues through the rich, role-playing experience afforded by the massively multiplayer online game world, *Star Wars: The Old Republic*. The user interacts with many very different personality types, represented by both primary and secondary avatars, many of which are simulated extraterrestrials.

Keywords Anthropology · Psychiatry · Culture · Personality · Online games · Computer games · *Star Wars*

1 Introduction

A classic perspective in the social sciences, sometimes called *Culture and Personality Studies*, uses the same conceptual framework to describe the predominant character of a society and the modal personality of an individual within it, often applying terminology taken from psychiatry, but also describing

---

The views expressed in this essay do not necessarily represent the views of the National Science Foundation or the United States.

W. S. Bainbridge (✉)
IIS Division, National Science Foundation, 4201 Wilson Blvd., Arlington, VA 22203, USA
e-mail: wbainbri@nsf.gov

D. A. Vakoch (ed.), *Extraterrestrial Altruism*, The Frontiers Collection,
DOI: 10.1007/978-3-642-37750-1_18, © Springer-Verlag Berlin Heidelberg 2014

human differences in terms of ideal types that were developed within the arts and religion (Opler 1959; Kaplan 1961). When Max Weber (1930) described *The Protestant Ethic and the Spirit of Capitalism*, he argued that an "elective affinity" existed between Protestant religion and capitalist economics, which at least for one period of human history defined both Protestant cultures and Protestant personalities. Both required austerity, deferred gratification, and rational calculation, an ethic not very different from that of the fictional Vulcan civilization in the *Star Trek* mythos, and indeed social historians have argued that Protestantism promoted scientific thinking (Westfall 1958; Merton 1970). Weber was a Protestant, and each ethnopsychiatric theory is to some degree an expression of the modal personality of the culture in which the theorist works (Edgerton 1966).

This chapter will examine a number of alien cultures from the *Star Wars* mythos, as expressed through the online virtual world *Star Wars: The Old Republic* (SWTOR), basing the analysis on a culture and personality framework informed by psychiatric and psychoanalytic traditions but not limited to them. The goal is not to critique a particular work of fiction, but to use it as a context for setting out principles that could well apply to real extraterrestrial cultures. Despite the scientific pretentions of psychology and psychiatry, personality typologies tend to be merely refined versions of popular beliefs, such as the Big Five dimensions often described in their OCEAN variant: Openness, Conscientiousness, Extraversion, Agreeableness, and Neuroticism (Wiggins 1996). Being high on two of these dimensions might encourage altruism. Conscientiousness is an unusual willingness to complete assigned practical tasks, including obligations to other people. Agreeableness supports cooperation, but also refers to pleasant communications that may not be related to tasks. While the research on the Big Five employs complex statistical techniques like factor analysis, the fundamental concepts of conscientiousness and agreeableness derive from popular culture rather than from any discoveries of neuroscience.

The *Star Wars* mythos was developed over more than a third of a century by many authors and media creators, and based on a century of science fiction literature and diverse branches of ordinary popular culture (Bainbridge 1986; Rinzer 2007). Thus it collected many cultural stereotypes of different human personalities, but found ways to present them as the temperaments of extraterrestrials. If humans are an unspecialized species, exhibiting much internal variation, then each type of person might represent a type of alien society. Since many of the concepts are psychiatric, they tend to describe different ways of being incapable of altruism.

For example, the gigantic slug named Jabba the Hutt, whom Princess Leia strangles to death in *Return of the Jedi*, is a slothful glutton whose position in society is comparable to that of a Mafia gang leader. Criminologists are the scientific specialists best able to analyze his pathology, and in conversations with psychiatrists they might bandy about the term *psychopath*. However, the traditional theory about gangs is that they emerge in societies where social order is breaking down (Thrasher 1927). In SWTOR, the Republic and Empire war with each other to rule the galaxy, thereby wrecking any existing stability. In consequence, many star systems suffer social disorganization, where gangs may be the only source of law and order. In this fictional galaxy, the Hutts dominate a planet and its moon, but have a base on a third world as well, perhaps because their psychopathology is adaptive given their chaotic environment (Searle 2011).

At the opposite extreme are the spiritual and highly organized Voss. They survive on a planet of the same name, surrounded by an apparently very different hostile species called the Gormak. The Voss are totally content within their own culture, apparently lacking any crime or mental illness, unless we were to diagnose their entire society as *catatonic*. They have no interest in forming alliances with other intelligent species, and wish they could totally ignore them. The Gormak, in contrast, are violent, seek to develop or appropriate the technology they need to leave their planet, and interact with other species by attacking them. On one level, the Gormak and Voss represent two polar opposite responses to stress: fight (aggression) or flight (withdrawal). Eventually, after dealing extensively with both species, the player learns that they originally were one species, similar to the future division of humanity described by H. G. Wells (1895) in *The Time Machine*, when *Homo sapiens* separated into angelic but oblivious Eloi and brutish but technological Morlocks. Thus, the Hutt, Voss and Gormak species can be seen as accentuations of human personality types, living in societies that are extensions of those personalities.

2 Ethnopsychiatry

Many readers might expect here an emphasis on the theories proposed by Sigmund Freud, but in fact Freud was far from the first influential figure in this area, and some of his contemporaries contributed more than he, such as Carl G. Jung, who proposed the *introversion-extraversion* dimension (Jung 1923; Bishop 1995). A landmark was the 1872 book *Die Geburt der Tragödie* (*The Birth of Tragedy*) by Friedrich Nietzsche, which offered a typology of three culture types that were equivalent to personality types: Apollonian, Dionysian, and Buddhist. Most influential for later writers is the Apollonian-Dionysian dichotomy, which Nietzsche derives from his reading of ancient Greek culture. Named after the competing Greek gods, Apollo and Dionysus, these two archetypes represent opposite modes of response to human existence. The Apollonian is cool, rational, classical, and when it does not speak in grammatical sentences, expresses itself through the visual arts. The Dionysian is hot, lustful, romantic, and when it does not roar with animal noises, expresses itself through music and dance.

Nietzsche argued that Apollonians were individualistic, Dionysians were collectivist, and Buddhists were neither. One way to map these concepts is to say that Apollonian cultures emphasize the individual and de-emphasize the group, while Dionysian cultures de-emphasize the individual and emphasize the group. Buddhist cultures de-emphasize both the individual and the group, which suggests that Nietzsche had in mind two cross-cutting dichotomies. This implies there should be a fourth category that emphasizes both the individual and the group. In his later work

Nietzsche (1885) struggled and failed to conceptualize this fourth type, which might be called the Zarathustran after his most influential book, *Also Sprach Zarathustra*. In another very influential work, *Patterns of Culture* (1934), anthropologist Ruth Benedict adapted Nietzsche's two main categories, but asserted opposite connections to social relations, considering Dionysians to be individualistic in their concentration on their own emotions, and suggesting that Apollonians were more collectivist because of their strict adherence to group norms.

Academic interest in the culture and personality perspective peaked in the 1950s, and began to decline in the following decade, as the various social sciences went their separate ways. This divorce was signified by the breakup of the Harvard Department of Social Relations in 1970, which had combined personality psychology and cultural anthropology with social psychology and sociology (Bainbridge 2012a). This boom and bust was associated with a similar pattern of popularity for psychoanalysis, which began fragmenting early in its history but splintered very visibly after its peak of popularity in the late 1950s (Brown 1967; Maddi 1968; Finkel 1976). This is not to say that psychoanalytically-oriented psychotherapy is an ineffective form of treatment for some neurotics, and many variants of it remain popular today (Bainbridge 2012b). In our secular society, psychotherapy has taken over some of the former functions of religion, including helping people construct coherent concepts of themselves and providing hope that can energize effective action to improve their lives (Bakan 1958; Frankl 1962).

The social sciences suffer recurrent fads, just as other sectors of culture do, so the culture and personality perspective could well revive. However, working against it is the trend toward human economic globalization, and the realization that different races do not differ significantly in those aspects of their genetics relevant to neurology. Thus here we must rely on rather old publications for much of our intellectual guidance, while nonetheless looking toward the future, when we might actually gain solid information about alien civilizations beyond the Earth.

Extraterrestrial civilizations do not presumably breed with each other, and are separated by vast distances, so their cognitive and cultural characteristics may fit the culture and personality perspective far better than the only slightly different societies that share the planet Earth do. Freud claimed that his concepts, like the Oedipus complex, which is based on a rivalry between a son and his father for the love of his mother, were naturally derived from the universal structure of human families. But anthropologist Bronislaw Malinowski (1927) argued that there in fact existed at least two very different family structures among human societies, patrilineal emphasizing fathers and matrilineal emphasizing mothers, with quite different consequences for adult personalities. Extraterrestrial intelligent species may have a variety of family structures, one per species, and thus the learned component of personality laid down in childhood could be quite different across species, an outcome intensified by the effect of genetic differences. Advanced extraterrestrial cultures would have already completed their equivalent of globalization long ago, which implies that each may have settled on a core set of norms and values, expressed through a modal personality type as well as in the institutions of the society.

3 *Star Wars* Avatars

Star Wars: The Old Republic is a massively-multiplayer online (MMO) role-playing game that debuted on December 20, 2011 with about 1.7 million players, and is the most recent major reconceptualization of the *Star Wars* mythos. MMOs are virtual worlds in which a player can explore and carry out missions for hundreds of hours, experiencing exotic environments as if they were real and getting deeply into the psychology of the player's avatar, a character in the world that provides the vantage point and focus of action. Several earlier MMOs depicted life on other planets, including *Anarchy Online, Entropia Universe, Tabula Rasa*, and *Star Wars Galaxies,* which shut down after eight years on December 15, 2011 (Bainbridge 2011). SWTOR depicts 17 planets and innumerable intelligent alien species, which compete and cooperate in a galaxy "far, far away," where interstellar travel is much easier than in our own galaxy. One open debate in the computer game literature concerns whether deep role-playing reflects the psychopathology of the players, a retreat from objectively unsatisfactory lives, or a creative synthesis between ordinary life and new communications technologies (Williams, Kennedy, and Moore 2011).

SWTOR represents a new *theatrical* trend in MMOs that tells especially rich pre-defined stories, including many cutscenes or short computer-generated movies in which the player's avatar interacts with other characters for whom recorded voice actors speak, including sometimes aliens who talk strange languages. This trend toward well-written but thus rather constraining stories draws on some traditions from solo-player computer games, and the creator of SWTOR, BioWare, had earlier produced a very successful one called *Star Wars: Knights of the Old Republic*. This theatrical trend emphasizes deep role-playing, and thus its MMOs delve more deeply into the psychology of the players, making SWTOR the ideal example for this chapter.

I ran four avatars all the way to the original top level 50 of experience, which given my intensive documentation of ethnographic observation took about 250 h for each. Two of them were human, one was a part-human cyborg, and the fourth belonged to the alien Zabrak species. I also ran several other avatars more briefly to study diversity. Avatars that represent the player in this gameworld interact with two main kinds of computer-operated characters: ordinary non-player characters and companions. Non-player characters, or NPCs, serve many functions in MMOs, including vendors in markets where virtual goods can be bought, teachers from which the player's avatar learns new skills, nameless enemy warriors who exist only to be killed, and named characters who play often complex roles in the story. Companions in SWTOR are semi-autonomous secondary avatars that in battle can be directed by the player in the same manner as the primary avatar, but also have their own personalities and stories.

The following exploration in exopsychology begins with an in-depth analysis of my primary avatars, with a special emphasis on the Zabrak alien. We start with human avatars to set out general principles, and use the cyborg to illustrate the possible transformation of personality that technology may cause at high levels of any society's development. This chapter also features companions, because they are secondary avatars and thus illustrate similar points, while displaying different personality types, largely dependent upon their species.

Despite the debates about the extent to which human personality is rooted in our biological nature and our family structures, much social science considers humans to be abstract cognitive systems, capable of developing any conceivable form of personality. An alternate view that was influential in the development of socio-biology is that insect societies are very different from human ones, perhaps pos-sessing a hive mind that could be said to be intelligent, although each individual might be stupid (Wilson 1971; 1975). It is possible that individualism might emerge in the evolution of an advanced hive culture, analogous to the hypothe-sized past evolution of humanity from an organic *community* form of organization, to a more rationalized *society* form of organization (Tönnies 1957).

Among mammals, some species are more sociable than others; for example tigers do not live in prides but lions do. So it is possible to imagine extraterrestrial intelligences outside the bounds of human variation, either more social or less social than we are. But *Star Wars* is not unreasonable in believing that most intelligent species could be understood as exaggerations of characteristics pos-sessed by at least a few humans. For reasons of movie-making cost, and to help unsophisticated audiences identify with the characters, aliens are depicted physi-cally as humans having just a few strange cosmetic features.

4 Primary Avatars

In exploring an online virtual world, one needs to work through an avatar, and over the years I have developed an approach in which the characteristics of the avatar are carefully chosen to maximize the intended research payoff (Bainbridge 2010a; b; 2013). To achieve maximum research results in this science-fiction world, my main avatars were based on four classic science fiction authors whose own main characters represented four distinct philosophies, and which are described in Table 1.

The primary goal and basis of morality were assigned by me, on the basis of my readings of the authors, rather than by SWTOR, but these characteristics do har-monize with the stories written for these classes of characters by the game designers. A major ethical dimension for magic-wielding characters in the *Star Wars* movies is orientation toward the Dark Side of the Force, which involves acting upon one's most powerful emotions, versus the Jedi "light side," which seems devoid of the compassion we usually associate with altruism, and appears to be a metaphor for a very different school of psychiatry, namely Zen Buddhism as adapted for western cultures (Herrigel 1953; Suzuki, Fromm, and De Martino 1960; Watts 1961; Shoji 2001). Avatars in SWTOR earn points with both sides

Table 1 Four research avatars in *Star Wars: The Old Republic*

Avatar:	Burroughs	Heinlein	Asimov	Bester
Based on:	Edgar Rice Burroughs	Robert A. Heinlein	Isaac Asimov	Alfred Bester
Faction:	Republic	Republic	Empire	Empire
Class:	Jedi Knight	Trooper	Agent	Sith Inquisitor
Primary Goal:	Exploration	Obey rules	Puzzle solving	Power
Basis of morality:	Personal code of honor, love for intimate friends	Total loyalty to one's assigned group	Intelligent problem-solving	Redress of personal grievances
Light points	5700	6277	7558	1700
Dark points	150	1382	7557	6050
Alien companions (species)	Rusk (Chagrian)	Aric Jorgan (Cathar)	Kaliyo Djannis (Rattataki)	Khem Val (Dashade)
	Scourge (Sith)	Tanno Vik (Weequay)	Vector (Human/Killik)	Ashara Zavros (Togruta)
		Yuun (Gand)		Xalek (Kaleesh)

through the decisions they take, which often concern the life or death of other characters.

This chapter focuses on how alien cultures might possess different orientations to altruism, but both Burroughs and Heinlein are humans. Both are heroes of the Galactic Republic who often help strangers, but are troubled by the corruption they see in the Republic's government and thus are not entirely committed to it. Neither author's characters engage in altruistic acts out of sympathy for victims of injustice, but rather because the situation of the victim violates their internalized sense of how the world should be organized. The story of each ends with his marriage to a woman, represented in SWTOR as a secondary avatar with her own complex story, who reinforces his own moral philosophy.

Burroughs (1917) is modeled primarily on the character of John Carter from *A Princess of Mars* by Edgar Rice Burroughs (Lupoff 1976). Carter had been a Confederate officer during the U.S. Civil War, and possessed a personal code of honor but had become an outcast and wanderer, lacking any home on Earth. On Mars he made close friendships with individual members of two very different intelligent species, Tars Tarkas of the Green Martians, and Dejah Thoris of the Red Martians. Only through love for them as individuals does he become committed to the well-being of their peoples.

My Burroughs avatar was more interested in exploring the planets he visited than completing the Republic's missions, and he fell in love with a secondary avatar Jedi named Kira Carsen, who had fled as a girl from the Empire and was a loyal but rather impulsive woman. Since romance violates the Jedi code of quasi-Buddhist emotional detachment, they kept their affair secret for a long time, thereby separating themselves psychologically from the Jedi Order. Once they both reached the top experience level, they retired from Republic service to live on

Tatooine, the planet where both Luke Skywalker and Darth Vader would grow up centuries later, and that was patterned on the Mars where John Carter married Dejah Thoris. While favoring both the Republic and the Jedi, and willingly performing missions for them, their love was bounded within their own personal relationships, rather than being what Christian theologians call *agape*, or the altruistic love of all sentient beings.

The heroes of Robert Heinlein's (1948; 1959; 1961; 1966) novels seem to be politically right-wing, but torn between two alternative varieties, libertarian individualism as in *The Moon is a Harsh Mistress* and *Stranger in a Strange Land*, or dutiful militarism as in *Space Cadet* and *Starship Troopers* (Panshin 1968). As scripted by SWTOR, my Heinlein avatar begins as the newest recruit to Havoc Squad, a successful and highly cohesive SWAT team, but near the very beginning the other members defect from the Republic, and Heinlein is assigned the difficult task of bringing them back, dead or alive. This presents an extreme dilemma, because loyalty requires a decision about whom one will be loyal to, and Heinlein can never quite convince himself that the Republic deserves his fealty. Eventually, he marries a secondary avatar named Elara Dorne, who quit the Empire as an adult woman and has an exceedingly bureaucratic, rule-based temperament. As with Burroughs, the marriage intensifies Heinlein's own moral code, which emphasizes law and order rather than altruistic benevolence, and completes the development of his loyalty to the rule-based Republic.

Thus, my Burroughs and Heinlein avatars represent two different ways of being human, both of their personalities fitting some stereotype of American culture, differing in degree largely based on past experiences and current situation. Their moral codes stress both bonds to other individuals, prior to group loyalty, but Burroughs differed from Heinlein in keeping his own moral judgment, rather than subordinating it any social organization to which he belongs. For my two Imperial avatars, I decided to go further in the direction of alien characters. Bester is a member of the humanoid but distinctively alien Zabrak species, while Asimov is a cyborg, partly human and partly robot. Isaac Asimov, of course, is famous for his three laws of robotics, the nearest thing to a code of ethics:

1. A robot may not injure a human being or, through inaction, allow a human being to come to harm.
2. A robot must obey the orders given to it by human being, except where such orders would conflict with the First Law.
3. A robot must protect its own existence as long as such protection does not conflict with the First or Second Laws (Asimov 1950, 7).

This is a passive moral code that subordinates a robot to human beings but does not give it a positive motivation to help humans on its own initiative. That alternative was explored in the novel *The Humanoids* by Jack Williamson (1950), in which overtly altruistic robots stifle human freedom in the obsessive attempt to do good. In his first two full novels about robots, *The Caves of Steel* and *The Naked Sun*, Asimov (1954; 1957) postulated that individual humans followed rules every bit as strict as his Three Laws of Robotics.

Economics and behavioral psychology, especially in the variant called *game theory*, assume not only that humans follow strict rules comparable to a computer program, but also that any intelligent being would follow exactly the same rules (Skinner 1938; Von Neumann and Morgenstern 1944; Homans 1974). In this school of thought, human cooperation is not derived from any instinct of altruism, but is a form of enlightened self-interest. As shown by Robert Axelrod (1984) in *The Evolution of Cooperation*, even a very simple artificial intelligence program can learn to cooperate, if it and a similar exchange partner interact repeatedly, learning to give in order to receive. *Homo economicus*, as this model of humanity is often called, is incapable of altruism, but also incapable of sadism. Every action taken with respect to another intelligent being is intended to result, sooner or later, in a favorable economic return. If someone is obviously not able to reward SWTOR characters like Asimov, it will simply never occur to him to help them. Thus, he is like a cyborg, whose humanity has been replaced by a program. Game theory may be correct when it says that any intelligent extraterrestrial would behave in exactly the same way, or it may apply only to extraterrestrials whose native propensities have withered away, after a hundred generations of rational civilization.

Bester was based on Alfred Bester (1953; 1956), the most highly-regarded science fiction author to be deeply influenced by psychoanalysis, and who won the first Hugo award given by the science fiction community each year, for his 1953 novel, *The Demolished Man*. Current science fiction scholars tend to be aware that this novel draws on Freud's psychoanalytic ideas, but they seem unaware that his second great novel, *The Stars My Destination*, similarly draws upon the power-oriented psychoanalysis of Alfred Adler, the first of Freud's main disciples to defect from his authority. Indeed, each of these novels achieves its greatness through a synthesis of one concept from parapsychology and one from psycho-analysis. *The Demolished Man* combines telepathy with Freud's Oedipus complex, while *The Stars My Destination* combines teleportation with Adler's inferiority complex. My understanding of Alfred Bester was increased when I met him in person, and his prominence in the history of science fiction is marked by the fact that he was posthumously the basis for a central character in the television series, *Babylon 5*, which was the subject of a trilogy of novels (Keyes 1998; 1999a; b).

Just as my avatar Burroughs was based on one particular character created by the novelist, Bester was based on Gully Foyle from *The Stars My Destination*. Donald Palumbo (2004) begins an essay about the novel by comparing it with the original *Star Wars* trilogy, then diagnoses Foyle as a victim of *monomania*, consumed by obsessive revenge. Fiona Kelleghan's (1994) diagnosis is *claustrophobic paranoia*, in which the mind itself is a *space of imprisonment*. The novel's Wikipedia (2013b) article explains Foyle's personality:

> It is the study of a man completely lacking in imagination or ambition, Gulliver Foyle, who is introduced with "He was one hundred and seventy days dying and not yet dead..." Foyle is a cipher, a man with potential but no motivation, who is suddenly marooned in space. Even this is not enough to galvanize him beyond trying to find air and food on the wreck. But all changes when an apparent rescue ship deliberately passes him by, stirring

him irrevocably out of his passivity. Foyle becomes a monomaniacal and sophisticated monster bent upon revenge. Wearing many masks, learning many skills, this "worthless" man pursues his goals relentlessly; no price is too high to pay.

Notice that Foyle undergoes a sudden personality change, from one kind of alien to another. He begins as a cipher who is incapable of altruism as well as every other form of intense emotional attachment. He is alien in the sense of being alienated, like the protagonists of existentialist novels of the period such as *Malone Dies* by Samuel Becket (1956) and *The Stranger* by Albert Camus (1946). Logically, some members of any real extraterrestrial society could also be alienated in this sense, precisely because it is a universal form of inhumanity. Then, Foyle is seized by a passion for revenge, when the commander of a passing ship fails to act altruistically and rescue him. More than that, he literally takes a leap into the dark, gaining two remarkable powers that illustrate the same principle on different levels of abstraction. First, where earlier he was a cipher, the lowest of the low, suddenly he becomes rich and adopts a new identity as an aristocrat, a leap up the status pyramid that may exist in most societies, human or otherwise.

The second power Foyle gains is indeed the ability to teleport to a new destination. The novel assumes that humans have developed the psychic ability to teleport from one point on the Earth to another, but only if they have first visited the destination to learn its mental coordinates. "Look before you leap" is the principle of the culture. Currently, this is a common convention in MMOs, including SWTOR, where one must first walk to a transportation hub before being able to teleport to it. Indeed clicking on any website hyperlink takes you to a new website, but only because its address was programmed into the web page from which you started. But Foyle can teleport himself to entirely new destinations. He is alien, free from the limiting assumptions of the prevailing culture, capable of doing things a normal member would find impossible. The relationship to altruism is that both before and after his sudden personality conversion, Foyle was incapable of identifying with other people, and thus lacks the intimate social relations that were the basis of morality for Burroughs and Heinlein. If he were not blinded by fierce obsession, he might coolly develop the exchange normality of Asimov.

Foyle passionately hates other people, especially those who he blames for failing to help him when he was stranded in space. His goal is not merely to destroy them, but to leap over them, because he suffers from an *inferiority complex*. There is an old joke about a patient who asks his psychoanalyst, "Doctor, do I have an inferiority complex?" "No," replies the doctor, "you really are inferior." Actually, people with inferiority complexes may or may not actually be inferior, but they deeply resent the fact that other people seem to have power over them. While Freud emphasized sexual feelings in his version of Psychoanalysis, his early follower and first rival, Alfred Adler (1927; 1928), emphasized the will to power, which he felt was as important an instinct in humans as the sex drive. Two key concepts of psychoanalysis may have come from the neurotic conflict between Freud and Adler, who were projecting their own internal problems. Freud wanted to dominate his followers, so he invented the Oedipus complex to explain away their disagreements with him, their intellectual father, while Adler suffered from his own inferiority complex with respect to Freud.

In creating a Bester avatar, I needed to select an appropriate story line, which is done by choosing one of four starting classes. Among the options for an Imperial character, the Sith Inquisitor was ideal. The avatar of this class has been a slave, but his masters noted that the Force was strong in him, so in preparation for the coming war against the Republic they send him for training as a Sith, the Dark Side equivalent of a Jedi. Most apprentices died during the severe trials on the Sith training planet, which helps make the personalities of the survivors even more shrewd and hostile than they were to begin with. This seemed a perfect background and ideal training to build a powerful inferiority complex.

Presumably, a few members of any real extraterrestrial civilization could develop inferiority complexes, and the proportion of them would be high—even the norm—in highly conflicted societies, such as those with high innate aggressiveness within a world of limited natural resources. This could also happen in those presumably rare cases in which a planet evolves two intelligent species, with one dominating but not exterminating the other. For Bester, I chose a species well-designed to play the intended role, the Zabraks. In the *Star Wars* movies, the prime example is Darth Maul in *The Phantom Menace*, who is not only violent but holds himself under exquisite control, and has deep red skin with horns, looking for all the galaxy like a devil (Wikipedia 2013a; Wookieepedia 2013b). The *Star Wars* wiki described his species:

> Zabrak were proud, strong, and confident beings. They believed that nothing was truly impossible, and strove to prove skeptics wrong at every turn. Some Zabraks carried themselves with an air of superiority, frequently discussing the achievements of their people with pride that could border on arrogance. As warriors or adventurers, Zabrak tended to be dedicated, intense, and extremely focused (Wookieepedia 2013f).

A Zabrak who had been forced to be a slave, however, would face ambivalence in his dedication, even worse than the three other avatars had experienced. The fact that Bester was a Zabrak and thus not a pure-blood Sith, which was a distinct species of extraterrestrial, meant he also faced racial discrimination during his training. However, he had not begun life as a cipher like Foyle, and conceivably could recover to some degree from his inferiority complex. His therapist, to use the term as a metaphor, was Ashara Zavros, a woman of the Togruta species from the planet Shili, who had been in training to become a Jedi, and must with difficulty negotiate her relationships between the Republic and the Empire when she becomes Bester's companion. Companions from a dozen different intelligent species can extend the scope of our psychiatric analysis of alien intelligences.

5 Alien Companions

Companions are secondary avatars that work closely with the primary avatars but have their own personalities and cultures, as reflected in the distinctive life stories they possess that shape some of the game's missions. One source of very explicit data about the cultures of some alien societies is the value-orientation of the 40 major companions in SWTOR. Over the course of a series of adventures ascending the ladder of experience, each of the eight classes of primary avatars sequentially gains five secondary avatars unique to that class, and can select one of the available secondaries at any time to be the one helping complete the current mission. Each of the five companions in a given class has certain combat abilities, but also has a personal moral code of likes and dislikes that in the case of aliens seems to reflect their native culture. A companion is a more effective assistant to the extent that it has gained *affection* toward the primary avatar, through approving the primary's words and deeds, and through gratitude over gifts received from the primary. A dozen of the companions who cooperated with my research avatars represent alien cultures, and are listed in Table 2. The data in the table came originally from online SWTOR databases (swtor.wikia.com, www.wikiswtor.com, www.wstor-spy.com), verified through direct familiarity with these characters.

The Togruta woman, Ashara Zavros, became Bester's companion after he deceived her into believing he had saved her life, and their relationship was a complex mixture of selfish calculation and gradual bonding. Because it was not based on truth, psychiatrists might call it a *folie à deux*, a two-person mental disorder, yet in this case it made them more normal rather than pathological—or pathological only with respect to Sith cultural standards. An appropriate socio-logical term could be *pluralistic ignorance*, a condition in which each person believes he or she is the only deviant, when in fact the functioning of society requires members to share illusions. For psychoanalysts, the chief example was religion, which they considered a *shared neurosis* or even a *shared psychosis* (Freud 1928; Roheim 1955). Ashara had been in training to become a Jedi within the Republic, but had too much independence to accept all Jedi teachings, an issue she and Bester sometimes discussed. One of the *Star Wars* wikis describes her species:

> The Togruta were a carnivorous humanoid species from the planet Shili. The race exhibited head-tails which were similar to those of Twi'leks. Togruta were also distin-guished by montrals, large hollow horn-like projections from the top of their heads that gave the species a form of passive echolocation. In order to protect themselves from dangerous predators, and to hunt their own prey, Togruta banded together in tribes and relied on their natural pigmentation to disrupt and confuse slow-witted beasts. Togruta worked well in large groups, and individualism was seen as abnormal within their culture, although it was also a necessary quality in leaders (Wookieepedia 2013d).

Twi'leks are similar in appearance to humans but with a variety of skin colors including blue and green, and posses two huge tails growing out of the backs of their heads that house special cognitive and communication functions. Both genders have

Table 2 Personalities of a dozen alien companions

Name	Species	Likes	Dislikes	Favorite Gift
Ashara Zavros	Togruta	Rational choices, secrets of the force, fighting bullies	Random cruelty, fighting Jedi	Republic memorabilia, cultural artifacts, military gear, weapons
Vette	Twi'lek	Anti-authority behavior, protecting the weak, treasure and getting paid	Bullying, killing innocents, kissing up	Cultural artifacts, underworld goods
Kaliyo Djannis	Rattataki	Disrespecting authority, casual violence, anarchy for the fun of it	Self-sacrifice for the greater good, sincerity, obedience, patriotic spirit and being taken advantage of	Luxuries, underworld goods, weapons
Lord Scourge	Pure-blood Sith	Using power against the weak, power, anger, revenge and spite	Greed, acts of mercy, Jedi and Republic authorities	Technology, Imperial memorabilia, trophies
Xalek	Kaleesh	Following the Sith Code, fighting overwhelming odds, brevity	Mercy, weakness, talking	Military gear, weapons
Aric Jorgan	Cathar	Efficiency, duty, the Republic military, honesty	Failure, excuses, callous sacrifices	Military gear, weapons
Sergeant Rusk	Chagrian	Killing Imperials, protecting the Republic, motivating others to fight	Avoiding fights, weakness, disrespecting authority	Republic memorabilia, weapons, military gear
Tanno Vik	Weequay	Ruthlessness, mercenary behavior, mocking authority and everyone else, blowing things up	Kindness, self-sacrifice	Weapons, underworld goods
Khem Val	Dashade	Killing Force users, displays of strength, making foolish people unhappy	Weakness in any form, not killing Force users	Cultural artifacts, weapons
Qyzen Fess	Trandoshan	Killing powerful enemies, encouraging others to defend themselves, danger, honor	Killing the weak, mercenary work, sparing powerful enemies	Weapons
Yuun	Gand	Mysteries, respect for unusual people or beliefs, patience, self-restraint	Unnecessary violence, chaos, rudeness, recklessness, bragging	Technology, cultural artifacts, trophies
Vector	Human who joined to the Killik	Diplomacy, helping people, exploring alien cultures	Greed, cruelty, prejudice, anti-alien sentiment	Imperial memorabilia, cultural artifacts

acute seduction skills and are often valued by slave-owning species like the Hutt. One was Oola, the dancing girl for Jabba the Hutt in *Return of the Jedi*, appearing also as a dancing NPC in the earlier MMO, *Star Wars Galaxies*. A *Star Wars* wiki says: "Twi'leks preferred to 'ride the storm' rather than 'defeat it,' as a proverb goes, and avoided to take a stand on any issue" (Wookieepedia 2013e). If they lack a solid core of conviction, and habitually use seduction to deal with other people, then they might be diagnosed as suffering from *histrionic personality disorder*. This is a somewhat vague psychiatric diagnosis, not far from the older categories of *hysteria* and *multiple personality neurosis*, but milder.

In the history of psychoanalysis, hysterics were the primary early focus of Freud's research, and through their emotive role-playing they may have seduced him into believing that their characteristics were universal flaws of mankind, rather than the symptoms of one particular disorder (Breuer and Freud 1936). Role-playing is central to MMOs, but we may wonder in the case of Twi'lek companions how much of their behavior is consciously simulated to elicit an emotional response from the particular primary avatar, rather than being a core characteristic of the Twi'lek companion. Seductiveness may seem like submissiveness, but it seeks to develop a successful relationship between two beings on the basis of one person satisfying the other's sensual desires, and thus is akin to sexual altruism but with qualities of subterfuge.

The table indicates that Kaliyo dislikes "self-sacrifice for the greater good," which is a definition of altruism. The *Star Wars* mythos does not present a clear picture of Rattataki culture, but it appears to have been a chaotic and violence-ridden adaptation to a hostile planetary environment and long isolation from other cultures. Anthony F. C. Wallace (1959) argued that societies existing in a chaotic environment tend to develop control-oriented psychotherapies, whereas those experiencing high stability tend to develop psychotherapies that emphasize liberation from control. However, Kaliyo has not sought therapy and acts out the sociopathic nature of her own background, even cuckolding an Imperial Agent character like Asimov, while acting as his companion:

> Kaliyo Djannis prizes her freedom and will lie, murder and blackmail in order to ensure that she is in control of a situation and able to indulge her vices. Known to pursue lengthy vendettas to redress grievances. Possesses a track record of expertly manipulating employers, lovers and associates (agents should not be fooled by attempts at seduction). As with many mercenaries, her loyalty cannot be purchased, but her services can be—if only temporarily (SWTOR Spy 2013).

One of the clearest representative of the Dark Side of the Force is Lord Scourge, the fifth companion gained by a Jedi Knight, who is a Sith and thus diametrically opposed to some of the Jedi's own values. He values power over other people but not greed, thus conceptualizing power as a goal in itself, rather than as a tool for gaining economic advantage. Technically, *sadism* implies inflicting pain to achieve erotic pleasure, but Scourge seems more stern than sensual (de Sade 1953). Normally, we would not expect a Sith to become the assistant to a Jedi, but Lord Scourge was part of a plot to assassinate the Emperor, superficially motivated by concerns that the Emperor sought to destroy all life in the galaxy, but perhaps also by Lord Scourge's wish to become emperor himself. Now he has fled the Empire, and became an unreliable ally of the Republic. His values are quite apparent, so he is not a wolf in sheep's clothing. He is more accurately a wolf in fox's clothing who follows The Sith Code:

Peace is a lie, there is only passion.
Through passion, I gain strength.
Through strength, I gain power.
Through power, I gain victory.
Through victory, my chains are broken.
The Force shall free me (Wookieepedia 2013a).

Xalek is a laconic Kaleesh. Introverted yet aggressive, this species believed members could become gods after death, following the same code as the Sith, but concealing their passions rather than expressing them. *Autism* would be too strong a diagnosis, but it not far from the mark. Despite their phonetic similarity, autism and altruism are negatively correlated.

Aric Jorgan belongs to the cat-faced Cathar species, somewhat calm in demeanor and well-controlled. He had worked with the original Havoc Squad and was quite surprised by their defection, apparently assuming that all his fellow soldiers exercised as much self-control and submission to duty as he did. Despite belonging to different species, he and Heinlein are similar in this respect. If asked to define altruism, they might offer "duty" and "loyalty" as synonyms, and might not spontaneously mention doing unselfish good to strangers in their definitions.

Rusk rebelled against the pacifism of his Chagrian background, becoming a most aggressive warrior for the Republic. As a psychiatric symptom, this is *reaction formation*, a defense mechanism in which a person responds to unacceptable emotions by exaggerating their opposites. This means that the person will not be very competent in handling the emotions that dominate behaviors, partly because of the internal conflict, and partly because the person has not learned the skills needed to implement the superficial emotion through an integrated personality. Thus Rusk takes far too many risks in military action, and as a low-level officer constantly endangers his troops and displeases his commanders, while some politicians respond positively to his false bravado.

Tanno Vik is a *sociopath*. His species, the Weequay, are clan-based and apparently were easily exploited by other species whose societies were not so heavily based upon social solidarity. However, he grew up on the streets of Nar Shaddaa, an urban moon dominated by the Hutt criminal cartels, so he did not have the advantage of traditional social support. In her book about Japanese society, *The Chrysanthemum and the Sword*, Ruth Benedict (1946) said Japan was a *shame culture* in which the threat of immediate social ostracism enforces morality. If an individual is taken out of such a society, all support for morality is lost and the individual becomes a sociopath. In a contrasting *guilt culture*, such as Protestant societies, an individual internalizes a moral code and will follow it even in the absence of continuing social support.

Khem Val belongs to the Dashade, who were nearly exterminated in an earlier war between the Sith and Jedi. Immune to the Force, they were repelled by both its Light and Dark sides. He was held in suspended animation for many years, until released by Bester. Now he serves Bester, but warns that at any opportunity he may kill Bester and any other Force-capable characters in the vicinity, such as Ashara Zavros. Not having much information about Dashade culture prior to the disasters, we must speculate he suffers from *post-traumatic stress disorder*.

Qyzen Fess belongs to a reptilian hunter species. Just as they hunted wild game in earlier centuries, they are now avid gamers who compete as lone individuals across the galaxy, believing a goddess called the Scorekeeper will reward them for victories but despise themselves if they ever are defeated. Given the fact that SWTOR is a game, many of the characters are dedicated to winning, but Qyzen Fess sank into depression when he lost, and only revived by the idea of flying away with the low-level Jedi Consular I used to investigate him. This suggests he has what Henry Murray (1955) called an *Icarus complex*, obsessed with unrealistic ascension, flying high both figuratively and physically.

The two concluding companions in the table, Yuun and Vector, are in one respect the most alien because they are at least partly insects, yet among the most mentally healthy. Yuun's species evolved on a planet with an ammonia atmosphere, so he must always wear estranging breathing equipment. While the Gand originally had a hive mind, as they evolved intellectually a degree of individualism became possible, and a few rare individuals like Yuun developed their own identities, not like so many others in the table as the result of psychopathology, but through a healthy, modesty-tempered *achievement motivation* (McClelland 1961). Yuun is one of his people's rare "findsmen," or shamanistic trackers, and shamanism was often associated with *schizophrenia* in the culture and personality literature (Silverman 1967). Yet Yuun exhibits no symptoms of schizophrenia, despite being a social outcast within the Republic's forces, and he has applied his unusual mind to achieving excellence with technology. One of the *Star Wars* wikis explains the cultural background of this admirable person:

> Gands were considered by galactic society to be a very humble species, a trait resultant of their culture, which dictated that an individual's identity had to be earned. Accordingly, most Gands were self-deprecating and polite, and usually referred to themselves in the third person. First-person pronouns were reserved for the most legendary of Gands, as the usage presumed that one was so renowned that everyone knew one's name. However, an accomplished Gand often responded with humility if praised and even downplayed his or her achievements (Wookieepedia 2013c).

Vector was once a normal human being who accepted a difficult diplomatic mission to the Killik insect species that was native to the planet Alderaan, the world from which Princess Leia came and which was shattered to splinters before her very eyes in the original 1977 movie. But SWTOR takes place earlier in history than the movie, and much of the action on this world is designed to explore the feudal society established there by humans, which is reminiscent of the Martian society described in the novel by Burroughs, but also resembles the system of aristocratic houses described in Frank Herbert's *Dune* series (1965). The Killiks were a hive-mind species, native to Alderaan, who mostly had migrated to other star systems, leaving behind only a few isolated nests that existed on the fringes of human society. A recurrent theme in the Alderaan missions of SWTOR is a human voluntarily or involuntarily becoming a *Joiner*, transformed by powerful pheromones into a member of a Killik nest. Vector, as his name implies in biological terminology, has become a link between humanity and the Killiks, shifting rather gracefully back and forth between total collectivism and some measure of individualism. Except for the fact that his eyes had turned totally black, he appeared to be a normal human, and indeed he may have achieved an entirely new form of normality, transcending the culture from which he came.

6 Conclusion

A key assumption of SWTOR, and indeed of much science fiction, is that extra-terrestrial civilizations will interact through direct contact, which means they will be in competition for natural resources and in a position to war against each other. In contrast, much serious writing about interstellar communication assumes it will be carried out by the equivalent of radio, and neither side will be in a position to help or harm the other, except through the information it provides. However, in doing so they may follow the habits their particular civilization developed in the more mundane context of their home world.

Imagine these possibilities. The Voss detect signals from another civilization, but choose not to reply. The Hutt engage in lively interstellar commerce in such goods as musical recordings and technology designs, but they do so duplicitously, using their skills as confidence artists to achieve great profits as trusted but deceitful traders. The Sith also deceive the civilizations with which they exchange radio signals, but with the purpose of destroying them and becoming the sole civilization in the galaxy, for example by broadcasting plans for machines that will supposedly provide great prosperity but actually are planet-destroying weapons. The poor Twi'leks vacillate and wind up being the victims of the Hutt and Sith. And which civilization in the galaxy practices altruism? We would like to think it will be the humans.

References

Adler, Alfred. 1927. *Understanding Human Nature*. Greenwich, CT: Fawcett.
Adler, Alfred. 1928. *Individual Psychology*. Totowa, NJ: Littlefield, Adams.
Asimov, Isaac. 1950. *I, Robot*. New York: Grosset and Dunlap.
Asimov, Isaac. 1954. *The Caves of Steel*. Garden City, NY: Doubleday.
Asimov, Isaac. 1957. *The Naked Sun*. Garden City, NY: Doubleday.
Axelrod, Robert M. 1984. *The Evolution of Cooperation*. New York: Basic Books.

Bainbridge, William Sims. 1986. *Dimensions of Science Fiction*. Cambridge, MA: Harvard University Press.

Bainbridge, William Sims. 2010a. *Online Multiplayer Games*. San Rafael, CA: Morgan and Claypool.

Bainbridge, William Sims. 2010b. *The Warcraft Civilization*. Cambridge, MA: MIT Press.

Bainbridge, William Sims. 2011. *The Virtual Future: Science-Fiction Gameworlds*. London: Springer.

Bainbridge, William Sims. 2012a. "The Harvard Department of Social Relations." In *Leadership in Science and Technology*, edited by William Sims Bainbridge, 496–503. Thousand Oaks, CA: Sage.

Bainbridge, William Sims. 2012b. "The Psychoanalytic Movement." In *Leadership in Science and Technology*, edited by William Sims Bainbridge, 420–428. Thousand Oaks, CA: Sage.

Bainbridge, William Sims. 2013. *eGods: Fantasy Versus Faith*. New York: Oxford University Press.

Bakan, David. 1958. *Sigmund Freud and the Jewish Mystical Tradition*. Princeton, NJ: Van Nostrand.

Beckett, Samuel. 1956. *Malone Dies*. New York: Grove Press.

Benedict, Ruth. 1934. *Patterns of Culture*. Boston: Houghton Mifflin.

Benedict, Ruth. 1946. *The Chrysanthemum and the Sword: Patterns of Japanese Culture*. Boston: Houghton Mifflin.

Bester, Alfred. 1953. *The Demolished Man*. Chicago: Shasta.

Bester, Alfred. 1956. *The Stars My Destination*. New York: New American Library.

Bishop, Paul. 1995. *The Dionysian Self: C. G. Jung's Reception of Friedrich Nietzsche*. New York: W. de Gruyter.

Breuer, Josef, and Sigmund Freud. 1936. *Studies in Hysteria*. New York: Nervous and Mental Disease Publishing Company.

Brown, J. A. C. 1967. *Freud and the Post-Freudians*. Baltimore: Penguin.

Burroughs, Edgar Rice. 1917. *A Princess of Mars*. Chicago: A. C. McClurg.

Camus, Albert. 1946. *The Stranger*. New York: Knopf.

de Sade, Marquis. 1953. *The Bedroom Philosophers*. Paris: Olympia.

Edgerton, Robert B. 1966. "Conceptions of Psychosis in Four East African Societies." *American Anthropologist* 68:408–424.

Finkel, Norman J. 1976. *Mental Illness and Health*. New York: Macmillan.

Frankl, Viktor E. 1962. *Man's Search for Meaning*. Boston: Beacon Press.

Freud, Sigmund. 1928. *The Future of an Illusion*. New York: H. Liveright.

Heinlein, Robert A. 1948. *Space Cadet*. New York: Scribner.

Heinlein, Robert A. 1959. *Starship Troopers*. New York: Putnam.

Heinlein, Robert A. 1961. *Stranger in a Strange Land*. New York: Putnam.

Heinlein, Robert A. 1966. *The Moon is a Harsh Mistress*. New York: Putnam.

Herbert, Frank. 1965. *Dune*. Philadelphia: Chilton.

Herrigel, Eugen. 1953. *Zen in the Art of Archery*. New York: Pantheon.

Homans, George C. 1974. *Social Behavior: Its Elementary Forms*. New York: Harcourt, Brace, Jovanovich.

Jung, C. G. 1923. *Psychological Types*. New York: Harcourt, Brace.

Kaplan, Bert, ed. 1961. *Studying Personality Cross-Culturally*. New York: Harper and Row.

Kelleghan, Fiona. 1994. "Hell's My Destination: Imprisonment in the Works of Alfred Bester." *Science Fiction Studies* 21:351–364.

Keyes, J. Gregory. 1998. *Dark Genesis: The Birth of the Psi Corp*. New York: Ballantine.

Keyes, J. Gregory. 1999a. *Deadly Relations: Bester Ascendant*. New York: Ballantine.

Keyes, J. Gregory. 1999b. *Final Reckoning: The Fate of Bester*. New York: Ballantine.

Lupoff, Richard A. 1976. *Barsoom: Edgar Rice Burroughs and the Martian Vision*. Baltimore: Mirage.

Maddi, Salvatore R. 1968. *Personality Theories: A Comparative Analysis*. Homewood, IL: Dorsey Press.

Malinowski, Bronislaw. 1927. *Sex and Repression in Savage Society*. New York: Harcourt, Brace.

McClelland, David C. 1961. *The Achieving Society*. Princeton, NJ: Van Nostrand.

Merton, Robert K. 1970. *Science, Technology, and Society in Seventeenth-Century England*. New York: Harper and Row.

Murray, Henry. 1955. "American Icarus." In *Clinical Studies in Personality*, volume 3, edited by Arthur Burton and Robert E. Harris, 615–641. New York: Harper and Row.

Nietzsche, Friedrich. 1872. *Die Geburt der Tragödie*. Munich: Goldmann.

Nietzsche, Friedrich. 1885. *Also Sprach Zarathustra*. Stuttgart: Kroner.

Opler, Marvin K., ed. 1959. *Culture and Mental Health*. New York: Macmillan.

Palumbo, Donald E. 2004. "The Monomyth in Alfred Bester's *The Stars My Destination*." *Journal of Popular Culture* 38:333–369.

Panshin, Alexei. 1968. *Heinlein in Dimension*. Chicago: Advent.

Rinzer, J. W. 2007. *The Making of Star Wars*. New York: Ballantine.

Roheim, Geza. 1955. *Magic and Schizophrenia*. Bloomington, IN: Indiana University Press.

Searle, Michael. 2011. *Star Wars: The Old Republic Explorer's Guide*. Roseville, CA: Prima Games.

Shoji, Yamada. 2001. "The Myth of Zen in the Art of Archery." *Japanese Journal of Religious Studies*, 28:1–25.

Silverman, Julian. 1967. "Shamans and Acute Schizophrenia." *American Anthropologist* 69:21–31.

Skinner, B. F. 1938. *The Behavior of Organisms*. New York: Appleton-Century.

Suzuki, D. T., Erich Fromm, and Richard De Martino. 1960. *Zen Buddhism and Psychoanalysis*. New York: Harper.

SWTOR Spy. 2013. "Kaliyo D'jannis." Accessed January 3, 2013. http://www.swtor-spy.com/companions/kaliyo-djannis/13/.

Thrasher, Frederic M. 1927. *The Gang*. Chicago: University of Chicago Press.

Tönnies, Ferdinand. 1957. *Community and Society*. East Lansing, MI: Michigan State University Press.

Von Neumann, John, and Oskar Morgenstern. 1944. *Theory of Games and Economic Behavior*. Princeton, NJ: Princeton University Press.

Wallace, Anthony F. C. 1959. "The Institutionalization of Cathartic and Control Strategies in Iroquois Religious Psychotherapy." In *Culture and Mental Heath*, edited by M. K. Opler, 63–96. New York: Macmillan.

Watts, Alan. 1961. *Psychotherapy, East and West*. New York: Vintage Books.

Weber, Max. 1930. *The Protestant Ethic and the Spirit of Capitalism*. London: Allen and Unwin.

Wells, H. G. 1895. *The Time Machine*. New York: H. Holt.

Westfall, Richard. 1958. *Science and Religion in Seventeenth-Century England*. New Haven, CT: Yale University Press.

Wiggins, Jerry S., ed. 1996. *The Five-Factor Model of Personality: Theoretical Perspectives*. New York: Guilford Press.

Wikipedia. 2013a. "Darth Maul." Accessed January 3, 2013. http://en.wikipedia.org/wiki/Darth_Maul.

Wikipedia. 2013b. "The Stars My Destination." Accessed January 3, 2013. http://en.wikipedia.org/wiki/The_Stars_My_Destination.

Williams, Dmitri, Tracy L. M. Kennedy, and Robert J. Moore. 2011. "Behind the Avatar: The Patterns, Practices, and Functions of Role Playing in MMOs." *Games and Culture* 6:171–200.

Williamson, Jack. 1950. *The Humanoids*. New York: Grosset and Dunlap.

Wilson, Edward O. 1971. *The Insect Societies*. Cambridge, MA: Harvard University Press.

Wilson, Edward O. 1975. *Sociobiology*. Cambridge, MA: Harvard University Press.

Wookieepedia. 2013a. "Code of the Sith." Accessed January 3, 2013. http://starwars.wikia.com/wiki/Code_of_the_Sith.

Wookieepedia. 2013b. "Darth Maul." Accessed January 3, 2013. http://starwars.wikia.com/wiki/Darth_Maul.

Wookieepedia. 2013c. "Gand." Accessed January 3, 2013. http://starwars.wikia.com/wiki/Gand.
Wookieepedia. 2013d. "Togruta." Accessed January 3, 2013. http://starwars.wikia.com/wiki/
    Togruta.
Wookieepedia. 2013e. "Twi'lek." Accessed January 3, 2013. http://starwars.wikia.com/wiki/
    Twi'lek.
Wookieepedia. 2013f. "Zabrak." Accessed January 3, 2013. http://starwars.wikia.com/wiki/
    Zabrak.

Evolutionary Perspectives on Interstellar Communication: Images of Altruism

Alfred Kracher

Abstract At least some part of an interstellar message of altruism will be in the form of images. Designing such iconic messages will require a comprehensive dialogue among representatives of science, arts and humanities, and include non-Western traditions. As a basis for discussing the likely success (or failure) of communicating cultural concepts to extraterrestrials, we can first look at some examples how this has been done across cultures here on Earth. I use the example of religious icons on altruism, because it is an area where the repertoire of iconic messages has been studied in some detail. We can then address particular questions about using this approach in interstellar messages. How much terrestrial experience can help depends on how human-like the recipients of our message are thought to be. Some conclusions are: (a) Images have to be combined with other forms of communication. (b) Terrestrial experience can be valuable for alien recipients that are human-like, but even in this case the difference in evolutionary history is a significant barrier to understanding. (c) In the case of aliens that are not human-like it is unclear what the equivalent of altruism would be for them. (d) Sending multiple diverse messages increases our chances to be understood, but also to be misunderstood. (e) A serious effort at designing interstellar messages of altruism will have positive consequences for humanity even if no alien civilization acknowledges our transmission.

Keywords Aesthetic strategy · Art · Convergent evolution · Iconography · Imagination · Layered messages · Madonna images

A. Kracher (✉)
15837 Garden View Drive, Apple Valley, MN 55124-7006, USA
e-mail: akracher1945@gmail.com

D. A. Vakoch (ed.), *Extraterrestrial Altruism*, The Frontiers Collection,
DOI: 10.1007/978-3-642-37750-1_19, © Springer-Verlag Berlin Heidelberg 2014

1 Introduction

The only aliens we have so far encountered are the ones existing in our imagination. This is a problem for interstellar message composition, because our imagination is limited. Although the technical aspects of interstellar communication and the astronomical issues of habitability are obviously the domain of science, the form of imagination required for communication itself falls at least partly in the domain of art. Interstellar communication is therefore a field where a thorough and detailed conversation between artists and scientists is a fundamental necessity. To begin with, the demarcations between science, art, religion, etc., are the product of a particular historical process in our particular culture, and not even universal among humans. We certainly cannot expect such boundaries to apply in the same way among other intelligent beings, although they may have different demarcations, unintelligible to us. In as much as our own compartmentalization of thought constrains our imagination, it is a liability in our efforts to communicate.

Altruism is a theme that transcends these traditional boundaries. It spans biology, social science, morality, and much else besides, and it can stimulate thinking on the kind of general scale that will presumably be required if we ever were to communicate with extraterrestrial intelligence. Moreover communication about altruism within and between cultures here on Earth has been going on in many forms, by words, images and actions, which provides us with a pattern for thinking about interstellar communication.

This chapter discusses the exercise of our imagination by using a concrete example, the transmission of an image symbolizing altruism. My example are "Madonna images" (the terminology is explained below), pictures of mother and child, or more generally adults and children in a specific familial setting, that are familiar from both religious and secular context as paradigms of one form of altruism. Through considering what kind of responses these images evoke as intercultural communication, how an image can or even must be supplemented with other messages, what misunderstandings arise, etc., our imagination can create a phase space for potential messages that have some chance of being successful. (For a summary of central questions relevant to communicating altruism in interstellar messages, see Table 1.) By choosing pictures as the focus of this chapter I do not mean to imply that images are the *only* way to convey altruism. As I argue below, a range of forms is desirable, perhaps even necessary. The following section explores the constraints on imagination before putting it to use reflecting on the design of potential messages. The intent of my examples is not to propose specific pictures for interstellar messages, but to stimulate discussion about the specific issues and problems that arise from iconic communication with extraterrestrials.

Table 1 Communicating altruism in interstellar messages

1. Would aliens understand the concept of altruism?	
Yes	**No**
Biological aspects	
Evolution may converge on humanoid organisms because of physical constraints	a. Cyborgs with machine-like qualities b. Members of a Hive Queen society c. James Blish's Lithians
Cultural aspects	
Technological societies need cooperation of individuals	Cooperation may be self-evident to them and not need a special concept to explain it
2. By what means can we communicate altruism?	
Yes	**No**
Through mathematical principles	
Mathematical principles are independent of culture	The understanding of these principles could be intuitive rather than conscious A formal language by itself may not be sufficient to convey altruism
By images	
Images of altruism are (almost?) universally understood among terrestrial cultures	Like other forms of communication, images rely on a social context to be interpretable
Images are rich in information ("worth a thousand words")	The possibility of multiple interpretations makes images inherently ambiguous
3. Are images of infant care a comprehensible icon of altruism?	
Yes	**No**
As long as recipients are sufficiently human-like with regard to infants	If recipients are non-humanoid (see "biological aspects" above)
If there are no cultural obstacles	If recipients are iconoclasts

2 The Limits of Imagination

Art, science and religion all rely in some way on human imagination. To put imagination to work, as it were, with regard to interstellar messages, we have to be aware of its limitations. Imagination is limited in a twofold way (Kracher 2006, 336):

1. We do not and perhaps cannot imagine everything that is possible in our universe. This is well expressed by J. B. S. Haldane's famous aphorism that the universe may well be "not only queerer than we suppose, but queerer than we *can* suppose."
2. We can and do imagine things that are impossible or entirely useless in our real world. This aspect of the imaginary world is commonly called "fantasy," and in art it is subject to its own rules and constraints.

Creativity in both the artistic and scientific sense pushes against the first limit, and scientific analysis delineates a boundary toward the second. Art and science tend to play complementary roles in this and have different strengths and weaknesses. Since science is in general analytic and reductive, it can provide detailed constraints in specific areas. But it is much harder, based on science alone, to arrive at conclusions about extraterrestrials that have overall plausibility. In this case art, for example storytelling in the form of science fiction, offers a holistic complement. The success of a story depends on our ability to imagine it. No imagined world, no story. What makes a story successful is a kind of inner logic and plausibility. Even though a story does not have to be about something that is actually possible in real life, it can still be judged on this basis (Kracher 2006, 337). On the other hand, with regard to interstellar communications only those stories are helpful that do not violate known scientific principles. Freewheeling imagination becomes productive creativity in this enterprise by exploring the scientific constraints that give a realistic shape to the imagined objects.

By making a distinction between artistic and scientific imagination I also want to draw attention to how individual imagination is shaped by personal preferences and professional experience. This inevitably leads to individually different expectations about how our messages might be received. One simple example is the question of how to make images decodable in a way that aliens would find obvious. Scientists who professionally work with mathematics find it obvious to encode a black-and-white picture as a matrix of $P1 \times P2$ pixels, where $P1$ and $P2$ are prime numbers (this can of course be extended to more than two dimensions). That mathematics as such is universal is incontrovertible, but the assumption that any civilization advanced enough to receive our message would recognize the significance of prime numbers for decoding is not entirely certain. Oliver Sacks describes humans who seem to have an intuitive access to such mathematical principles while intellectually barely functional in other respects (Sacks 1970, 195–213). It is at least conceivable that this poorly understood phenomenon of intuition, which is associated with pathology in humans, is the basis for technology in extraterrestrials. For them prime numbers might not stand out as unusual in the way they do for human scientists.

3 Sending Messages

These considerations provide the framework for a discussion of communication proper, understood in the sense that at least some *mutual* comprehension is possible. The question *can we communicate* with extraterrestrials is not one that can be answered wholesale by yes or no. Even if we exclude from the definition of *communication* the possibility of merely being noticed (maybe because our powerful military radar signals are detected), there are any number of deliberate actions conceivable on our part that are potentially interpretable by a recipient. Just think of the fact that "sending a message" has become a political euphemism for inflicting violence. Here however we will limit discussion to messages that have the potential to convey altruism. After some brief remarks about the nature of suitable methods of communication we will consider message content in the following sections.

By far the most promising kind of signal for reaching an extraterrestrial culture is some kind of electromagnetic transmission. How closely message content is connected to transmission method can vary. For example, Vakoch (2008) suggests that transmission frequencies can themselves constitute a message. In addition to transmitting at content-carrying frequencies, conceivably several layers of content could be added to such a message in the manner suggested by Carl Sagan (1985) in his novel *Contact*. The point he makes is that the more complex and richer layers may be discovered only successively by the recipient, and partly on the basis of understanding the content of the simpler layers. I will refer to such multi-level messages with a wrapper-and-content metaphor, so that the simplest layer is the outer "envelope," the increasingly complex information being contained "inside" successive layers of wrappers.

Besides complexity the issue of ambiguity has to be considered. This is particularly important in the following discussion, because in this chapter I will focus on pictures. A picture may be worth a thousand words, but it may also be subject to a thousand interpretations, whereas words admit to only very few. We cannot foresee all possible interpretations of messages that we send, but we should try to consider as many as possible. How to achieve this is one main purpose of this chapter.

As a rule more complex messages like images admit to a larger scope of possible interpretations: what are the things that a picture shows? What is it trying to tell us? Are the depicted objects the important aspect, or is it how they act? If we want the recipient to focus on action rather than objects, a movie or at least a sequence of images would be preferable, but that may not always be feasible, at least not in an easily decodable way. Images are therefore inherently ambiguous. In this regard we need to distinguish between *intent* and *content*. Ambiguity of content may be inevitable, or we may even deliberately send ambiguous messages. But if the latter, we should nonetheless ourselves be clear that this is what we intend to do. For the interpreter of the message this raises of course the problem that the (potentially ambiguous) *content* of the message may not reveal the *intent* of the sender. In the case of altruism in particular it may be enough to convey that we do think about this issue, and that it is of sufficient importance to us to make it the subject of an interstellar message.

I do not think that we can know enough about our potential conversation partners to be able to express ourselves unambiguously, and based on principles of physics I hope that it does not matter. As long as we are only exchanging electromagnetic signals, the damage that we can do to each other is probably limited. As Randolph et al. (1997) point out, SETI by itself does not raise issues of planetary protection. Immediate physical contact is so much less likely than a long-distance exchange of messages that until we accomplish the latter it is pointless to worry about the former. Even so, it may be prudent to consider the ethical effects that a message might have on potential recipients. Alan Rubin (2002, 299–320) has compiled a list of potential human responses to first contact, as well as reasons for seeking it, and his compilation makes it clear that the aftermath of first contact is likely to be a complicated business. There is every reason to believe that the same would apply the recipients of our message.

The following discussion assumes recipients that are in a general sense humanoid, the products of an evolutionary process analogous to the one to which we owe our own existence. How similar or different evolution on another astronomical body might turn out to be is a highly controversial subject. Arguments that evolutions will converge on humanoid aliens have been made for example by Ronald Fox (1997) and Sjoerd Bonting (2003); a more general view of convergent evolution is presented by Simon Conway Morris (2003), who believes however that the evolution of intelligence is very rare. Stephen Jay Gould has argued against convergentism (e.g., Gould 1999). In the case of intelligent life, the development of culture will necessarily add another dimension of complexity to biological evolution. Because of these uncertainties we cannot be sure that aliens are sufficiently like us, but we have to start with some assumptions to organize our thinking. As I have argued elsewhere, when it comes to aliens a form of critical anthropomorphism is the only viable heuristic strategy (Kracher 2002). We start with a list of our own features and attributes, then try to assess how likely or unlikely it is that extraterrestrials would share them.

In reality aliens may of course not be very anthropomorphic. At least two such extraterrestrials are conceivable. One of these are intelligent machines or, perhaps more likely, cyborgs in which machine-like qualities dominate over organic ones. Although it is inherently interesting to consider the question what the equivalent of altruism might be for such a civilization, this would require a much more extensive analysis than is available so far. Communication with them may well turn out to be impossible in principle. The other non-humanoid population I call the "Hive Queen scenario," a species more akin to eusocial insects like ants and bees, in which the intelligence would be concentrated in a single individual (the Hive Queen) or a small "aristocracy." Following the technical term used in biology such a community could be called a "eusociality," but here I adopt "Hive Queen" borrowed from the *Ender* novels of Orson Scott Card. I am not sure that the required form of intelligence, specifically one that gives rise to technology, could actually evolve in a Hive Queen community, but it is a possibility that should be considered.

4 Altruism

Altruism is itself not an unambiguous concept. In biology it refers to a genetic predisposition of individuals to cooperate and can be observed in a large number of animal species. On our current understanding this trait is a prerequisite for the evolution of intelligence, and therefore we expect to share some form of it with extraterrestrials. In a moral sense, however, altruism is something that we strive to attain rather than possess automatically as a species. We may call it a virtue or moral ideal; it belongs to the "horizon of human conduct" (Kracher 2010) that guides our actions in a general way, and which I believe to be in some sense universal.

Such abstract ideals are hard to articulate even across human cultures, let alone in a message to beings unknown. However, when we consider how to go about communicating the concept, we are not entirely without examples. I rely here on an idea of Loyal Rue (1998; 2005), who has pointed out that virtues such as compassion, sacrifice, serenity, etc., have been promoted across diverse human cultures for a long time by the world religions, and that art has played a prominent role in this process. For Rue the act of proclaiming and encouraging these virtues is the primary function of religion, and he refers to art as part of the "aesthetic strategies" that religions use to promote their ideals.

One reason why images are particularly successful in conveying the idea of altruism is that they avoid the necessity of translation inherent in verbal communication. This is not simply a problem of literacy encountered by missionaries; after all, we assume that a technological extraterrestrial society would be literate. But there is a specific grammatical requirement in communicating *moral* altruism. If I am correct that moral altruism requires the deliberate transference of genetically based altruism to objects we can choose, e.g., treating adopted children as if they were biological offspring (Kracher 1996), then the language we use not only has to convey an *algorithmic* conditional (if–then–else), but a *counterfactual* condition (treat them/us as if they/we were kin). Since this is not merely a formal issue, I find it difficult to believe that any language-like message could successfully convey the concept of altruism on its own, without being integrated with other forms of communication. From a different starting point Paolo Musso (2011), who proposes to communicate cultural concepts by analogy, has likewise concluded that integrating formal language with iconic forms of communication will be necessary for successful communication.

Integration of language and images is what we actually find in Rue's examples of altruism as promoted by religions. The history of this form of disseminating moral ideas through art is very long, thousands of years in some cases. It is thus in principle possible to study the response of different human cultures to these symbolic expressions, to ask how successfully the intended ideas were conveyed, and to what extent the imagery was culturally contingent. Given that these messages have often been misinterpreted, we also need to ask where such communication may go wrong, either in transmission or reception.

The following examples come from the Christian tradition, partly because of familiarity. But it is also fair to say that out of the five world religions whose aesthetic strategies are examined by Rue (2005), Christian art focusses in a special way on acts of altruism (later I will discuss the cross-cultural implications of this). Story illustrations such a the Good Samaritan in the Gospel of Luke, or the legend of St. Martin sharing his cloak with a beggar, are presumably first of all directed to followers of the religion, to give them examples of conduct they should emulate. But it is clear that there is also a *missionary* aspect to these messages, an aim to advertise one's religion as one that values altruism. As such they are specifically intended to work across cultural boundaries, and their history can tell us something about cross-cultural message composition. We cannot be sure that what we learn from studying such images is applicable to extraterrestrials. Aesthetic strategies are after all designed to appeal to a genetic predisposition within the recipient. The fact that extraterrestrials, unlike fellow humans, do not share our evolutionary history makes it difficult to know how far the cross-cultural missionary experience is applicable to interstellar communication, but it is a place to start.

Perhaps Rue's most poignant example of pictorial religious messages are what I call "Madonna images," and they can serve as a useful paradigm of artistic portrayals of moral maxims. The next section explains their iconic content and their appeal as symbols of altruism. I choose this example not because I think this is necessarily the kind of picture we should transmit to aliens, but because it is a good example for bringing out the problems that need to be considered and discussed in the context of iconic interstellar messages. This is therefore not a proposal of *what* images to send, but *how to think about* the kind of images that are likely to promote mutual understanding.

5 Madonna Images

The label "Madonna images" for the type of pictures I use as example derives from the paradigmatic image of Mary with the baby or infant Jesus (Virgin and Child). However, as used here the label is meant to include several iconographic types, such as pictures that also show John the Baptist (Fig. 1), the Holy Family, Mary's mother, etc. In Western art such mother-and-child images, with or without extended family, have been popular for many centuries, not only in religious but also in secular forms. For example, practically all the known artistic compositions of Madonna iconography just mentioned can also be found in secularized versions in the oeuvre of the American Impressionist painter Mary Cassatt (1844–1926). And even though the subject may not be as common in other cultures, it seems that these images are at least *universally understood* across human cultures as icons of altruism. The point would have to be verified by anthropological studies, though I am confident that it is true.

Some of the appeal of these pictures ultimately derives from a genetic predisposition to act altruistically toward infants, a trait that is widespread in nature. There are several reasons to think that care of one's young is the best illustration of altruism we can muster. Among humans it is perhaps its least controversial manifestation, and it is a widespread trait in our imaginary panoply of aliens. First, the prolonged infancy of humans is a byproduct of the evolution of intelligence. A genetic predisposition for infant care is found in many animals and certainly not unique to humans. However, the particular form of human infant care with its intricate acquisition of language as well as other uniquely human features must have co-evolved with a high-level intelligence (e.g., Pinker 1994). Second, the development of culture, which is surely a prerequisite for the kind of technology that would be able to communicate with us, would seem to require teaching and learning across generations.

For these reasons, benign interactions between adults and infants may be as universal an icon of altruism as we can conceive. If it is really true that evolution

Fig. 1 Raphael (Raffaello Sanzio, 1483–1520), Virgin Mary with Jesus and John the Baptist, known as *Madonna del Prato*, c. 1506. Vienna: Kunsthistorisches Museum. Accessed January 4, 2013. http://upload.wikimedia.org/wikipedia/commons/5/51/Raffaello_belvedere_madonna.jpg

of intelligence has to converge on beings that are in many ways human-like, they would give birth to relatively helpless infants which, unlike insect larvae, none-theless rather resemble adults on a smaller scale. They would also share traits such as slow development and consequently the need for parental care, education, and perhaps other forms of cross-generational communication. Therefore there is some probability that extraterrestrial recipients would interpret Madonna images (or generally pictures of infants with parents or caregivers) as icons of altruism. Later I will return to the question how (im)probable this is.

6 Problems of Cultural Diversity

There is, however, another aspect to the popularity of this subject in terrestrial art. On closer inspection the many variations of the theme are also a clue that their message is not context-free. If one looks at collections of Madonna images from many periods and artistic styles it quickly becomes apparent how the art form is shaped by pervasive iconographic conventions. There is a reason for every change in style, and in many cases a reason for every detail in the picture. Typically it is the cultural context of the times that shapes the message, even though a viewer at a later time may be unaware of it and can no longer decipher the subtleties. If this partial loss of content can occur even within the same human culture in the space of a few hundred years, how much more difficult will it be to find an adequate representation for recipients who are non-human and probably in an evolutionary phase that is different from ours by millennia. Even with our best efforts at a context-neutral representation, sending pictorial messages is more problematic than it might appear. Once we design new images capable of electronic transmission over interstellar distances, technology inevitably becomes a new context that will create its own iconographic subtleties.

To illustrate the point I have chosen the particular picture in Fig. 1 to contain an implicit altruistic reference that is not entirely obvious, even though the picture comes from a mostly familiar cultural background. This is the fact that we see two children who are too close in age to be siblings. The Renaissance painter Raphael (Raffaello Sanzio), who created many versions of this particular subject, may have just liked it for its artistic symmetry. However to understand its iconic message requires knowledge of the Gospel of Luke, which informs us that Mary is here not only looking after her own child, but also that of her kinswoman Elisabeth (Luke 1:36–58).

I have chosen the painting in Fig. 1 to point out how difficult it may be for aliens to, as we say, "get the full picture" from any iconic communication, if we might miss a point even if it comes from within our own cultural tradition. A picture actually suitable for interstellar transmission would surely have to be much simpler and without such subtleties. Note also how our culture, and with it our iconic understanding, has changed since the times of Raphael, when viewers would have been more familiar with gospel stories. The roughly 500 years since then are not an unreasonable time to expect for only one or at most a few interstellar exchanges. It would take a long time if we needed to explain ourselves. The fact that within the number of generations since Raphael we have lost some understanding of our own messages warns us about the fragility of context. Among extraterrestrials, generations may be longer, but they cannot be too long or there is not enough time for intelligence to evolve; and cultural understanding may be more stable, but it cannot be too stable, or the transition to a high-technology culture would take too long.

Thus with the possible exception of very simple diagrams, images are not comprehensible without context. They are therefore not very suitable as stand-alone kind of messages but are likely to be most useful if they are combined with other kinds of information as part of a layered message. As a very simply suggestion how such a "layering" might work in my wrapper-and-content metaphor, consider a message like Vakoch's (2008) frequency-encoded family trees as an envelope wrapping specific images that show the family relationships schematically encoded in the frequencies. For example, his Fig. 2 might be associated with the image of a family of two parents and one child.

A second class of problems that can still be understood on the analogy of terrestrial intercultural relations is that not all kinds of images are considered acceptable by all cultures, either as a matter of general principle or for certain subjects. Some cultures are generally iconoclastic, especially when it comes to expressing religious ideas. But even among cultures with a rich pictorial history, Hinduism for example, the iconic types discussed above are not always a major artistic feature. These cultures may have their own preferred ways of depicting altruism implicitly, and persons brought up in the Western tradition may not always recognize them as such. This means that message design requires a thorough conversation not only between scientists and artists, but also between persons from different cultural backgrounds. It would seem important to gain an in-depth understanding of these cultural differences among humans as a basis for speculating about extraterrestrial cultures.

7 Limited Knowledge About Aliens

The problems of iconography raised in the preceding section can at least to some extent be understood and investigated through the study of intercultural communication. Even more difficult however are the problems arising from our ignorance of how anthropomorphic other intelligent beings might be. Should we rely on our assumptions that other intelligent beings are like us with regard to infant care? Although from an evolutionary psychology viewpoint this seems plausible, an array of science fiction leads us to imagine alternatives: James Blish's (1958) Lithians (*A Case of Conscience*) and Robert A. Heinlein's (1961) Martians (*Stranger in a Strange Land*) go through metamorphosis like terrestrial amphibians, and Orson Scott Card's (1985) insect-like aliens (*Ender's Game* and sequels) do not even share our sense of individuality.

Here the importance of science fiction in creating a plausible entire world through holistic imagination becomes obvious. Of course evolution of intelligence may not realistically be possible under the conditions portrayed in these novels. Civilization implies passing knowledge on across generations, and in a technological age the content would presumably change so fast that this can only happen by learning rather than the "hardwired" genetics that, for example, Blish envisions for his Lithians. Perhaps science fiction in this regard merely illustrates the principle that we can imagine more than is physically possible. But on the other hand, we are still being surprised about the infinite diversity and adaptability of life on our own planet, so we should not discount the possibility that evolution outside our own domain creates life forms that we would have considered improbable. Moreover, Blish was writing long before the ubiquity of personal electronic devices demonstrated the possibility of intergenerational learning without face-to-face contact. In humans this arises only after the early stages of childhood, but it is an alternative model for education without parental care in alien species.

Regardless of parental care or its absence, even a Lithian society would still need to be cooperative. Therefore other icons of altruism could possibly appeal to them, although it is not easy to imagine what they would have to be like. Our own conception of altruism relies heavily on kinship models, as shown by the pervasive metaphorical use of words like mother, brother, etc., in this context. It may be harder than expected to reconceptualize altruism without this association. What matters for message design is not so much what *we* think of an alien society, but whether *they* conceptualize altruism in a way similar to us and would therefore recognize the content of our message. Maybe extraterrestrials evolved cooperation in such a way that they are not even aware that it could be otherwise.

A reasonable strategy that partially answers these problems is to transmit not just one example of altruism but many, and to make them as diverse as possible. Even though we are now imagining a population that may be quite unlike our own, Rue's example from world religions still has some relevance. In the Earthbound aesthetic strategies, too, there is a multiplicity of examples and stories—I mentioned the Samaritan and St. Martin above, in addition to the Madonna images. And if we look at their reception we find that different cultures respond in different ways, some picking out, as it were, icons that are particularly congenial to their own traditions. Whether someone sharing his coat makes a lasting impression as example of altruism depends on the climate the audience lives in, so not all Earthly cultures respond to it the same way. We can similarly try to imagine a variety of extraterrestrial societies and send a variety of messages, each of which we hope to be appealing to different recipients. It is important, however, to acknowledge that this strategy has a downside: more images may increase our chances to be understood, but also to be misunderstood.

8 The Benefits of Imaginary Dialogue

It might seem by now that the analysis of image-based communication has brought us to a deep skepticism regarding the success of interstellar message composition. However, I actually want to conclude on a positive and optimistic note. There are compelling reasons for continuing the project of designing interstellar messages.

First, there is the familiar SETI paradigm that we have to practice how to talk in order to be able to listen. Receiving signals from alien civilizations is still a distinct possibility, and our chances of making sense of them are very much better if we have in advance thought through the issue of how we would ourselves construct a message.

Second, even if we never make contact, we learn something important about ourselves. Preparing for a conversation on a cosmic scale will force us to think in the most general terms. In this way even fictional aliens have their uses (Kracher 2006), but the ones that may really exist in our universe are uniquely relevant to our self-knowledge. If we never thought about how we would explain ourselves to them, we might never confront the contingency of our own civilization in quite the same way. Both aspects of SETI underline the necessity for the broadest possible dialogue among different cultures, academic disciplines, and diverse human abilities for the purpose of message design. Only a thorough understanding of both the scope and the limits of human diversity can prepare us for contact with an intelligence unlike our own.

There is merit in this effort, and it should not be underestimated. To return once more to our mission analogy: perhaps if we in our Western culture had spent more time getting to know ourselves and imagined possible encounters before we embarked on spreading our message across the globe, we might have been more cautious in our dealings with unfamiliar cultures. Our actual encounters might then have turned out more benign and less bruising than they did. Even if no human were ever to meet any extraterrestrials, the benefit of better self-knowledge is well worth the effort of preparing for the encounter. As for aliens preparing to encounter us, we can only hope that they have already gone through an analogous process of gaining self-knowledge.

References

Blish, James. 1958. *A Case of Conscience.* New York: Ballantine Books.
Bonting, Sjoerd L. 2003. "Theological Implications of Possible Extraterrestrial Life." *Zygon* 38:587–602.
Card, Orson Scott. 1985. *Ender's Game.* New York: Tor Books.
Conway Morris, Simon. 2003. *Life's Solution.* Cambridge, UK: Cambridge University Press.
Fox, Ronald F. 1997. "The Origins of Life: What One Needs to Know." *Zygon* 32:393–406.
Gould, Stephen J. 1999. *Wonderful Life.* New York: Norton.
Heinlein, Robert A. 1961. *Stranger in a Strange Land.* New York: G. P. Putnam.
Kracher, Alfred. 1996. "Genetic Evolution and Moral Choice." In *Investigating the Biological Foundations of Human Morality*, edited by James P. Hurd, 209–221. Lewiston, NY: Edwin Mellen Press.
Kracher, Alfred. 2002. "Imposing Order—The Varieties of Anthropomorphism." *Studies in Science and Theology* 8:239–261.
Kracher, Alfred. 2006. "Meta-humans and *Metanoia*: The Moral Dimension of Extraterrestrials." *Zygon* 41:329–346.
Kracher, Alfred. 2010. "Proliferation, Diversity and the One True Faith." *Studies in Science and Theology* 12:151–168.

Musso, Paolo. 2011. "A Language Based on Analogy to Communicate Cultural Concepts in SETI." *Acta Astronautica* 68:489–499.
Pinker, Steven A. 1994. *The Language Instinct.* New York: W. Morrow & Co.
Randolph, Richard O., Margaret S. Race, and Christopher P. McKay. 1997. "Reconsidering the Theological and Ethical Implications of Extraterrestrial Life." *CTNS Bulletin* 17(3):1–8.
Rubin, Alan E. 2002. *Disturbing the Solar System.* Princeton, NJ: Princeton University Press.
Rue, Loyal D. 1998. "Sociobiology and Moral Discourse." *Zygon* 33:525–533.
Rue, Loyal D. 2005. *Religion is Not about God.* Brunswick, NJ: Rutgers University Press.
Sacks, Oliver W. 1970. *The Man Who Mistook His Wife for a Hat.* New York: Touchstone.
Sagan, Carl. 1985. *Contact.* New York: Simon and Schuster.
Vakoch, Douglas A. 2008. "Representing Culture in Interstellar Messages." *Acta Astronautica* 63:657–664.

About the Editor

Douglas A. Vakoch, Ph.D. is Director of Interstellar Message Composition at the SETI Institute, as well as Professor of Clinical Psychology at the California Institute of Integral Studies. He serves as chair of both the International Academy of Astronautics (IAA) Study Group on Interstellar Message Construction and the IAA Study Group on Active SETI: Scientific, Technical, Societal, and Legal Dimensions. Through his membership in the International Institute of Space Law, Dr. Vakoch examines policy issues related to interstellar communication. His research spans the fields of psychology, anthropology, environmental studies, and space sciences, and his books include *Psychology of Space Exploration: Contemporary Research in Historical Perspective* (NASA, 2011); *Communication with Extraterrestrial Intelligence* (SUNY Press, 2011); *Ecofeminism and Rhetoric: Critical Perspectives on Sex, Technology, and Discourse* (Berghahn Books, 2011); *Feminist Ecocriticism: Environment, Women, and Literature* (Lexington Books, 2012); *Altruism in Cross-Cultural Perspective* (Springer, 2013); *On Orbit and Beyond: Psychological Perspectives on Human Spaceflight* (Springer, 2013); *Archaeology, Anthropology, and Interstellar Communication* (NASA, 2013); and *Astrobiology, History, and Society: Life Beyond Earth and the Impact of Discovery* (Springer, 2013).

D. A. Vakoch (ed.), *Extraterrestrial Altruism*, The Frontiers Collection,
DOI: 10.1007/978-3-642-37750-1, © Springer-Verlag Berlin Heidelberg 2014

309

About the Authors

William Sims Bainbridge, Ph.D. earned his doctorate in sociology from Harvard University in 1975, writing a dissertation on the history of the space program conceptualized as a social movement, and joined the Sociology Department of the University of Washington. In 1992, after a five-year return to Harvard, he joined the National Science Foundation to manage its sociology program, where he represented the social sciences on many computational programs, notably the Digital Library Initiative, before transferring in 2000 to NSF's Computer Science Directorate. After managing a variety of programs, such as artificial intelligence, and serving as Deputy Division Director, Dr. Bainbridge helped create the new program in human-centered computing. He is the author of 21 books, including *Dimensions of Science Fiction* and *Goals In Space*, as well as the editor of 10 volumes, most recently *Leadership in Science and Technology*.

Jerome H. Barkow, Ph.D. is Professor Emeritus of Social Anthropology at Dalhousie University in Halifax, Canada and Honorary Professor at the Institute of Cognition and Culture at Queen's University Belfast in Northern Ireland. He is an anthropologist with a career-long interest in evolution and human behavior. Dr. Barkow's particular focus is the adaptations that make it possible for human populations to edit local culture with each generation in a manner that, at least until recently, kept cultural information pools at least somewhat adapted to local conditions. His current research deals with the attentional mechanisms involved in socially mediated culture-editing. Among Dr. Barkow's publications is *Darwin, Sex, and Status: Biological Approaches to Mind and Culture*. He is a co-editor, along with Leda Cosmides and John Tooby, of *The Adapted Mind*, as well as editor of *Missing the Revolution: Darwinism for Social Scientists*.

Frank Drake, Ph.D. is Professor Emeritus of Astronomy and Astrophysics at the University of California, Santa Cruz, where he was previously Dean of the Natural Sciences Division. He is also Chairman Emeritus of the Board of Trustees of the SETI Institute, as well as former Director of the Carl Sagan Center for the Study of Life in the Universe at the SETI Institute. In 1960 Dr. Drake conducted Project OZMA, the first organized search for radio signals from extraterrestrial intelligence. The following year he devised the widely-known Drake Equation,

D. A. Vakoch (ed.), *Extraterrestrial Altruism*, The Frontiers Collection,
DOI: 10.1007/978-3-642-37750-1, © Springer-Verlag Berlin Heidelberg 2014

giving an estimate of the number of communicative extraterrestrial civilizations that we might find in our galaxy. In 1974 he constructed and transmitted the Arecibo interstellar message. Three additional messages were sent to outer space using the techniques and methods he developed: the Pioneer 10 and 11 plaques and the Voyager interstellar recording. He is author of *Is Anyone Out There?* (with Dava Sobel).

David Dunér, Ph.D. is Professor of History of Science and Ideas at Lund University, Sweden and researcher at the Centre of Cognitive Semiotics, Lund University. His research concerns the development of science, medicine, mathematics, and technology during the scientific revolution and onwards. His latest book is *The Natural Philosophy of Emanuel Swedenborg: A Study in the Conceptual Metaphors of the Mechanistic World-View*. Dr. Dunér studies the history of astrobiology, and he was guest editor of the special issue "The History and Philosophy of Astrobiology" of the journal *Astrobiology*. His astrobiological research includes "astrocognition," the study of cognition in space, detailed in his chapter "Astrocognition: Prolegomena to a Future Cognitive History of Exploration," which appeared in *Humans in Outer Space—Interdisciplinary Perspectives*. Dr. Dunér's work on the semiotic and cognitive foundations of interstellar communication appeared in the volume *Communication with Extraterrestrial Intelligence*.

Yvan Dutil, Ph.D. is a researcher in the Department of Mechanical Engineering at the École de Technologie Supérieure in Montréal. Over the years, his researcher interests and jobs have varied widely. With training in astrophysics, he drifted to remote sensing, then on to the design of various instruments for remote sensing by satellite. Later he taught in a college, worked as a science journalist, and has been a consultant in sustainability. Now his research interests are mostly concentrated on biophysical economics as a decision-making tool for energy policy, especially in the context of sustainable building. He is most known for his participation in the conception of the interstellar message sent from the Evpatoria radar facility in the Ukraine in 1999 and 2003. For this project, he introduced the use of noise hardening technique, optimum formatting, and functional redundancy in interstellar message design.

George F. R. Ellis, Ph.D. is Emeritus Distinguished Professor of Complex Systems in the Department of Mathematics and Applied Mathematics at the University of Cape Town. He obtained his doctorate at Cambridge University in 1964. Dr. Ellis is a relativist and cosmologist who has written on the philosophy of cosmology as well as the science and religion interface, and he is co-author with Stephen Hawking of *The Large Scale Structure of Space-Time*, and with Nancey Murphy of *On the Moral Nature of the Universe*. He is past President both of the International Society on General Relativity and Gravitation and of the International

Society for Science and Religion. Dr. Ellis is a Fellow of the British Royal Society, and in 2004 he was awarded the Templeton Prize.

Harold A. Geller, D.A. is Term Associate Professor and Observatory Director at George Mason University (GMU). From 2006 to 2008 he was Associate Chair of the Department of Physics and Astronomy at GMU. In 2008 Geller was awarded the GMU Faculty Member of the Year and he authored *Astrobiology: The Integrated Science*. In 2009 and 2010 he shared six Telly Awards for online educational videos in association with Astrocast.TV. In 2012 he became a Solar System Ambassador for NASA/JPL. Dr. Geller's past achievements include being an award-winning tour guide and lecturer with NASA/GSFC, President of the Potomac Geophysical Society, producer of two educational multimedia CD-ROMs, faculty at Northern Virginia Community College, doctoral fellow of the State Council of Higher Education for Virginia, and lecturer/operator at the Einstein Planetarium in the Smithsonian Institution's National Air and Space Museum. Dr. Geller has contributed to publications in education, astrophysics, astrobiology, and biochemistry.

Abhik Gupta, Ph.D. is Professor in the Department of Ecology and Environmental Science at Assam University, which is located at Silchar in Assam, India. As a zoologist and animal ecologist, he started his research by studying the ecology of hill streams in the mountains of Meghalaya, India, and later on the bioaccumulation and toxicity of heavy metals in aquatic ecosystems. Presently, he is also investigating with his students the issue of arsenic contamination of groundwater in North East India, which constitutes an important concern for human and ecosystem health. Other areas in which he has written include the intricacies of human altruism towards non-human animals, plants, and even non-living entities, as well as the role of nature religions in generating and nurturing this form of altruism in India. The ethical, ecocentric facets of traditional community conservation through sacred groves and other mechanisms remain yet another area of his interest.

Albert A. Harrison, Ph.D. is Professor Emeritus in the Department of Psychology at the University of California, Davis. He is the author or co-author of over 100 papers in a wide range of journals, and his books on space exploration and astrobiology include *Spacefaring: The Human Dimension*; *Living Aloft: Human Requirements for Extended Spaceflight* (with Mary Connors and Faren Akins); *After Contact: The Human Response to Extraterrestrial Life*; *Starstruck: Cosmic Visions in Science, Religion, and Folklore*; and *Civilizations Beyond Earth: Extraterrestrial Life and Society* (with Douglas A. Vakoch). He was a member of NASA's Space Human Factors Engineering Science and Technology Working Group and the International Academy of Astronautics Space Architecture Study Group. In December 2003, Dr. Harrison was principal investigator of a NASA-sponsored conference on new directions in behavioral health, and he edited a special supplement on this topic for *Aviation, Space & Environmental Medicine*.

Denise L. Herzing, Ph.D. is Research Director of the Wild Dolphin Project and Affiliate Assistant Professor in the Departments of Biological Sciences and Psychology at Florida Atlantic University. She has completed 28 years of her long-term study of the Atlantic spotted dolphins inhabiting Bahamian waters. In 2008 Dr. Herzing received a Guggenheim Fellowship. She is also a fellow with the Explorers Club as well as a scientific advisor for the Lifeboat Foundation and the American Cetacean Society, and she is on the board of Schoolyard Films. In addition to many scientific articles, she is the author of *The Wild Dolphin Project* and *Dolphin Diaries: My 25 Years with Spotted Dolphins in the Bahamas*. Coverage of her work with the spotted dolphins has appeared in *National Geographic, BBC Wildlife, Ocean Realm* and *Sonar* magazines. Her work has been featured in broadcasts on Nature, Discovery Channel, PBS, ABC, and BBC.

Tomislav Janović, Ph.D. is Assistant Professor in the Department of Philosophy and the Department for Communication Science, Center for Croatian Studies of the University of Zagreb, Croatia. His interests range from philosophy of mind and philosophy of science to ethics and communication. He has worked for the Croatian Ministry of Science and Technology's Division of Research Projects, at the Institute for Anthropology in Zagreb (as a Junior Research Fellow), and at the University of Zadar, Croatia (as a Senior Lecturer in philosophy). He is author or co-author of scientific articles in peer-reviewed journals and books, and he is currently working on a project on evolutionary naturalism and the problem of moral knowledge.

Adam Korbitz, J.D. in an attorney and writer in Madison, Wisconsin. In 2012, he launched CeleJure, a space law consulting practice, following a 20-year career practicing law and working in various positions related to politics and the development of public policy. Since 2008, he has written about the legal and policy implications of the Search for Extraterrestrial Intelligence and other space law topics and has presented papers at various national and international symposia and conferences. He is a member of both the Space Law Committee and the United Nations Law Committee of the American Branch of the International Law Association, and he also participates in NASA's Astrobiology and Society Focus Group. He earned his B.A. in 1985 and his J.D. in 1993, both from the University of Wisconsin-Madison.

Alfred Kracher, Ph.D. is retired Staff Scientist at Ames Laboratory, Iowa State University, and Affiliate Scientist of the Arkansas Center for Space and Planetary Sciences at the University of Arkansas. He studied the chemistry of meteorites and lunar rocks, first at the Museum of Natural History in his native Vienna, later at UCLA and the University of New Mexico. At Iowa State University he also worked on the analysis of advanced materials. Aside from cosmochemistry and materials science, his interests include philosophical questions pertaining to religion and natural science. In this field he has contributed chapters to *Investigating the*

Biological Foundations of Human Morality, Creation's Diversity, and *Is Religion Natural?* He is on the editorial board of the *European Journal of Science and Theology*.

Mark C. Langston is a computer expert with over 20 years of experience in various areas, including computer and network security, software development, and testing. He has a master's degree from the University of Chicago in the field of experimental cognitive psychology. He has published numerous peer-reviewed academic papers and chapters in that field, and he has also published and presented in various venues in computer science. His interests in computer security and psychology inform his views on and approaches to the complex act of communication, and in particular the challenges faced by the recipient of information, whether intended or unintentional.

Mark Lupisella, Ph.D. is an engineer and scientist at the NASA Goddard Space Flight Center. He has co-led the Human Spaceflight Architecture Team Cis-Lunar Destination Team and has been a member of the Near-Earth Asteroid Working Group. Previously, he worked on the Hubble Space Telescope, Mars planning, Exploration/Constellation Programs (emphasis on science integration), wearable computing, cooperative robotics, and areas of astrobiology such as planetary protection, artificial life, and broader societal issues of astrobiology such as ethics and worldviews. He co-founded the Horizons Project, which aims to improve humanity's ability to address long-term survival challenges. Dr. Lupisella recently co-edited *Cosmos and Culture: Cultural Evolution in a Cosmic Context* with previous NASA Chief Historian, Steven Dick. He has a B.S. in physics, an M.A. in philosophy (emphasis in philosophy of science), and a Ph.D. in behavior, ecology, evolution, and systematics, all from the University of Maryland at College Park.

George Michael, Ph.D. is Associate Professor at Westfield State University in Massachusetts. Previously, he was Associate Professor of Nuclear Counterproliferation and Deterrence Theory at the Air War College in Montgomery, Alabama. He received his doctorate from George Mason University's School of Public Policy. He is a veteran of both the U.S. Air Force and the Pennsylvania Air National Guard. Dr. Michael is the author of five books: *Confronting Right-Wing Extremism and Terrorism in the USA, The Enemy of My Enemy: The Alarming Convergence of Militant Islam and the Extreme Right, Willis Carto and the American Far Right, Theology of Hate: A History of the World Church of the Creator*, and *Lone Wolf Terror and the Rise of Leaderless Resistance*.

Alexander Ollongren, Ph.D. is Professor Emeritus of Theoretical Computer Science at the Leiden Institute of Advanced Computer Science, Leiden University. He is a member of the International Astronomical Union's Commissions on Celestial Mechanics, Galactic Dynamics, and Bioastronomy, as well as a member of the European Astronomical Society and the Dutch Astronomical Society. Dr. Ollongren is also a member of the International Academy of Astronautics (IAA) SETI Permanent Committee; the IAA Study Group on Interstellar Message Construction;

and the IAA Study Group on Active SETI: Scientific, Technical, Societal, and Legal Dimensions. He has published research papers on dynamical astronomy, celestial mechanics, theoretical computer science, formal languages, and automata. Over the past decade his research has focused on formal approaches to interstellar message design, culminating in his recent book *Astrolinguistics: Design of a Linguistic System for Interstellar Communication Based on Logic*.

Douglas Raybeck, Ph.D. is Professor Emeritus in the Department of Anthropology at Hamilton College. His research addresses topics ranging from nonverbal communication and psycholinguistics to physiological correlates of behavioral dispositions. He is familiar with future studies and has written a book on the topic, *Looking Down the Road: A Systems Approach to Future Studies*. He studies Malaysian culture and has published *Mad Dogs, Englishmen, and the Errant Anthropologist*, a book summarizing his fieldwork in Kelantan, Malaysia. Dr. Raybeck is co-editor of *Deviance: Anthropological Perspectives* and *Improving College Education of Veterans*, as well as co-author of *Improving Student Memory* and *Improving Memory and Study Skills: Advances in Theory and Practice*. He has been a Fellow at the National Institutes of Health and is past President of the Society for Cross-Cultural Research.

Holmes Rolston, III, Ph.D. is University Distinguished Professor and Professor of Philosophy Emeritus at Colorado State University and a founder of environmental ethics as a philosophical discipline. His books include *Philosophy Gone Wild*; *Environmental Ethics*; *Conserving Natural Value*; *Science and Religion: A Critical Survey*; and *Three Big Bangs: Matter-Energy, Life, Mind*. His most recent book is *A New Environmental Ethics: The Next Millennium for Life on Earth*. Dr. Rolston gave the Gifford Lectures, University of Edinburgh, 1997–1998, published as *Genes, Genesis and God*. Advocating environmental ethics, he has lectured on seven continents. He is featured in Joy A. Palmer's *Fifty Key Thinkers on the Environment*. He won the Templeton Prize in 2003, an award worth well over a million dollars, and more than a Nobel Prize, awarded by Prince Philip in Buckingham Palace.

Index

A

abstractions, 252, 256, 257
achievement motivation, 290
Active SETI, 94, 233
    debate regarding risk, 111, 112, 114–124
    and risk analysis, 118–124
acts of speech, 180, 254
adaptation, 146, 147
addressee, 182–185, 187
Adler, Alfred, 283, 284
adversatives, 255
aesthetic strategy (Rue), 301, 306
affection, 286
Africa, 73
Age of Reason, 6
aggression, 57, 58
agonistic behavior, 49, 57–59, 150, 161, 162
agreeableness, 276
alarm calls, 195
Alex (African Gray parrot), 199
Alexander, Richard D., 40, 42
Algol 60 report, 254
Alien (film), 223
aliens. *See* extraterrestrial intelligence
alloparental care, 149
Alpha Centauri, 69, 70
altruism, 15, 19, 38–40, 42, 43, 45, 54, 57, 59,
    65, 72, 75, 79–82, 84–87, 89, 131,
    136–138, 141–143, 148, 149, 158,
    161, 162, 177, 179
    accidental, 140
    advanced, 17, 93, 95, 96, 99–104, 106
    beacon, 137, 139
    biocultural, 96, 97, 99, 101–103, 106
    biological, 96, 97, 99
    class, 219
    communicating, 44–46, 201, 228, 257–259,
        272, 296, 301–304
    content, 137

    cosmic framework, 4
    definition of, 4, 11, 272
    and economics, 236, 237, 272
    ethical, 232, 235–237
    extraterrestrial, 112, 113, 116, 118–122,
        125, 211–222, 232–239, 243, 244,
        246
    inference of, 18
    in-group, 216, 218
    interspecies, 17, 18
    kinship-based. *See* kin selection
    multigenerational, 98
    reciprocal, 10–12, 17, 18, 40, 41, 45, 96,
        161, 179, 191, 215, 217
    religion in, 221
    social capital and, 12
    terrestrial, 211–222
    tribal, 215–217, 219
    types of, 17
    universal, 136
    values in, 216, 219, 221
ambiguity, 299
American culture, 282
American legal realism, 245
amino acids, 68
amplification process, evolutionary, 41–43,
    268, 269
Anarchy Online (online role-playing game),
    279
Andrén, Mats, 159
Andromeda galaxy, 70, 159, 227
anecdotes, data and, 14
animal rights, 245
anthropocentrism, 18, 82, 211, 212
anthropomorphism, 171n1, 171n2, 178, 300,
    305
    critical, 201
Anti-Catastrophe Principle, 117–122
antimatter, 69

Apollonian culture, 277
archaeology, 9
Archer, Michael, 29
Arecibo Message, 26
Aristotle, 262
Armageddon, 224
Arrow, Kenneth, 271
art, 44–46, 295–298, 301, 302, 304
artificial intelligence, 245
Asimov, Isaac, 282
astrobiology, 142, 174
astrocognition, 145, 147
astrolinguistics, 251, 256
atmosphere, 157, 158
atomic table, 213
atomic weapons, 15. *See also* nuclear weapons
attention, 141–145, 148, 152–154, 158–161
    joint, 142, 145, 158, 159, 185n23
Au Lushan revolt, 9
authoritarian regimes, 4, 5, 7–10, 14. *See also*
        totalitarian regimes
    science and, 7, 268, 269
autism, 289
availability heuristic, 115, 116
avatars, 20
Axelrod, Robert, 138
Axtell, Roger, 139

B
Babylon 5 (television series), 139
background assumptions, 186
Bainbridge, William Sims, 20
Barkow, Jerome H., 16, 38, 41n3, 42, 173n4
Barrow, John, 31
Batson, Daniel, 179n15
Battleship (film), 26
Becket, Samuel, 284
behavior, 254, 255
    communicative, 180, 184
    cooperative, 169, 179
    mindlike, 171, 171n1, 172
belligerence, 74–75
Benedict, Ruth, 278, 289
benefits, 256, 257
benevolence, 80
Benford, James, on threat of alien invasion,
        118, 119, 119n2, 120
Berger, Peter, 226
Bester, Alfred, 283, 285
Big Five personality dimensions, 276
bilateral insula, 85
Billingham, John, on threat of alien invasion,
        118, 119, 119n2, 120
bioaltruism, 6

biochemistry, 68, 72–74
biocultural coevolution, 143, 147, 148
biodiversity, 155, 158, 162
bioengineering, 30
biomarkers, 158
biophilia, 17, 79, 80, 83, 86, 88
BioWare, 279
black holes, 28, 32
Blish, James, 297, 305, 306
bonding, 197
bootstrapping problem (Wittgenstein), 180
Borda, Jean Charles, 269
Bracewell, Ronald, 32
Brinck, Ingar, 159
British empire, 30
Brockhurst, Michael A., 43n4
Buddhist culture, 277
Burroughs, Edgar Rice, 281
Buss, David, 40n2, 43

C
Calculus of Constructions with Induction
        (CCI), 256, 257
Campbell, David, 6
Campbell, Donald T., 42
Camus, Albert, 284
Card, Orson Scott, 300, 305
caring capacity, 93, 95, 97–99, 103, 105, 106
Carrigan, Richard, 132, 133
carrying capacity, 95, 98
Cashdan, Elizabeth, 42
catatonia, 277
Categorical Imperative, 218, 234, 238–240,
        242
Catholic sovereigns, 72
celegistics, 19, 239, 242, 244, 245
    definition of, 239
central Indian tribes, 89
chemical propellants, 69
Chesterson, G. K., 221
chimpanzees, 74, 214, 215
China, 9
Chingoiron, 86
choice, 226
citizenship, 217
civilizations, 67, 141, 142, 145, 146, 149–155,
        157, 158, 161, 162
    post-biological, 16, 149
    See also longevity of civilizations
civilizing process, 6
clarification, 259
claustrophobic paranoia, 283
Clement, Hal, 38n1
Clever Hans (horse), 198

climate change, 158
code, 132–134, 136, 139, 169, 170, 180,
     180n16, 181, 183–185, 187
  model of communication, 180n16
cognition, 141–143, 145–151, 153–158, 161,
     172
  distributed, 142, 150, 154–156
  evolution of, 143, 145–151, 154, 157
cognitive abilities, 141, 142, 145–149, 151,
     155, 157, 161, 170, 179n15
cognitive dissonance, 73
cognitive science, 141–143, 145
Cohen, Jack, 171n3, 173n4, 174, 175, 175n6,
     175n7, 186n25, 187
Cold War, 6, 31, 76
collaboration, 142, 150, 154, 155
collective intelligence, 104
collective mind, 104
collective security, 5, 8, 9
collectivism, 291
Columbus, Christopher, 16, 26, 31, 65, 66,
     70–73, 118
Columbus, Ferdinand, 72
Commandments, Ten, 218
commensurability, 169, 170, 186. *See also*
     representational systems,
     commensurability between
common ground, 252, 253
communicants, 169, 170, 171n1, 173, 177,
     178, 181, 182, 182n18, 184–187
communication, 13, 14, 17, 18, 20, 141–146,
     148–158, 161, 162, 191
  channel, 184
  coded, 180
  codeless, 181
  as consensus, 136, 137
  conventional, 182–184
  images and, 20
  intentional, 182, 185, 188
  interspecific, 144, 191, 196, 202
  intraspecific, 194
  linguistic, 182
  mutual, 199
  non-conventional, 180, 181, 183
  pre-verbal, 184
  process, 169, 180, 180n16, 182, 183n20,
     185
  prosodic, 203
  symbolic, 171, 172, 179–181
  theory of human, 194
  two-way, 197
  verbal, 184
  See also interstellar communication
communicative motive, 181, 187

communicative system, complex, 151–155
communicators, 181, 182, 182n18, 183, 185
community, 280
compassion, 275
competition, 4, 226
complexity, 141, 142, 146–156, 158, 159, 161
  communicative, 141, 142, 146, 149–156,
     158
  social, 141, 142, 147–151, 154–156
computer viruses, 131–135, 138, 139
conceptualization, 142, 154, 157
Condorcet, Marie Jean Antoine Nicolas
     Caritat, 269
conflict
  intergroup, 226
  resolution, 225
conscientiousness, 276
consciousness, 171n1, 177, 177n10, 224–226
constraints, 170, 174, 175, 185n23, 186, 255
constructive logic, 251, 253
Contact
  (film), 223, 228
  (novel), 27, 137, 228
content
  explicit, 169, 181, 182, 182n19, 185, 187
  implicit, 169, 181, 185
  informational/representational, 169, 185,
     187
  shared/common, 169, 182, 185, 186
context, 136–138, 140, 169, 177, 182, 183,
     183n20, 185–187, 195
  definition of, 183, 183n20
  explicit, 185
  implicit, 185
  verbal, 182
contextual knowledge, 182, 185, 186
convergence, 81.   *See also* genetic and me-
     metic convergence
Conway Morris, Simon, 39
cooperation, 38, 40–43, 79, 81, 84, 96, 154,
     155, 161, 169, 173n4, 175, 178,
     182, 185, 193, 203, 215, 216
  in-group, 57
  out-group, 57
Coq Development Team, 253
Corollary of Negentropic Equality, 240
Correy, Lee. *See* Stine, G. Harry
cosmic evolution, 5
Cosmides, Leda, 40, 176
Crick, Francis, 30
critical legal studies movement, 245
Cronin, Helena, 44
Crosby, Alfred, 76
cross-species signals, 194

cultural encounters, 161
cultural learning, 173, 180
cultural relativity, 139
cultural transmission, 198
culture, 38, 42, 53, 54, 141–149, 151,
        153–157, 161, 216–220, 226, 262
  -bearing species, 38
  definition of, 38
  lag, 15
  and personality studies, 275
Cusack, Thomas, 9

D
data, anecdotes and, 14
deception, 181, 187, 203
decoding, 169, 170, 172, 180, 180n16, 182n19,
        185, 187
  passive, 194
deep structures, 252
defection, 85
demarcations between disciplines, 296
democracy, 5, 7, 8, 14, 20, 267–269
  definition of, 8
  science and, 7
democratic peace, 7–9
Dennett, Daniel, 171n1
Denning, Kathryn, 15
deoxyribonucleic acid (DNA), 30, 72, 74
DeScioli, Peter, 40
destructive behavior, 150, 155, 161
destructive capacities, 224, 225
Diamond, Jared, 26
Dick, Steven, 75
difference, appreciation of, 224
Dionysian culture, 277
diplomacy, 245
directed panspermia, 30
diseases, 73
division of labor, 6
DNA. *See* deoxyribonucleic acid
dodo bird, 73, 74
dogs, 72
dolphins, 74, 192, 200
dopamine, 81, 86
Drake, Frank, 26, 34, 39
Drake Equation, 39, 66, 67, 119, 145, 153, 176
Dunér, David, 7, 18, 185n23
Dunér, Sten, 160
Dutil, Yvan, 20
duty cycles, 71
Dyson, Freeman, 28, 33
Dyson sphere, 28

E
Earth, 142–144, 147, 149, 150, 157, 158, 169,
        174–176, 184
  analogue, 142, 143, 157, 158
ecological refugees, 89
economics, 11
The Economist (periodical), 16
ecophilia, 86–88
ecosystem people, 85
ecstasy, 226
efficiency, 184, 263
egoism, extraterrestrial, 112, 119–122
Einstein, Albert, 26, 27, 69
elective affinity, 276
electromagnetic leakage radiation, viii, ix, 74,
        158
electromagnetic waves, 142, 145
elephants, 193
Ellis, George F. R., 19, 227
emotions, 143–146, 169, 181
empathic stance, 178
empathic understanding/recognition, 169,
        179, 179n15, 180, 181, 185,
        187, 188
empathy, 18, 141–144, 148, 161, 169, 170,
        177, 178, 178n12, 179, 179n14,
        179n15, 181, 182, 185, 187, 193
Enceladus, 124
encephalization, 147, 148
encoding, 169, 170, 180, 180n16, 181, 182,
        182n19, 185, 196
energy, 70
engineering, 69, 70
Enlightenment, 6
Entropia Universe (online role-playing game),
        279
Entropic Censorship Rule, 241
environment, 141, 145–149, 151, 153–159,
        162, 171–173, 180n16
  physical, 141, 145–148, 156–159
  social, 141, 145–148, 171, 183n20
environmentalism, 219, 220
envy, freedom from, 263, 267
epistemology, 212
  realism, 212
  universalism, 213, 214
equitability, 263, 267
equity
  democracy and, 20, 262–272
  theory, 11, 12, 262–268
E.T. (film), 223
ethical egoism, 235, 243, 244

ethics, 142, 151, 154–156, 161
  in animals, 214, 215, 221, 271
  culturally relative, 218, 219, 262, 263
  Darwinian, 215, 218
  deep, 225–227
  definition of, 19
  deontological, 219–221
  Earthbound, 214–218
  environmental, 99, 101, 219, 220
  extraterrestrial, 227, 228
  "formal", 97
  inclusive global, 218, 220
  kenotic, 19, 225–228
  machine, 16
  naturalized, 217
  normative, 82
  realist view of, 227
  religious, 218, 219, 221
  and technology, 224, 226
  universal, 19, 212, 220–222
  and value, 219–221
ethnocentrism, 42, 86
  origins of, 42
  unit of, 42
ethnopsychiatry, 277, 278
ethology, 199
ETI. See extraterrestrial intelligence
Europa, 124
European Constitution, 113
Europeans, 73
events, 254, 255
evolution, 11, 13, 50–59, 142–149, 151, 154,
          156, 172, 173, 175–177, 178n11,
          184, 193
  biological, 38–44
  convergent, 39, 147, 268, 269, 297, 300,
          303
  cosmocultural, 17, 93, 95, 102, 103, 106
  cultural, 142, 148, 149, 153, 154
  Darwinian, vii, 98, 103, 226, 268, 269
  processes of, 41–43
  See also amplification process,
          evolutionary
evolutionary biology, 49
evolutionary psychology, 40, 40n2, 43, 176,
          226
existential risks, 117, 119
existentialist novels, 284
exoplanets, 157, 158, 174, 174n5, 176
exopsychology, 279
experience, shared, 141–145, 148, 151, 152,
          157, 160–162
extelligence, 171, 173n4, 175, 175n7, 178,
          180, 188

external norms, 82, 83
extra-linguistic information, 252
extrasolar planets, 123, 124
extraterrestrial intelligence (ETI), 49, 50,
          56–60, 65, 66, 68, 70, 71, 73, 74,
          79–81, 84, 86–89, 112, 116,
          118–125, 131–133, 141–146, 148,
          149, 153, 156, 157, 159, 160, 169,
          174, 180, 183, 232–239, 241–246
  characteristics of, 18
  evolution of, 300, 302, 305
  and morality, 220–222
  possible machine nature of, 237, 243
  technological level of, 224
  unknown ethical and legal views of, 232,
          236, 243–246
extraterrestrial life, 173, 174, 178
extraterrestrial mind, 170, 171n1, 172–177
extraterrestrials, 39, 41–43, 50, 52, 54, 57–59,
          170, 271
  hostile, 39, 43, 44
extraterrestrial society, 169, 170, 172–175,
          177, 178, 185, 268
extraversion, 276, 277
extrinsic value, 79, 82, 83

F
factor analysis, 276
fairness, 11, 12, 41, 45, 219, 262, 263, 271,
          272
Falger, V.S.E., 42
Fasan, Ernst, 19, 234, 235, 237–245
f_c, 153
Fehr, Ernst, 138
Fein, Daniel, 40
Feldman, Marcus, 75
Fermi, Enrico, 67
Fermi Paradox, 32, 67, 119, 161, 176
Ferris, Timothy, 7
Finney, Ben, 76
First European "Seas at Risk" Conference,
          final declaration, 114
First Principle of Metalaw. See Interstellar
          Golden Rule
Fischbacher, Urs, 138
fitness, inclusive. See inclusive fitness
flexibility, cognitive, 141, 145–148, 150, 156
folie à deux, 286
folklore, 85
The Force, 280
forgiveness, 225, 226
Forward, Robert, 34
Fossey, Dian, 200, 201
Fouts, Roger, 198

freedom, 219, 220
Freitas, Robert A., 19, 239–244
Freud, Sigmund, 278, 283, 284
Freudenthal, Hans, 19, 136
fundamentalists, 224, 225

G
Galactic Club, 10
game theory, 12–14. *See also* Prisoner's
    Dilemma Game
Gandhi, Mahatma, 228
Gärdenfors, Peter, 144, 151, 154
gazelles, 196
Gelarden, Ian, 43n4
Geller, Harold, 16
gene-culture coevolution, 81
General Theory of Relativity, 27
genes, 85
    selfish. *See* selfish genes
genetic and memetic convergence, 89
gesture, 181, 182, 188
Gliese 581 (star), 34
God, universal nature of, 227
Goddard, Robert H., 7
Godfrey-Smith, Peter, 75
gold, 72
Gold Beach, Oregon, 11
Golden Rule, 10, 12, 15, 19, 99, 101, 218–221,
    226, 233
Goldilocks zone, 30, 31
Goldstein, Joshua, 10
Goodall, Jane, 200, 214, 215
goodness, 255
goods, 257
gorillas, 200, 268
great apes, 196
Grice, Paul, 183, 184n22, 187
guilt culture, 289
Gupta, Abhik, 17
Gurven, Michael, 44

H
habitability, 141, 143, 156–158, 296
habitable zones, 67, 157
habitats, 75
Hafner, Everett M., 159
Haley, Andrew G., 100, 233–235, 237, 238,
    240–242, 242n1, 244, 245
    and stark concept of absolute equity, 233,
        235, 237, 240
Haley's Rule. *See* Interstellar Golden Rule
handicap principle, 43, 44
harmful ETI hypothesis, 65, 66, 68, 76
Harrison, Edward, 31

Hart, Michael, 76
Hawkes, Kristen, 44
Hawking, Stephen, on threat of alien invasion,
    16, 25, 26, 29, 65, 66, 68, 71, 112,
    118, 119, 156, 162
hawkish attitude, 84
Heinlein, Robert A., 282, 305
Herzing, Denise, 18, 200
High Resolution Microwave Survey (HRMS),
    125
Hill, Kim R., 44
history, cultural, 144, 155, 161
histrionic personality disorder, 288
Hitler, Adolf, 7
    Hive Queen society, 297, 300
Homo
    habilis, 55
    neanderthalensis, 39
    sapiens, 39, 55, 68, 74, 75, 81, 144, 154,
        213, 218, 221, 277
Hubble Space Telescope, 66
human
    -animal kinship, 83
    -heartedness, norm of, 12
humanitarian revolution, 6
humans, cognitive powers, 213
hunter-gatherers, 79, 85
hysteria, 288

I
Icarus complex, 290
iconic communication, 296, 301, 304
iconic messages, 295, 302, 304
iconoclasm, 297, 305
iconography, 302, 304, 305
icons, 20
idealist position, international relations, 5
imagination, 296–298, 305
imitation, 144, 149, 152, 153, 171
inclusive fitness, 11, 84
Independence Day (film), 29, 223
index, 159, 161
Indians. *See* Native Americans
individualism, 225, 291
infant care, 302
inferiority complex, 284, 285
information, 142, 144, 145, 147, 149, 151, 153,
    155, 161, 169, 181–183, 183n20,
    184, 185, 187
    transmission of, 72, 196
in-groups, 42, 44, 57, 86, 88, 226
inherent value. *See* intrinsic value
innovation, 153
insects, 74

insider treatment, 5, 6
instrumental value. *See* extrinsic value
intelligence, 38–41, 43, 49–59, 141–149, 151,
		153, 154, 156, 157, 159–161,
		173n4, 175, 178, 186, 186n24, 188
	amplifying processes, 41
	definition of, 38
	human, 211–213, 218–222
	machine, 300
	non-Darwinian, 105
	origins of, 41
	practical, 171
	theoretical, 172
	See also extraterrestrial intelligence
Intel Pentium 4, 133
intent. *See* intention
intention, 96, 131, 135, 136, 138–145, 148,
		151, 152, 155, 161, 169, 170, 175,
		178, 179, 179n15, 180–183,
		183n21, 184, 184n22, 185, 187, 201
	communicative, 169, 170, 180, 182–185,
		187, 188
	informative, 169, 170, 182–185, 187
	model of communication, 183, 184
	social, 177–181, 183, 184
intentional act, 177, 178, 184
intentional agent, 170, 181, 187
intentional being, 169
intentional stance, 171n1
intentional state, 179, 180
intentional system, 171n1
intention-like (mental) state, 178, 179, 181
interaction, 142, 143, 146, 147, 149–159, 161
	intersubjective, 142, 143, 146, 151,
		157–161
	interspecific, 195, 196
internal norms, 82
international relations, 245
Internet, 6, 42
interpretation, 183n20, 251, 255, 259
interstellar communication, 38, 44–46,
		141–143, 145, 146, 149, 151–154,
		156, 161, 169, 170, 171n1,
		172–178, 185n23, 186, 187
	cognitive foundations of, 143, 151
Interstellar Golden Rule, 233, 235, 236, 241
interstellar messages, 80, 142–144, 156, 169,
		170, 172, 173, 185, 187
	construction of, 142, 143
	emission of, 187
	reception of, 185
interstellar signals, 169, 170, 172, 173, 177,
		181, 183, 185, 187
	emission of, 172, 173

reception of, 169, 170, 172, 173, 186, 187
interstellar travel, 16, 66, 69, 70, 74
	readiness for, 225
intersubjectivity, 18, 141–146, 148, 150–152,
		154, 156–162, 178n12
intrinsic value, 79, 82, 83, 85
introversion, 277
intuitionism, 253
isomers, 68

J
Jainism, 101
Janović, Tomislav, 18
Jedi, 280, 281
Jung, Carl G., 277
justice, 11, 17, 41, 45, 218–221, 262

K
Kagan, Jerome, 214
Kaku, Michio, 27, 29, 32, 33
Kant, Immanuel, 99, 218
Kanzi (bonobo), 197
Kardashev scale, 28, 33
Kelleghan, Fiona, 283
Kelly, Walt, 76
kenosis, 225–228
	communicating, 228
Kepler space observatory, 39, 67, 123, 174n5
Kerr, Benjamin, 75
keyboards, 197
kin selection, 40, 42, 81, 85, 96, 192, 226
King, Martin Luther, 228
kinship, 192
knowledge, 252, 255, 259
	contextual, 182, 185, 186
	implicit, 181
	non-genetic transmission of, 171
	social construction of, 212, 219
Koko (gorilla), 198
Korbitz, Adam, 17, 19
Kracher, Alfred, 20
Krasnikov, Sergei, 27
k-selection, 84, 175
Kummer, Helmut, 214
Kurzban, Robert, 40

L
L. *See* longevity of civilizations
laboratory studies, 196
Lagrange points, 244
Lana (chimpanzee), 197
Landes, William, 236, 237
landscapes, 211
Langston, Mark, 18

language, 179, 180, 197
language (*cont.*)
   acquisition, 151
   human, 149, 151
Latin, 254, 255
Law
   of the Rights of Mother Earth, 99
   universal, 19, 186
Laws of Robotics (Asimov), 242, 282
layered messages, 305
legal interpretivism, 245
LeVine, Robert A., 42
Lewis, C. S., 223
life, 142, 146–148, 154, 158, 161, 212–216
   carbon-based, 175, 213
   silicon-based, 175
   -supporting environments, 173
   -supporting planets, viii, 176
light, 213
LINCOS, 253–256, 259
Lingua Cosmica, 19, 252–259
linguistics, 143, 148, 151
Lithians, 297, 305, 306
Lively, Curtis M., 43n4
Living Systems Theory, 245
Loewen, James, 71, 72
longevity of civilizations, 15, 155
loss aversion, 115, 116
love, 218
   suffering, 221
Lupisella, Mark, 11, 17
Lysenko, Trofim, 7

M
M13 (globular star cluster), 26
Maastricht Treaty, 113
Machiavellianism, 57–59
machine ethics, 16
MacIntyre, Alasdair C., 219
Madonna images, 20, 296, 302, 304, 306
Malinowski, Bronislaw, 278
Manhattan project, 155
Mao Zedong, 7
marine mammals, 192
marriage, 217
Mars, 31, 124
Mars Needs Women (film), 30
Martin, Saint, 301, 306
mate choice, 43, 44
mature civilizations, 84
maximin, 117, 119–122
Mayr, Ernst, 147
McCain, John, 11
meaning-intention, 182

mean time between failures (MTBF), 70
meme-complexes, 80, 87
memes, 80, 81, 85, 87, 88, 178
memory, 155, 161, 254
mesolimbic dopamine system, 85
messages, 169, 180, 180n16, 181, 182,
      182n19, 183–185, 187
   content, 169, 184, 187
   See also interstellar messages
Metalaw, 19, 100, 233–243, 245, 246
   and altruism, 236, 237
   criticism of, 238, 243–245
   definition of, 233–235, 239
   future directions for development of,
      243–245
   later development of, 239–242
   and machine intelligence, 237, 243
metaphors, 299, 305
meta-representation, 172, 186, 186n24
Michael, George, 16
Michaud, Michael, 34
Mideast Slave Trade, 9
Milky Way, 66, 67, 112, 232
Miller, Geoffrey, 44
Miller, James Grier, 5
minds, 142–144, 146–148, 151–155, 157,
      171n1, 172, 173, 175, 176, 179n15,
      181, 186, 187
   human-like, 171, 171n1
   See also other minds
Ming Dynasty, 9
Minsky, Marvin, 186
Mirror Self Recognition, 193
misunderstanding, 296
modalities
   of communication, 194
   of human behavior, 255
Modelski, George, 8
Mongol conquests, 9
monomania, 283
Moon, 75, 76
morality, 40, 226, 227, 251, 256, 257, 259
Moral Landscape, 100
moral realism, 227
Morran, Levi T., 43n4
Moss, Cynthia, 201
motivations, extraterrestrial, 29–32, 50, 54,
      56–59, 112, 124
MTBF. *See* mean time between failures
multi-level systems, 251, 255, 259
multiple personality neurosis, 288
Murphy, Nancey, 227
Murray, Henry, 290
music, 44–46, 259

N

NASA. *See* National Aeronautics and Space
    Administration
National Aeronautics and Space Administra-
    tion (NASA), 26, 44, 69, 75, 125,
    174n5
Native Americans, 25, 26, 31, 72, 73
NATO. *See* North Atlantic Treaty
    Organization
natural law theory (jurisprudence), 234,
    237–240, 242, 245
natural meaning (Grice), 184n22, 187
natural selection, 186, 268
natural sign, 187
nature
    appreciation, 88
    religions, 17, 79, 87, 88
    worship, 80, 85, 87
Nazi Germany, science in, 7
negative entropy. *See* negentropy
negentropy, 240, 243
nepotism, 40
neurotics, 278, 286
NEW LINCOS, 253, 256–259
Nietzsche, Friedrich, 277, 278
Nim (chimpanzee), 198
non-Darwinian intelligence, 105
non-human organisms, 82, 194
non-player characters, 279
non-zero sum game, 5, 12, 13. *See also* zero
    sum game
normative ethics, 82, 262, 263
North Atlantic Treaty Organization (NATO),
    10
northeast India, 86, 89
Nowak, Martin, 215
nuclear weapons, 10. *See also* atomic
    weapons
nucleotides, 68
Nungoiron, 86

O

Oakley, Kenneth Page, 41
Obama, Barack, 113
oblique speech, 255
Oedipus complex, 278, 284
Office of Commercial Programs, 76
O'Gorman, Rick, 42
Ollongren, Alexander, 19
O'Neill, Gerald, 32
"one world-one science" argument (Rescher),
    186
operating system (OS), 134, 135
Operation Overlord, 32

opportunity benefits, 119, 122
Orcas, 199
original position (Rawls), 218, 220
other minds, 169, 176, 179n15
out-group problem, 95, 98, 100
out-groups, 54, 57, 58, 86, 226

P

pacification process, 6
pacts, defensive, 12
Palumbo, Donald, 283
pantheism, 87
paranoid views of extraterrestrial life, 4, 5, 15
Parrish, Raymond C., 43n4
Passive SETI, 119, 119n2, 122. *See also*
    Search for Extraterrestrial
    Intelligence
patriotism, 217
Patterson, Francine, 198
peace, 4
    democratic, 7–9
Pepperberg, Irene, 199
periodic table of elements. *See* atomic table
Perry, Gardner, III, 8
personification, 255
persons, 254, 255
phenomenal property, 172, 177n10
phenomenology, 144
physics, 69, 172, 175
Piaget, Jean, 146
Pinker, Steven, 4, 6, 9, 10
Pioneer plaque, 44, 159
Plan 9 from Outer Space (film), 31
Plato, 212
pluralistic ignorance, 286
political risks, to SETI, 116, 124, 125
pollution, 155, 158
Population I stars, 68
Posner, Richard, 236, 237
postmodernism, 212
post-traumatic stress disorder, 290
prairie dogs, 195
Precautionary Principle, 17, 18, 46, 112–121,
    124
    and Active SETI, 115–121, 124
    alternatives to, 117, 118
    as applied in Europe, 113–115
    as applied in United States, 113–115
    criticism of, 114–117
    definition of, 113, 114
    strong version, 113–115
    weak version, 113–115
predation, 17
    co-, 41–43

predation (*cont.*)
  self-, 42
  theories of, 41–43
Predator (film), 29
prejudice, reduction of, 6
Premack, David, 197
Preston, Stephanie, 144
primates, 191, 198
Principal Thermoethic, 240
principle
  of economy (Minsky), 186
  of sparseness (Minsky), 186
Prisoner's Dilemma Game (PDG), 12, 13, 82,
      131, 138, 140. *See also* game
      theory
probability, 68, 71, 176
  neglect, 115, 116
progress, 28, 269
  ethical, 220, 225, 228
  technological, 173, 173n4, 225
promise keeping, 19, 214, 220
pronoid views of extraterrestrial life, 4, 5, 15
propellants, 69
property, respect for, 220
Protestantism, 276
Proxima Centauri, 27, 132
Proxmire, William, 125
pseudo-intentional processes, 184
psychoanalysis, 278, 284
psychology, 11, 38, 39, 41
  extraterrestrial, 16
  See also evolutionary psychology
psychopaths, 276
psychotherapy, 278
Putnam, Robert, 6

Q
Queen of Spain, 72

R
Raphael, 303, 304
Rare Earth hypothesis, 31
Rawls, John, 218
Raybeck, Douglas, 17
reaction formation, 289
realist position, international relations, 5
reality, 146, 152, 157–159
receiver, 169, 185, 187
reciprocity, 5, 12, 79, 81, 83
  indirect, 161
reconciliation, 225
redundancy, 251
refrigerators, 70, 71
Relativistic Heavy Ion Collider, 31

relativity theory, 69, 70. *See also* General
      Theory of Relativity; Special
      Theory of Relativity
reliability, 70
religion
  American, 6
  intermarriage and, 6
  nature and, 17
religions, 226–228, 296, 297, 301, 306
Rendell, Luke, 42
representation (internal/mental), 170, 171,
      171n2, 172, 180, 180n16, 184,
      186n24
representational content, 169, 187
representational systems, 169, 172, 184, 186,
      188
  commensurability between, 169, 170, 186,
      188
reproductive value, 40
reputation, 13
Requirement, 72
Rescher, Nicholas, 178, 186
resources
  struggle for, viii, 224
  and technology, 224
respect, 218, 220–224
revolution, democratic and scientific, 7
Reynolds, Vernon, 42
ribonucleic acid (RNA), 68
Ridley, Matt, 43n4, 44
rights, 201, 214, 219. *See also* Universal
      Declaration of Human Rights
Rights Movement, 6
Rio Declaration, 113
risk
  analysis, 115, 118–124
    and Active SETI, 118–124
  communication, 117, 233
  human preference for familiar vs. unfa-
      miliar, 116
  human preference for voluntary vs. invol-
      untary, 116, 117
  perception, 116
    cultural differences in, 114
  reduction, 46
  situations of vs. uncertainty, 117, 118, 122
RNA. *See* ribonucleic acid
robotics, 245
rocketry, 7
Rolston, Holmes, III, 19
Romantic Movement, 87
Rousseau, Jean-Jacques, 269
r-selection, 84, 175
Rue, Loyal, 301, 302, 306

rule-making, social, 97
rules, sharing, 20
Rumbaugh, Duane, 197
Ruse, Michael, 215
Russett, Bruce, 8
Russia, 10

S
Saad, Gad, 44
Sacks, Jonathan, 224
sacred groves, 85, 89
sadism, 283, 288
Sagan, Carl, 26–28, 32, 125, 136, 137, 159. *See also* Contact
SAIC. *See* Science Applications International Corporation
Samaritan, Good, 301, 306
Sarah (chimpanzee), 197
Savage-Rumbaugh, Sue, 197
schizophrenia, 290
Schmidt, Olivia G., 43n4
Schwarzenegger, Arnold, 31
science, 153, 154, 156, 157, 159, 174, 295–298
    authoritarian regimes and, 7
    democracy and, 7
    fiction, 5, 20, 174, 298, 305, 306
    in Nazi Germany, 7
    realism in, 212
Science Applications International Corporation (SAIC), 76
Search for Extraterrestrial Intelligence (SETI), viii, ix, 4, 15, 25, 33, 34, 39, 49, 50, 60, 94, 111, 118n1, 123, 125, 132–134, 136, 141–143, 145, 146, 154, 156, 174, 191, 202, 203, 223, 232, 233, 239, 243–245
Searle, John, 179, 179n14, 180, 186
secondary avatar, 279, 286
Second Law of Thermodynamics, 33
security, collective, 9
seduction, 288
self-consciousness, 175
self-defense, 224, 228
selfhood, 95, 101, 102
self-interest, 4, 11, 221
selfish genes, 17, 192, 216
selfishness, 72, 75, 221, 272
    beacon, 137, 139
    content, 137, 139
    trap, 94, 100
self-sacrifice, 225, 228
semantics, 151
semiotics, 156, 161
sender, 169, 182, 183, 185, 187

senses, 147, 153, 155, 157
sensory impressions, 157
sensory systems, 199
sentient species, 192
SETI. *See* Search for Extraterrestrial Intelligence
sexual selection, 43, 44
shamanism, 86
shame culture, 289
Shannon, Claude, 180n16
shared knowledge problem, 186
shared neurosis, 286
sharing rules, 20, 182
Shermer, Michael, 4, 9
Shostak, Seth, 15
Sieve
    of Eratosthenes, 67
    of Fermi, 67
signal co-opting, 195
signals, 142, 151, 153, 156, 158, 180n16, 181, 182, 184, 186
    content, 169, 170, 183, 187
    See also interstellar signals
signs, 149, 152, 153, 158, 161, 162, 180, 184, 185
    iconic, 152, 153, 161
Singer, Peter, 214
The Sith Code, 289
slave trade, 73
Sobel, Dava, 39
Sober, Elliott, 216
sociability, 141–143, 148, 156, 161
social behavior, 49, 52, 54, 56–58, 142, 144, 146–153, 155, 156, 159, 160
social-brain hypothesis, 148
social capital, 12
social choice theory, 20, 262
social complexity hypothesis, 151
social coordination, 173n4, 178
social exchange, 40
    theory, 11
socializing, 142, 146, 149, 151
social learning, 171, 202
Social Relations, Department of, 278
social sciences, 278
social systems, complex, 141, 142, 147–151, 154–156
societal impact, 132
societies of states, 5
society, 169, 172, 173, 173n4, 185, 280
    despotic, 150, 269
    egalitarian, 150
    See also authoritarian regimes; extraterrestrial society; totalitarian regimes

sociobiology, 226
sociopaths, 289
Solaris, 148
Sonesson, Göran, 153
Soviet Union, 76
space travel. *See* interstellar travel
Spaniards, 72
spatial behavior, 49, 55, 56
Special Theory of Relativity, 26, 27, 69
species, 74, 176
spectroscopy, 158
speed of light, 70
Sperber, Dan, 183n20, 183n21, 186n24
spheres, viii
spices, 72
Sputnik, 76
stages of action, 255
Stalin, Joseph, 7
Star Trek (mythos), 276
Star Wars (mythos), 275–294
Star Wars Galaxies (online role-playing
    game), 279
Star Wars: Knights of the Old Republic (online
    role-playing game), 279
Star Wars: The Old Republic (online
    role-playing game), 20, 279
Stewart, Ian, 171n3, 173n4, 174, 175, 175n6,
    175n7, 186n25, 187
Stine, G. Harry, 19, 75, 241, 242, 242n1, 243,
    244
Stoll, Richard, 9
strong typing, 256
subjectivity, 144
submissiveness, 288
sugars, 68
Sunstein, Cass N., 113–117, 121
surface structures, 252
survival, 146–150, 155, 161, 202
    of the fittest, 4
    human, 228
sustainability, 141, 143, 154, 155, 158, 161
symbols, 148, 152–155, 157, 158, 161, 181,
    186

T
Tabula Rasa (online role-playing game), 279
Tarter, Jill, 33
Tattersall, Ian, 41
tea gardens, 89
technological adolescence, 15
technology, 141–146, 149, 150, 153–156, 158,
    161
    advanced, 38, 40, 141–143, 146, 149, 150,
        153–156, 158, 161

and ethics, 224, 226
high, 38, 39, 173, 175, 178
history of, 142, 150, 153–155
and resources, 224
role in "caring capacity", 97, 98, 103
Templeton, Sir John, 225
Ten Commandments. *See* Commandments,
    Ten
Terrace, Herbert, 198
terraforming, 31
territorialism, 218
thermodynamic ethics. *See* Thermoethics
Thermoethics, 240, 241, 243
Three Laws of Robotics. *See* Laws of Robotics
time
    difference, 133
    dilation, 70
The Time Machine (novel), 277
Tipler, Frank, 31
Titan, 124
Tit for Tat, 13, 14
Tomasello, Michael, 152, 178, 179n14,
    180n17, 181, 182, 185n23
Tooby, John, 40, 176
tool use, and eye-hand coordination, 49, 54, 55
"To Serve Man" (television episode), 29
totalitarian regimes, 8, 173n4. *See also*
    authoritarian regimes
Total Recall (film), 31
totemism, 87
transhumanism, 245
Transmissions from Earth Working Group, 34
travel, interstellar. *See* interstellar travel
tribal groups, 85
tribalism, 219
Trivers, Robert L., 216
Trojan horse, 18
trust, 12
truth-telling, 19, 214, 220
The Twilight Zone (television series), 29, 74

U
UFOlogy, 13
Umwelt, 200
uncertainty, situations of vs. risk, 117, 118,
    122
understanding, 297, 299, 300, 302, 304, 305,
    307
United Nations, 10
United States, 10
Universal Declaration of Human Rights. *See
    also* rights, 97, 99, 218, 221
universal intent, 214, 218
universality, 254, 262

universal signals, 203, 204
universals
    psychological, 169, 174–176, 179, 181,
        186, 186n25
    vs. parochials, 174, 175
use value, 86
utilitarianism, 221

V
Vakoch, Douglas, 13, 157, 178, 223, 228
value, and ethics, 219–221
Van den Berghe, Pierre, 42
verification, 250
viewpoints, 212
Vine, Ian, 42
violence
    decline of, 15, 97
    television, 228
viruses. *See* computer viruses
visitors from space, 221
von Neumann, John, 32
Voyager
    recording, 44, 159
    spacecraft, 27

W
Waal, Frans de, 144, 214
Wallace, Anthony F. C., 288
Walsh, Anthony, 40
war, 4, 9, 10, 269
    decline of, 15
    democracy and, 8, 269
The War of the Worlds (novel), 30
Washoe (chimpanzee), 197
Weaver, Warren, 180n16
Weber, Max, 276
Wells, H. G., 277

Wendt, Alexander, 5
White House, Office of Information and Regulatory Affairs, 113
Whiten, Andrew, 39
Whorf, Benjamin Lee, 136
Wiio's First Law, 140
Williams, George, 216
Williamson, Jack, 282
Wilson, David Sloan, 216
Wilson, Deirdre, 183n20, 183n21
Wilson, Edward O., 83, 86, 87, 215
Wittgenstein, Ludwig, 180, 180n17
wolf children, 195
world
    government, 225
    states, 5
wormholes, 27, 29, 30
worst-case scenario, 116–119, 121
"Wow!" signal, 33

X
xenophobia, 14, 17, 18, 39, 43, 46, 138
xenoscience, 174–176

Y
Yerkish, 197

Z
Zahavi, Amotz, 43
Zampino, Edward, 69, 70
Zarathustra, 278
Zen Buddhism, 280
zero sum game, 224. *See also* non-zero sum game
Zlatev, Jordan, 143, 144, 159
Zone of Sensitivity, 242, 242n1
zoo hypothesis, 161